Springer Tracts in Modern Physics
Volume 121

Editor: G. Höhler
Associate Editor: E. A. Niekisch

Springer Tracts in Modern Physics

Volumes 90–106 are listed on the back inside cover

107 **Electronic Excitations in Condensed Rare Gases**
By N. Schwentner, E.-E. Koch, and J. Jortner 1985. VIII, 239 pages

108 **Particles and Detectors** Festschrift for Jack Steinberger
Edited by K. Kleinknecht and T. D. Lee 1986. 91 figs. X, 291 pages

109 **Metal Optics Near the Plasma Frequency**
By F. Forstmann and R. R. Gerhardts 1986. 30 figs. VIII, 132 pages

110* **Electrodynamics of the Semiconductor Band Edge**
By A. Stahl and I. Balslev 1987. 42 figs. IX, 215 pages

111 **Surface Plasmons on Smooth and Rough Surfaces and on Gratings**
By H. Raether 1988. 113 figs. X, 136 pages

112 **Tests of the Standard Theory of Electroweak Interactions**
By C. Kiesling 1988. 86 figs. X, 212 pages

113 **Radiative Transfer in Nontransparent Dispersed Media**
By H. Reiss 1988. 72 figs. X, 205 pages

114 **Electronic Transport in Hydrogenated Amorphous Semiconductors**
By H. Overhof and P. Thomas 1989. 65 figs. XIV, 174 pages

115 **Scattering of Thermal Energy Atoms from Disordered Surfaces**
By B. Poelsema and G. Comsa 1989. 74 figs. VIII, 108 pages

116 **Optical Solitons in Fibers**
By A. Hasegawa 1989. 22 figs. X, 75 pages

117 **Determination of Hydrogen in Materials** Nuclear Physics Methods
By P. K. Khabibullaev and B. G. Skorodumov 1989. 38 figs. X, 87 pages

118 **Mechanical Relaxation of Interstitials in Irradiated Metals**
By K.-H. Robrock 1990. 67 figs. VIII, 106 pages

119 **Rigorous Methods in Particle Physics**
Edited by S. Ciulli, W. Thirring, and F. Scheck
1990. 21 figs. VIII, 220 pages

120* **Nuclear Pion Photoproduction**
By A Nagl, V. Devanathan, and H. Überall
1991. 53 figs., Approx. 270 pages

121 **Current-Induced Nonequilibrium Phenomena in Quasi-One-Dimensional Superconductors**
By R. Tidecks 1990. 109 figs. IX, 341 pages

---

* denotes a volume which contains a Classified Index starting from Volume 36

Reinhard Tidecks

Current-Induced Nonequilibrium Phenomena in Quasi-One-Dimensional Superconductors

With 109 Figures

Springer-Verlag Berlin Heidelberg GmbH

Dr. Reinhard Tidecks

I. Physikalisches Institut der Universität Göttingen, Bunsenstr. 9
D-3400 Göttingen, Fed. Rep. of Germany

Manuscripts for publication should be addressed to:

Gerhard Höhler

Institut für Theoretische Kernphysik der Universität Karlsruhe, Postfach 6980
D-7500 Karlsruhe 1, Fed. Rep. of Germany

Proofs and all correspondence concerning papers in the process of publication should be addressed to:

Ernst A. Niekisch

Haubourdinstraße 6, D-5170 Jülich 1, Fed. Rep. of Germany

Library of Congress Cataloging-in-Publication Data – Tidecks, Reinhard, 1949 – Current-induced nonequilibrium phenomena in quasi-one-dimensional superconductors/Reinhard Tidecks. p. cm. – (Springer tracts in modern physics; v. 121) Includes bibliographical references (p.) Includes index.
1. Superconductivity. 2. One-dimensional conductors. I. Title. II. Series. QC1.S797 vol. 121 [QC611.97.E69] 530 s – dc20 [537.6'23] 90-10393

ISBN 978-3-662-15043-6 ISBN 978-3-540-46720-5 (eBook)
DOI 10.1007/978-3-540-46720-5

Originally published by Springer-Verlag Berlin Heidelberg New York in 1990.
Softcover reprint of the hardcover 1st edition 1990

2157/3140-543210 – Printed on acid-free paper

Preface

The present work reviews the research activities of the past two decades in the field of current-induced nonequilibrium phenomena in quasi-one-dimensional superconductors from the early experiments to the most recent developments.

The basis of this book is our own experimental work on the investigation of the current-induced breakdown of superconductivity in monocrystalline filaments (whiskers) close to their critical temperature. At the critical current these samples enter a dissipative phase-slip state and, thus, represent a complex problem of nonequilibrium superconductivity.

The observations are discussed in comparison with results on long thin film microbridges and related specimens investigated by other groups. In addition, a detailed introduction is given into the theoretical work done in this field. Basic topics such as the time-independent Ginsburg-Landau theory and the structure of the BCS ground state and its quasiparticle excitations are discussed. Then all the ingredients of a nonequilibrium state in a superconductor are introduced in detail, followed by a description of the available models for a phase-slip center. Finally, the development of the time-dependent Ginsburg-Landau theory is reviewed. The experiments are interpreted within the framework of the theories. In some cases, our own model calculations are presented which describe the experimental observations.

More than five hundred different articles are discussed, dealing with experimental and theoretical work, and several still unpublished experimental results are included. One aim of the present work is to bring the different experiments and the experiments and the theories together. For this purpose we always try to point out the essential content of the theories and to present the theoretical results in a version which can be directly qualitatively and quantitatively compared with the experiments.

The intention of this work is to be a guide for readers who want to inform themselves about the research in this field of nonequilibrium superconductivity or who plan own research activities. The discussion always starts with the basic experimental and theoretical facts, leads to the most recent results, and gives many further references for more detailed information. Experimental techniques and yet unresolved problems are also presented. For readers already working in this field, this work may serve as a 'handbook' containing all important theoretical and experimental facts (or references to find them). Finally, for readers who have worked on problems of current-induced nonequilibrium superconductivity for a longer time, the present article may serve as a comprehensive summary.

It is hoped that this work will fulfill the intended purposes, to be basic enough for an introduction, explicit enough for the specialist, enjoyable for the experienced scientist, and that it will reflect and transmit some of the fascination I have found during my work on this subject.

Finally, may the reader forgive the somewhat german English. To enhance the timeliness of this manuscript, the publisher has agreed to sacrifice some linguistic beauty and style.

Göttingen, August, 1990 Reinhard Tidecks

Contents

1. Introduction

The most striking feature of superconductivity is the vanishing of the electrical resistivity for direct currents. The destruction of the superconducting state by an overcritical transport current is, therefore, one of the fundamental problems in superconductivity research.

At the critical current a superconducting sample generally does not simply enter the fully normal conducting state. The building up of an intermediate state structure, flux flow, and heating effects lead to a complicated transition behaviour [1 - 4]. To reduce the complexity of studies of the transition behaviour, so-called 'quasi-one-dimensional superconductors' have been investigated. These are thin filaments with two dimensions which are less than or comparable to the temperature-dependent Ginsburg-Landau (GL) coherence length $\xi(T)$ and the magnetic penetration depth $\lambda(T)$. In such a sample the order parameter and, thus, the density of Cooper pairs, should be nearly constant over the conductor's cross-sectional area and the transport current should be equally spread over the whole cross-sectional area (see refs. 2 and 5 for a review of the GL theory).

Since $\xi(T)$ and $\lambda(T)$ grow if the temperature T approaches the critical temperature, T_{c0}, an experimental realization of a quasi-one-dimensional superconductor is possible. Suitable samples are 'microbridges' (thin, evaporated metallic films) and 'whiskers' (perfect monocrystalline filaments with very small diameter) at some millikelvins below their critical temperature. Temperatures close to T_{c0} have the additional advantage of a reduced Cooper pair density, leading to a low critical current density. Together with the tiny cross-sectional area of a thin filament the resulting critical currents are so small that Joule heating is of minor importance. This is especially true for clean monocrystalline wires.

Such a quasi-one-dimensional superconductor has been expected to be the most simple system one can imagine for the investigation of the current-induced breakdown of superconductivity. However, measurements of voltage-temperature (V-T) transition curves at fixed currents I and V-I characteristics at fixed temperatures T of microbridges and whiskers at the transition temperature T_c and the critical current I_c, respectively, do not show the direct transition into the fully normal conducting state expected by the GL theory. Instead, a large transition width from the first onset of voltage to the completely normal state has been observed. In this transition region the voltage increases in a series of regular voltage steps.

This phenomenon has been investigated experimentally as well as theoretically. The result is that a single voltage step is related to the appearance of a localized phase-slip center at which phase-slip processes at Josephson frequency occur and the voltage developed is governed by the relaxation of nonequilibrium quasiparticle excitations produced by the phase-slip process. The observed transition width, which is large even for such a homogeneous filament as a whisker, is caused by a stabilization of superconductivity by the phase-slip centers. It can be shown that the quasi-one-dimensional superconductor between the superconducting and the fully normal conducting state is a system of interacting phase-slip centers and, thus, represents a complicated problem of nonequilibrium superconductivity.

Numerous nonequilibrium phenomena have been studied in superconductors, and the reader may be referred to some recent reviews for an overview [6-9]. This book concentrates on the current-induced nonequilibrium effects in quasi-one-dimensional superconductors. It is based on our own experimental work performed with monocrystalline filaments (whiskers). The results will be discussed together with those of measurements on long microbridges and related specimens. To explain these experiments, phase-slip models and the time-dependent Ginsburg-Landau (TDGL) theory will be considered. The theoretical results used will be explained in detail in a separate chapter containing a comprehensive review of the theoretical work. Some experiments cannot be understood within the framework of the existing theories. These are measurements of the electrochemical potentials of Cooper pairs and quasiparticles as well as the behaviour of the intrinsic hysteresis. For this purpose, phenomenological models have been developed which are able to explain our observations.

While the central chapters of this book deal with the dissipative phase-slip state, equilibrium properties such as the critical current and the critical temperature are also discussed and compared with theory. The result is that the nonequilibrium phenomena develop at overcritical currents. We also report the determination of the thermal boundary resistance between the whiskers and the surrounding helium ('Kapitza resistance') which is important for an estimate of Joule heating effects in our samples. Without this information, no separation between thermal and intrinsic hysteresis would be possible.

In a chapter about tunable weak links we study how the properties of a phase-slip center depend on the strength of superconductivity at the locus where the phase-slip center develops. The experimental arrangement enables us to control continuously the characteristic properties of a phase-slip center in a quasi-one-dimensional superconductor.

A final chapter dealing with ongoing work gives information about our current research activities. Several unpublished and still unexplained experiments are reported.

The appendix contains additional details about some of the problems discussed in the text. Among other information, a comprehensive summary of experimental and theoretical results for the inelastic electron-phonon scattering time is given for a number of different superconducting materials.

The article starts with two chapters describing the experimental methods. Because both, the growth and preparation of samples and also the low-temperature techniques are somewhat unconventional, they are briefly discussed.

The international MKSA unit system has been used throughout.

The author is pleased to thank G. v. Minnigerode for his encouragement to write this book, for helpful suggestions concerning the design of the low-temperature techniques, and for stimulating discussions. Special thanks are due to J. D. Meyer for introducing me into the research field. The author would like to thank all his collaborators of the past years. The first one was G. Slama, followed by T. Werner, U. Schulz, B. Damaschke, and X. Yang. Thanks are also due to the precision mechanics workshop of the author's institute for the construction of the low-temperature equipment.

In addition, the author would like to thank L. Kramer and R. Rangel for many helpful discussions concerning the time-dependent Ginsburg-Landau theory. During the preparation of this work numerous discussions with B. Damaschke and X. Yang were quite productive. The author also thanks X. Yang for a critical reading of the manuscript. The accurate typing of the final manuscript by Ms. C. Meyer is gratefully acknowledged and also S. Dina's assistance in handling the computer text processing system.

Special thanks to those institutions which financially supported the project: For some years two of my collaborators (U. S. and B. D.) received support by the *Graduiertenförderung des Landes Niedersachsen.* This research project was also supported by the *Akademie der Wissenschaften, Göttingen* and the *Deutsche Forschungsgemeinschaft, Bonn*.

This work was submitted as habilitation thesis to the University of Göttingen in May 1990 to acquire the venia legendi for physics.

2. Monocrystalline Filaments (Whiskers)

2.1. Crystal Growth

Whiskers are perfect monocrystalline filaments with diameters in the order of 1 μm and a typical length of 1 mm [10]. The spontaneous growth of whiskers (as for instance observed on tinned iron plates) is very slow but can be accelerated by applied pressure and heating [10-13].

The whiskers for the present work are grown by a "squeeze technique". A detailed description of this method is given in ref. 14. The principle is as follows: The raw material is evaporated on thin washers made of iron sheets. Thereafter a screw is sticked through a sandwich of several washers. The washers are compressed together by tightening the nut. The sandwich is smoothened by the cutting tool of a turner's lathe and polished with abrasive paper. Then whiskers begin to grow from the small layer between the iron sheets. In most cases the growth can be accelerated by a warming up of the sandwich in an inert gas (^{4}He) atmosphere.

Using this technique whiskers are grown from Sn and <u>Sn</u>-In alloys [15], In and <u>In</u>-Sn alloys [16,17], the In-Pb system [14], and from Zn and <u>Zn</u>-Ag alloys [18]. Also J.D. Meyer used a similar method for his pioneering investigations with Sn whiskers [19].

2.2. Handling

The handling of a whisker is watched by a stereo lightmicroscope. Using a needle, a straight-grown whisker is picked from the screw and put into a contact to connect the filament with the current and voltage leads. Usually squeeze contacts are used, where the whisker is held by two electrically isolated metal blocks soldered with the electrical leads [14,19]. The whisker is put into a groove at the top of the blocks and fastened by squeezing the material near the ends of the whisker (Fig. 1a). The kind of metal used for the contact blocks depends on the material of the whisker to be investigated. For filaments of Sn, <u>Sn</u>-In, Zn, and <u>Zn</u>-Ag, Wood's metal has been used throughout [15,20]. Whiskers of In-Pb alloys can be contacted with lead [14]. For crystals of pure or nearly pure Pb, Wood's metal leads to a better metallic contact with the whisker [14]. In the case of pure In whiskers, the

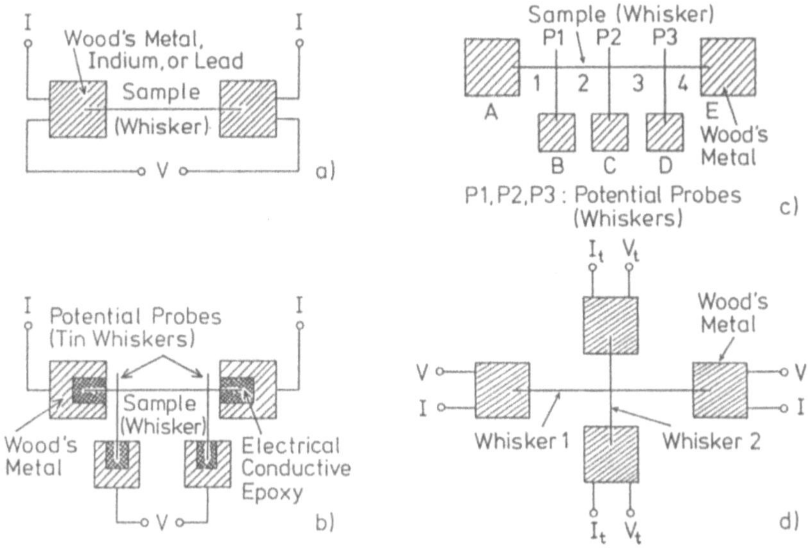

Fig. 1: *Contacts. (a) Squeeze contacts, (b) epoxy contacts, (c) multi-potential-probe contacts, (d) tunable-weak-link contacts. Here, V and I are the voltage and current, respectively. In sketch 'd', V_t and I_t denote the voltage and current of the 'controlling' (or 'tuning') whisker.*

use of indium is of advantage [16,17]. In all cases the material of the squeeze contact is superconducting during the measurements.

Some investigations demand "multi-potential-probe contacts" (Fig. 1c) with several additional whiskers as potential probes [21,22]. Other experiments need two whiskers crossing each other (Fig. 1d) with a metallic contact at the crossover [23]. In these cases the whiskers are held by squeeze contacts. The metallic connection between the whiskers can be made by bringing them into touch with each other and then allowing a discharge current to flow across the touching region.

For In and In-Sn whiskers also expoxy contacts were used which are normal conducting during the measurements. In this case the voltage was probed by a pair of superconducting tin whiskers [17].

2.3. Characteristic Properties

The investigations of our samples with a scanning electron microscope (SEM) show that the whiskers have complicated cross-sectional areas which, however, do not change over the whole length of the whisker. Fig. 2 shows an SEM picture of a pure Zn whisker lying on a copper grid as used in the transmission electron microscope. Several other SEM pictures of whiskers grown from Zn [18], In [14], Sn [24], and Pb [14] as well as pictures of contact

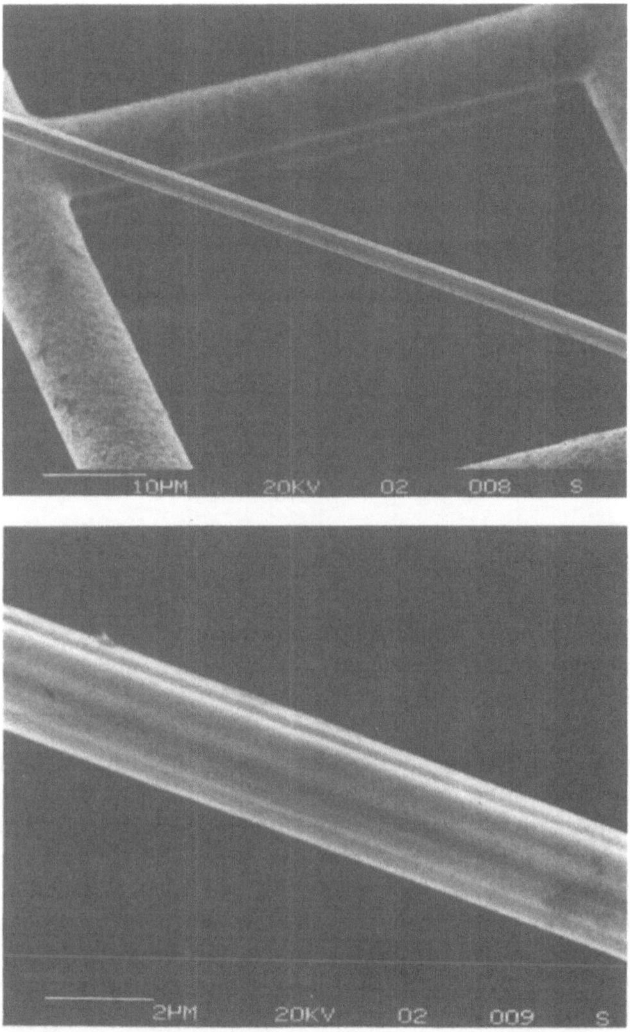

Fig. 2: *SEM picture of a Zn whisker lying on a copper grid (upper picture). The lower picture shows a part of the same sample at a higher magnification.*

arrangements [14, 24] as taken with the SEM and the light microscope are given in the cited literature.

The length L of a sample is usually measured with a light microscope [14, 15]. In some cases an SEM was used in order to enhance accuracy. This is especially important in the case of very short whiskers [20]. The cross-sectional area A is determined from the resistance at room temperature R_{298K}. The electron mean free path ℓ at low temperatures can then be calculated from the residual resistance R_n if the product $\rho_n \cdot \ell$ is known, where ρ_n is the residual resistivity [14, 15].

For anisotropic materials, the material parameters depend on the crystallographic orientation of the elementary cell of the material of the whisker relative to the direction of the applied current which flows parallel to the axis of the whisker. For indium we neglect the orientation dependence, because its face centered tetragonal elementary cell is only slightly anisotropic. This is not allowed for strongly anisotropic materials such as tetragonal tin and hexagonal zinc. For these materials the growth direction of the sample has to be determined.

This can be done by the evaluation of electron diffraction patterns taken in a transmission electron microscope (TEM) [15,18]. For the case of zinc whiskers we published a detailed description of this method [18]. These investigations show that our squeeze-grown Zn whiskers have one preferred growth direction. This is similar to the case of Sn, where four preferred growth directions are known [15].

For the determination of the crystallographic orientation by electron diffraction in a TEM, the whisker has to be removed from the contact, so that the sample is destroyed. Therefore, we have also developed a nondestructive method, where the orientation is determined from a comparison of critical current measurements with the GL theory [15,20]: As predicted by the GL theory, it is $I_c^{2/3}(T_c) \sim (T_{c0} - T_c)$ a straight line, where T_c is the transition temperature at the critical current I_c. For every preferred growth direction, we calculate the slope $d\,j_c^{2/3}/d\,T_c$ and the electron mean free path ℓ with the experimental data. Here $j_c = I_c/A$ is the critical current density. The orientation which gives the best agreement with the theoretical value $d\,j_c^{2/3}/d\,T_c(\ell)$ is taken as the whisker orientation.

For whiskers which were removed from the contact and investigated by electron diffraction, we found agreement between the nondestructive method and the TEM results. However, the nondestructive method can only be successful, if the sample has an orientation for which material parameters are available from the literature.

The material parameters which characterize the superconducting properties of our sample are the critical temperature, T_{c0}, as extrapolated from the $I_c^{2/3}(T_c)$ straight line, the BCS coherence length, ξ_0, and the London penetration depth, $\lambda_L(0)$, so that the GL coherence length, $\xi(T)$, and the magnetic penetration depth, $\lambda(T)$, can be calculated for arbitrary electron mean free path [25]. Values for ξ_0 and $\lambda_L(0)$ for the different materials are given in the cited literature [14-16,20,25].

A special problem is the characterization of alloy whiskers [14,15,20,25]. For small alloy contents the system can be regarded as a clean matrix with impurities. Then the resistance at room temperature can be split into a temperature dependent part and the temperature independent residual resistance using Mathiessen's rule [26], the product $\rho_n \cdot \ell$ and $\lambda_L(0)$ can be approximated by the value of the matrix, the change of ξ_0 with T_{c0}, only, has to be considered, and the impurity concentration can be obtained from the residual resistance ratio [15,20]. In the other case, where samples with concentrations spread over the whole alloy system are examined, the problem is much more involved. For the In-Pb alloy system [14] we collected all data available in the literature concerning T_{c0}, the resistance at 0°C and 20°C (see also Fig. 6 of ref. 27), $\rho_n \cdot \ell$, and the energy gap at zero temperature $\Delta(0)$, and plotted them as a function of the lead concentration, c_{Pb}. From the results for the gap we calculated ξ_0 for the whole alloy system. For $\lambda_L(0)$ we assume a step-like change from one to the other value of the pure materials at $c_{Pb} = 50$ at%. The key to the material parameters of the sample is then the value of T_{c0} which we get information from about the concentration c_{Pb}. Interpolating our results for the electron mean free path, we get an ℓ-c_{Pb} plot for the whole alloy system [25]. Calculating the GL parameter $\varkappa = \lambda(T)/\xi(T)$ for all samples gives an information about the concentrations which are the borders between type I and type II superconductors [25].

A more explicit discussion concerning the determination of material parameters is given in the cited literature together with several tables summarizing sample properties. The problem of characteristic parameters of samples has to be very carefully dealed with, because they are needed for a comparison of experimental results with theoretical predictions. This, however, is the basis for an understanding of the phenomena observed.

3. Low Temperature Techniques

3.1. General Features

For real samples the requirements of a quasi-one-dimensional superconductor are usually only satisfied at temperatures of some millikelvins below their critical temperature. The width of the superconducting to normal transition at fixed current is typically a few millikelvins. For an investigation of step-like structures within this transition special cryostats are needed which allow a precise measurement of temperatures relative to T_{c0} and which enable us to adjust fixed temperatures with a very high stability. For a good thermal contact between the sample and the cooling medium, the whiskers should always be immersed in a ^{4}He bath.

To perform measurements in the whole temperature range between 0.45 K and 7.5 K, we developed and constructed three different cryostats. These are a ^{4}He bath cryostat (1.4 K - 4.2 K), a ^{4}He overpressure cryostat (3.0 K - 7.5 K), and a ^{3}He cryostat with attached superfluid ^{4}He bath (0.45 K - 1.4 K). The last two cryostats are equipped with a top-lading system for rapid sample mounting and change. In all cryostats the temperature is measured by a carbon resistance thermometer being part of a sensitive bridge circuit. The temperature stabilization works on the principle of cooling and countercurrent heating. The temperature stability is at least $\pm 3 \cdot 10^{-5}$ K, for all temperatures between 0.45 K and 7.5 K.

The current can be set at a fixed value or can be continuously adjusted. The current source is electronically stabilized, so that measurements are performed at impressed current. The voltmeters used are slow, measuring time-averaged values even if voltages are oscillating at high frequencies. The earth magnetic field is shielded and the sample is protected against electrical noise by low-pass filters.

In the following sections the basic ideas and components of the three cryostats are briefly explained. Further details are given in the cited literature. Whiskers of Sn, <u>Sn</u>-In, In, and <u>In</u>-Sn are measured in the ^{4}He bath cryostat, samples of Zn and <u>Zn</u>-Ag in the ^{3}He cryostat, and those of Pb, <u>Pb</u>-Bi, and In-Pb alloys in the ^{4}He overpressure cryostat. Moreover, there are some very recent measurements of the hysteresis of pure In whiskers and of their behaviour in a high frequency radiation field which were carried out in the ^{4}He overpressure cryostat.

3.2. ^{4}He Bath Cryostat (1.4 K - 4.2 K)

In this construction we use a liquid nitrogen shielded standard glass Dewar vessel with a narrow bottom. The sample is immersed in the helium bath, surrounded by a brass can with an electric heater at its end. At the top of this sample chamber the carbon resistor thermometer is placed. Connections to the external electronic devices are made of thin copper wires. The sample is protected against noise by low pass filters.

Temperatures below the normal boiling point of liquid helium can be adjusted by vapour pressure reduction. The carbon resistor thermometer is part of a Wheatstone bridge. By comparison with a precision resistance decade the resistance of the thermometer and, thus, the temperature can be measured. The out-of-balance signal of the bridge can be plotted as a temperature axis. To stabilize the temperature, the bridge is adjusted to the desired value and the out-of-balance signal then regulates the power dissipated in the heater. The stability of the temperature is monitored by plotting the bridge signal as a function of time. Calibration points for the thermometer are obtained from a ^{4}He vapour pressure table and the critical temperatures of suitable superconductors as determined by an inductive method using the Meissner-Ochsenfeld effect. In the case of measurements at fixed temperatures the temperature of the liquid helium bath was stabilized to about $\pm 2\cdot 10^{-5}$ K.

For additional details see refs. 15 and 28. The construction is similar to the cryostat as described in refs. 19 and 29.

3.3. ^{4}He Overpressure Cryostat (3.0 K - 7.5 K)

The overpressure cryostat [30] also works on the principle of cooling and countercurrent heating for the adjustment of highly stabilized temperatures. However, the cryogenic system is quite different from the bath cryostat and allows the performance of measurements above 4.2 K. The sample is mounted in a copper cell containing pressurized ^{4}He. The cell is weakly coupled to a liquid helium bath giving rise to cooling. The temperature can be adjusted by an electric heater. The typical temperature stability is $\pm 3\cdot 10^{-5}$ K. For fast sample mounting the cryostat is equipped with a top-loading system.

The sketch in Fig. 3 shows the core of the cryostat. This insert is mounted in a liquid nitrogen shielded standard glass Dewar vessel. Between the copper pressure cell and the outer liquid helium bath there is an insolating vacuum. The copper cell is connected with the head of the cryostat by a stainless-steel tube. The wire of an electric heater is wound around the lower part of the copper cell. A copper rod ('cold link') leads from the lower part of the copper cell to the vacuum can. The sample and a carbon resistor thermometer are placed in the sample chamber situated at the bottom of the top-loading system. The sample is connected with the

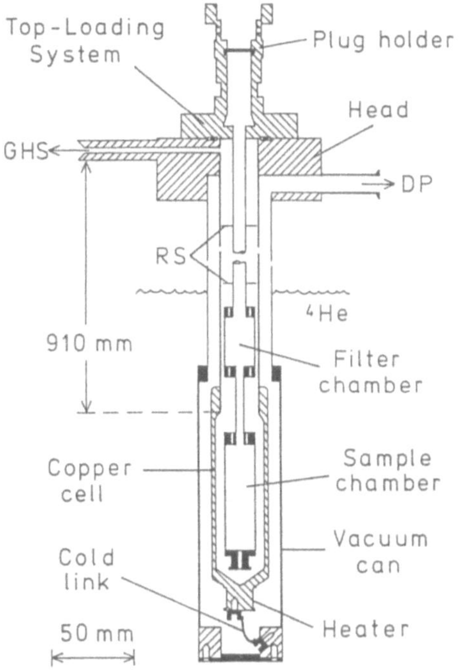

Fig. 3: *Sketch of the ^{4}He overpressure cryostat. GHS: gashandling system, RS: radiation shields, DP: diffusion pump*

external electronic devices by thin copper wires leading to a central plug at the top of the loading system. The electric low-pass filters are situated in the filter chamber.

For a regulation of the temperature a sufficient inertia of the system is needed. This inertia is given by the heat capacity of the pressurized helium. At low temperatures this heat capacity is much larger than that for an equal volume of copper. Due to the real gas behaviour the heat capacity of the ^{4}He does not simply grow with increasing pressure. Very high values of the pressure are not needed, because they do not lead to larger heat capacities as already being present at our working pressure of 0.8 MPa (8 bar).

For a working pressure of 0.8 MPa no condensation of helium can occur for temperatures below the critical point. Thus, no second phase can establish in the copper cell. The different thermal properties of the two phases would cause a reduction of the temperature stability.

Finally, the cold link is coupled to the lower end of the copper pressure cell, where the heater is, and so the temperature is regulated at this point. Due to the high thermal conductivity of copper which is several orders of magnitude greater than that of pressurized ^{4}He, the temperature is uniform throughout the copper cell. Thus, we apply an 'integral heating method' and avoid turbulences around the sample. The carbon resistor thermometer is placed close to the sample, because the thermal conductivity of the ^4He is very small.

3.4. ^3He Cryostat with Superfluid ^4He Bath (0.45 K - 1.4 K)

For measurements below 1 K a ^3He cryostat has been constructed with a sample space which can be filled with superfluid ^4He through a thin capillary [31]. Again the cryostat works on the principle of cooling and countercurrent heating. The cooling tendency is generated by vapour pressure reduction of the ^3He bath. A top-loading system allows to feed the sample from outside the cryostat directly into the sample space. A lambda tight low-temperature seal is used to avoid film-flow from the sample space into the top-loading system.

A sketch of the lower part of the cryostat and the top-loading tube is shown in Fig. 4. It is comfortable to mount the insert into a nitrogen shielded standard glass Dewar vessel filled with ^4He at 4.2 K. The low temperature part of the cryostat is separated from the outer bath by an insulating vacuum. Two cooling stages are needed to generate temperatures below 1 K.

The first one is a continuously operating cooling stage (cold plate) at 1.4 K. It consists of a copper plate with a sickle-shaped evaporation room. From the outer bath ^4He is sucked into the evaporation room through a flow impedance by a mechanical pump. One purpose of the cold plate is to be a heat sink which reduces the heat-flow into the measuring stage. Furthermore, it is the top of the concentric ^3He and ^4He condensation chambers, moving away the heat generated during the condensation procedure.

The second stage is the measuring stage consisting of the sample space which in the upper part is surrounded by a concentric ^3He evaporation room. The entrance of the sample space is the female part of the lambda-tight seal of the top-loading system. The male seal is part of the top-loading tube. The sample is placed in the sample chamber situated at the bottom of the tube. The carbon resistor thermometer is placed at the top of the sample chamber. The electrical leads are connected with a central plug at the loading system.

Before loading, the male seal area is coated by a lambda-tight cement made of flake soap and glycerine. By a seal heater the seal area can be warmed up, so that a sample change is possible, even if the cryostat remains at low temperatures.

Essential for the functioning of the cryostat is that a very thin capillary is used as the filling tube between the condensation chamber and the sample space. The reason is that superfluid film-flow from the sample space through the filling tube to the condensation chamber cannot be avoided. Here the film evaporates and a certain part of the gas would recondense into the sample space through the filling tube, if the gas could not be removed by an

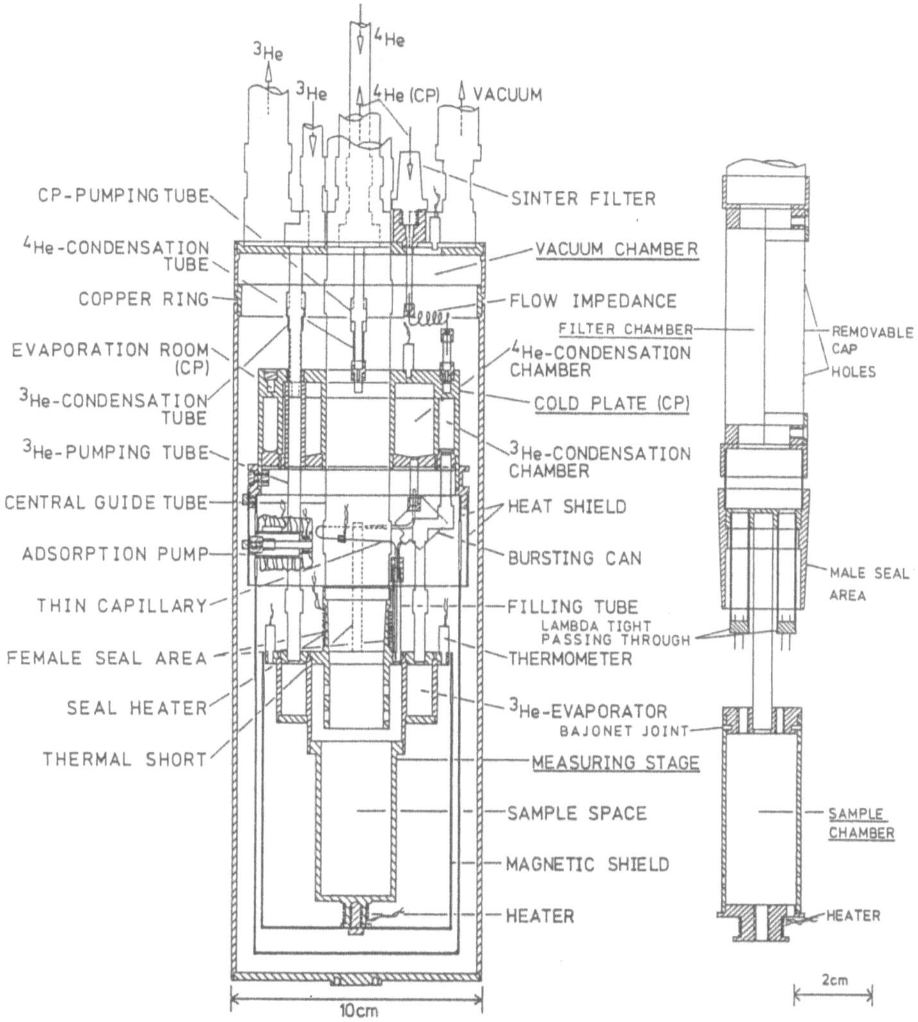

Fig. 4: *Core of the ^{3}He cryostat with attached superfluid ^{4}He bath, together with the lower part of the top-loading system*

external pump. This convection would lead to an enormous heat load of the measuring stage. An effective removal of the gas with an external mechanical pump is, however, only possible if the cryogenic pump effect of the measuring stage is severely handicapped by a small diameter of the filling tube.

With a quantity of about 15 cm^3 of ^{3}He which can effectively be used for evaporation cooling, a measuring time of more than 6 h can easily be reached. The temperature can be adjusted with a stability of $\pm 5 \cdot 10^{-6}$ K without any problem.

4. Basic Experimental Observations

For a current-carrying thin wire the GL theory predicts a sudden transition from the superconducting to the normal state at the transition temperature and the critical current, respectively. The experimental result is, however, quite different. As well V-T transition curves at fixed currents I as also V-I characteristics at fixed temperatures T show a wide transition with a series of regular voltage steps.

The phenomenon has been observed for long microbridges made of tin [32], indium [33], and aluminium [34, 35] as well as for whiskers made of tin [19, 36 - 38], tin-indium alloys [15, 39], indium [16, 17], the indium-lead alloy system [40], zinc [20, 41], and zinc-silver alloys [20]. First reports on the effect are given in refs. 36 - 38. The first systematic investigation of the step-like structure had been performed with tin whiskers by J.D. Meyer during the years 1969 - 1973. He obtained the following basic results [19]:

For small measuring currents very close to the critical temperature, T_{c0}, the V-T transition curves show the typical shape of a fluctuation governed phase transition [38, 42 - 44] without any voltage steps. The transition is very small (~ 1 mK), as expected for a homogeneous filament. For higher measuring currents the transition width rises and voltage steps build up which become more distinct at larger fixed currents (Fig. 5).

For sufficiently low temperatures, voltage steps are also observed in the V-I characteristics at fixed temperatures. These characteristics show a large transition width from the first onset of voltage to the normal state. The lower parts of V-I characteristics for several fixed temperatures are shown in Fig. 6. The step structure becomes more distinct for lower temperatures, that means larger values of $\Delta T = T_{c0} - T$ and critical currents, respectively.

Outside the fluctuation region the current I_c associated with the first onset of voltage across the sample follows the prediction of the GL theory for the critical current of a thin wire. Thus, I_c is called the 'critical current of the whisker' and the voltage step structure has to be regarded as an overcritical phenomenon.

Between the voltage jumps the characteristics are straight lines. At least for the first few steps the differential resistance is a multiple of the differential resistance $(dV/dI)_1$ after the first voltage jump. As will be discussed in the next chapters, this indicates that higher voltage steps are a repetition of the phenomenon of a localized phase-slip center leading to the first step.

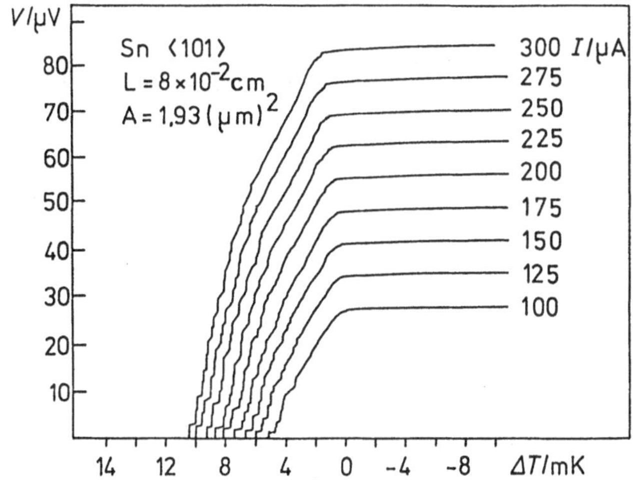

Fig. 5: *V-T transition curves of a tin whisker at different fixed currents I as measured by J.D. Meyer [19]. Here, $\Delta T = T_{c0} - T$, L the length of the sample, A the cross-sectional area. Furthermore, the crystallographic orientation is given.*

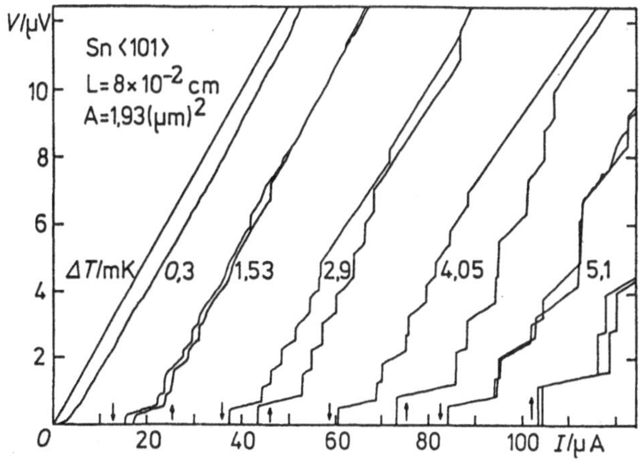

Fig. 6: *V-I characteristics (increasing and decreasing current) for several fixed temperatures $T = T_{c0} - \Delta T$, as measured by J.D. Meyer [19].*

The first voltage step represents the case of an isolated phase-slip center. Therefore, its properties are of special interest: The height of the first voltage jump $V_1(I_c)$ depends linearly on I_c. For tin whiskers the slope $(dV/dI)_1$ is nearly independent of temperature. Nevertheless, the voltage is not simply generated by a region becoming normal conducting at the critical current. The back extrapolation of the characteristic after the first voltage jump shows a zero voltage intercept of I_0 which is not zero but a certain part of the critical current. For tin whiskers the ratio I_0/I_c is nearly independent of temperature.

For not too low currents the transition is hysteretic. This hysteresis grows with decreasing temperature, that means with increasing critical current. While V_1 grows linearly with I_c, the jump-back voltage shows a saturating behaviour.

To explain these estonishing experimental results two basic questions have to be answered: The first one concerns an explanation of the phenomenon of the single voltage step. The second one has to deal with the problem, how a very homogeneous sample such as a whisker can have a large transition width.

A great deal of experimental and theoretical work concerning this subject has been done since the discovery of the phenomenon. In the following chapter an introduction is given to the theories which are considered for an interpretation.

5. Overview of Theories

In this chapter a comprehensive survey of the theoretical work is given in the framework of which we will try to understand the phenomena observed. The survey starts with the time independent Ginsburg-Landau (GL) theory and the Bardeen-Cooper-Schrieffer (BCS) theory. Concerning the GL theory, special attention is drawn to the characteristic lengths of the superconducting state and to the result for the critical current of a quasi-one-dimensional superconductor. Regarding the BCS theory, the structure of the BCS ground state and its quasiparticle excitations are explained.

Then all ingredients for a nonequilibrium state of a superconductor are introduced: Charge imbalance is one of them. Its relation with the electrochemical potentials in a superconductor is discussed as well as its relaxation by inelastic and elastic processes. Moreover, the energy mode may be excited and also collective excitations may appear. A special section deals with the dynamics of charge imbalance and the derivation of the charge imbalance wave equation.

Next, there is a detailed explanation of the models for a phase-slip center: The Rieger-Scalapino-Mercereau (RSM) model concentrates on the dynamics of the superconducting order parameter. The Skocpol-Beasley-Tinkham (SBT) model adds the nonequilibrium quasiparticles and their relaxation during diffusion. The Kadin-Smith-Skocpol (KSS) model considers the dynamics of charge imbalance in the vicinity of the phase-slip center.

Furthermore, a brief review is given of the time-dependent Ginsburg-Landau (TDGL) theory from the first attempts to the most recent results. The TDGL theory is based on the microscopic theory of superconductivity. The application of the TDGL theory to solve the problem of the breakdown of superconductivity in a quasi-one-dimensional superconductor is discussed in much detail.

The chapter ends with a summary and some final remarks. The problem that the theories to describe the nonequilibrium state are usually only valid for weak-coupling superconductors and the question of what to do in the case of a comparison with experiments on strong-coupling superconductors are discussed in these remarks.

5.1. Ginsburg-Landau Theory

In the phenomenological GL theory a space dependent 'macroscopic wave function' $\psi(\underline{r})$ of the superconducting state is introduced as complex order parameter. The square of the absolute value of this wave function is equal to $n_s(\underline{r})$, the local density of superconducting charge carriers *1. The central result of the theory is a system of coupled differential equations for $\psi(\underline{r})$ and the vector potential, $\underline{A}(\underline{r})$, of the magnetic field [2, 5]. In international units these equations read [25]

$$0 = (1/2m')(-i\hbar\underline{\nabla} - e'\underline{A})^2\psi + \alpha\psi + \beta|\psi|^2\psi \qquad (1)$$

$$\underline{j_s} = (e'\hbar/2im')(\psi^*\underline{\nabla}\psi - \psi\underline{\nabla}\psi^*) - (e'^2/m')|\psi|^2\underline{A} \qquad (2)$$

where [1]

$$\alpha = -B_{cth}^2/\mu_0 n_{s0} \qquad (3a)$$

$$\beta = B_{cth}^2/\mu_0 n_{s0}^2 \qquad (3b)$$

Here: $e' = -2e$ with $e > 0$ is twice the electron charge; $m' = 2m$ is twice the electron mass; $\underline{j_s}$ is the supercurrent density; B_{cth} is the thermodynamical critical magnetic field; $n_{s0} = |\psi_0|^2$ is the density of the particles described by ψ in the absence of currents or magnetic fields; $\hbar = h/2\pi$, where h is Planck's constant; and $\mu_0 = 4\pi\cdot 10^{-7}\,\mathrm{Vs\,A^{-1}m^{-1}}$.

The differential equations of the phenomenological GL theory were derived as a rigorous limit of a many-particle theory by Gorkov [5, 46, 47]. Thus, microscopic expressions for ψ, α, and β are obtained by comparing the result for the GL equations of the many-particle theory with that one of the phenomenological approach. The results of the microscopic theory for arbitrary mean free path ℓ for the conduction electrons (non-magnetic impurities) are given in eqs. (46) and (73) of ref. 47. The comparison mentioned yields *2

---

*1 This normalization is usually used in the literature. More rigorously, the order parameter should vary between 0 and 1. This can be achieved by the normalization [45], $|\psi(\underline{r})|^2 = n_s(\underline{r})/n_0$, where n_0 is the maximum possible density of superconducting particles (as present at T = 0 K), and leads to similar differential equations.

*2 Note that in ref. 47 the electron charge is denoted by e and by (−e) in the present work. Furthermore, in eq. (64) of that work a cross at the operator Π seems to have not been printed. Finally, $\sum_{\nu=0}^{\infty}(2\nu+1)^{-3} = 7\zeta(3)/8$ has been introduced into the definition of χ, as given by eq. (65) of that work.

$$\psi(\underline{r}) = (\xi_0(T_{c0})/\hbar)((7\zeta(3)/6)\ m'N_0\chi)^{1/2}\boldsymbol{\Delta}(\underline{r}) \tag{4}$$

$$\alpha(T) = (12/7\zeta(3))(\hbar^2/2m'\xi_0^2(T_{c0})\chi)((T-T_{c0})/T_{c0}) \tag{5a}$$

$$\beta = (18/7\zeta(3)\pi^2\chi^2 N_0)(\hbar^2/2m'\xi_0^2(T_{c0})kT_{c0})^2 \tag{5b}$$

Here, $\boldsymbol{\Delta}(\underline{r})$ is the space dependent order parameter of the microscopic theory, $\xi_0(T_{c0}) = \hbar v_F/2\pi k T_{c0}$, $\zeta(3) = \sum_{\nu=1}^{\infty} \nu^{-3} = 1.202...$, and $\chi(\tilde{\rho}) = (8/7\zeta(3))\sum_{\nu=0}^{\infty}(2\nu+1)^{-2}(2\nu+1+\tilde{\rho})^{-1}$, with $\tilde{\rho} = \xi_0(T_{c0})/\ell$. A very useful approximation of the function χ is given in eq. (13.32) of ref. 46, namely $\chi(\tilde{\rho}) \approx (1+0.853\tilde{\rho})^{-1}$. Furthermore, $N_0 = m^2 v_F/2\pi^2\hbar^3$ is the number of electronic states (in a free electron model) for one spin direction per volume and energy interval at the Fermi energy [26]. Here, v_F is the Fermi velocity and k is Boltzmann's constant.

There are two characteristic lengths in the GL theory, the GL coherence length $\xi(T)$ and the magnetic penetration depth $\lambda(T)$. While $\xi(T)$ is the characteristic decay length for a disturbance of $\psi(\underline{r})$ from ψ_0, $\lambda(T)$ is the typical decay length of the magnetic field.

From the phenomenological theory the GL coherence length is given by [2,5], $\xi^2(T) = -\hbar^2/2m'\alpha$, with α from eq. (3a). An expression for arbitrary ℓ can be obtained by inserting $\alpha(T)$ from the microscopic theory. The result is the same as given by Lüders and Usadel [46] who derive the microscopic expression for the linearized GL equation for metals containing impurities (see chap. 13 of ref. 46). Thus, it is

$$\xi(T) = 0.74\chi^{1/2}\xi_0(T_{c0}/(T_{c0}-T))^{1/2} \tag{6}$$

Here, we have introduced the 'BCS coherence length'

$$\xi_0 = \tilde{\gamma}\hbar v_F/\pi^2 k T_{c0} \tag{7}$$

where $\tilde{\gamma} = 1.781...$ is Euler's constant. Because $\xi_0(T_{c0}) = 0.882\xi_0$, the approximate form for χ given above yields

$$\chi = (1+0.752\xi_0/\ell)^{-1} \tag{8}$$

In the 'Gorkov version' of the microscopic theory, the coherence length $\xi_0(T_{c0})$ and the BCS coherence length ξ_0 can be interpreted as the typical decay length of the probability density for a Cooper pair in a weak coupling pure superconductor at $T = T_{c0}$ and at $T = 0\,K$, respectively (see Appendix 2 of ref. 14 for a detailed discussion). Although $\xi_0(T_{c0})$ is the more 'natural' decay length in the vicinity of the critical temperature, the BCS coherence length ξ_0 has been introduced to get results as usually given in the literature.

From the phenomenological theory, the magnetic penetration depth is given by [5], $\lambda(T) = (m'/e'^2\mu_0|\psi_0|^2)^{1/2}$, where $|\psi_0|^2 = -\alpha/\beta$. An expression for arbitrary electron mean free path is obtained by inserting $\alpha(T)$ and β from the microscopic theory. In agreement with eq.(75) of ref. 47 it results

$$\lambda(T) = 0.5^{1/2}\, \lambda_L(0)\, [T_{c0}/\chi(T_{c0}-T)]^{1/2} \tag{9}$$

where

$$\lambda_L^2(0) = 3/2e^2 \mu_0 N_0 v_F^2 \tag{10}$$

is the square of the 'London penetration depth'.

An important quantity is the GL parameter $\varkappa$, defined by [5]

$$\varkappa = \lambda(T)/\xi(T) \tag{11}$$

so that with eqs. (6) and (9)

$$\varkappa = 0.956\, \lambda_L(0)/\xi_0 \chi \tag{12}$$

The parameter $\varkappa$ is temperature independent and decides whether our samples are type I ($\varkappa < 1/2^{1/2}$) or type II superconductors ($\varkappa > 1/2^{1/2}$) [5]. This is of importance for our alloy whiskers, where ℓ changes with the alloy content, so that the type of superconductor may change [25].

The most important result for the investigation of the present work is the calculation of the absolute value of the critical current density j_c for a quasi-one-dimensional superconductor [2, 19], that means a thin wire with transverse dimensions which are small compared to $\xi(T)$ and $\lambda(T)$. The ideas for this calculation are for instance outlined in ref. 19 for the case of a homogeneous filament. Neglecting the vector potential in eqs. (1) and (2), introducing the phase $\varphi(\underline{r})$ of the order parameter so that $\psi(\underline{r}) = |\psi| \exp(i\varphi(\underline{r}))$, and identifying $(\hbar/m')\underline{\nabla}\varphi(\underline{r})$ with the superfluid velocity leads to an expression for j_s as a function of $|\psi|^2/|\psi_0|^2 = n_s/n_{s0}$ with a maximum value at $n_s/n_{s0} = 2/3$. This maximum value is the critical current density given by

$$j_c = 8e B_{cth}^2(T_c)\, \xi(T_c)/3\sqrt{3}\hbar\mu_0 \tag{13}$$

Here, T_c is the transition temperature at current density j_c. To eliminate the thermodynamical critical field we use

$$B_{cth}(T)\, \xi(T)\, \lambda(T)\, 2e = \hbar/2^{1/2} \tag{14}$$

This equation can be obtained by converting eqs. (2.33) and (2.34) of ref. 5 into international units.

Then the critical current density can be calculated for arbitrary electron mean free path by inserting the expressions for $\xi(T)$ and $\lambda(T)$. The result is

$$j_c^{2/3} = K_{GL}^{2/3}(T_{c0}-T_c)/T_{c0} \tag{15}$$

where

$$K_{GL} = 0.52\,\hbar\,\chi^{1/2}/\mu_0\, e\,\xi_0\,\lambda_L^2(0) \tag{16}$$

Thus, $j_c^{2/3}(T_c)$ is predicted to be a straight line with the zero current intercept T_{c0} and the slope

$$d\,j_c^{2/3}/d\,T_c = -K_{GL}^{2/3}\,(1/T_{c0}) \tag{17}$$

The presentation of the GL theory in the present chapter is far away from being complete. However, we introduced the quantities and results which are of basic importance for a discussion of our experiments.

5.2. BCS Ground State and Quasiparticle Excitations

An important ingredient of a nonequilibrium superconducting state are nonequilibrium quasiparticles. Within the framework of the Bardeen-Cooper-Schrieffer (BCS) theory of superconductivity we will therefore briefly discuss the properties of a quasiparticle excitation of the superconducting ground state. For reviews and further information about the BCS theory see refs. 1, 2, 5, 45, and 48 - 55.

In the BCS theory the superconducting ground state consists of electron pairs ($\underline{K}\uparrow, -\underline{K}\downarrow$), so called 'Cooper pairs', where $\underline{K}$ denotes the wave number vector and $\uparrow, \downarrow$ the spin direction. There is an attractive interaction between the electrons in the pair by an exchange of virtual phonons. The physical idea is that the first electron polarizes the crystal lattice by attracting positive ions and repulsing other electrons and that the second electron is moving through the polarized region.

The graph symbolizing the contribution of effective electron-electron interaction via virtual phonons in the Hamilton operator of the problem is drawn in Fig. 7. The effective interaction potential, $V_{eff}(\underline{K}, \underline{q})$, summarizes contributions of the interaction between electrons and of electrons with 'real'

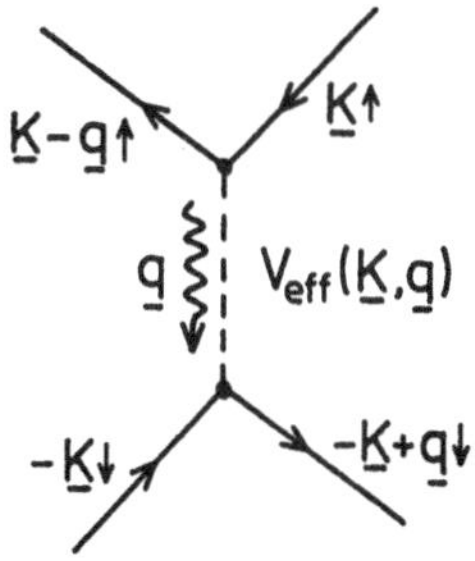

Fig. 7: *Formation of a Cooper pair ($\underline{K}\uparrow, -\underline{K}\downarrow$) by an exchange of a virtual phonon with wave number vector $\underline{q}$. Here, $V_{eff}(\underline{K}, \underline{q})$ is the effective interaction potential (see the text).*

phonons. Because the sound velocity is much smaller than the Fermi velocity of the electrons we have a time retarded electron-electron contact interaction occurring at the locus of the emission of the virtual phonon.

The BCS theory approximates the effective interaction potential by a constant as long as the wave number vectors involved in the scattering process are related to an unperturbed plane-wave energy $\eta_{\underline{K}}$ of free electrons less than $\hbar\omega_D$ above or below the chemical potential μ_F ('Fermi energy'). For all other cases the interaction potential is set to zero. Here, ω_D is the Debye frequency[26].

The many particle wave function of the superconducting ground state is approximated by BCS using the expression

$$\Psi_{BCS} = \prod_{\underline{K}} (u_{\underline{K}} + v_{\underline{K}} c^*_{\underline{K}\uparrow} c^*_{-\underline{K}\downarrow}) \Omega_0 \tag{18}$$

Here, $c^*_{\underline{K}\uparrow}$ and $c^*_{-\underline{K}\downarrow}$ are 'creation operators', creating an electron with wave number vector $\underline{K}$, spin up and $-\underline{K}$, spin down, respectively. Here, Ω_0 is the vacuum state with no particles present. The coefficients $u_{\underline{K}}$ and $v_{\underline{K}}$ are given by

$$|u_{\underline{K}}|^2 = (1/2)\ (1 + \varepsilon_{\underline{K}}/E_{\underline{K}}) \tag{19}$$

$$|v_{\underline{K}}|^2 = (1/2)\ (1 - \varepsilon_{\underline{K}}/E_{\underline{K}}) \tag{20}$$

where

$$\varepsilon_{\underline{K}} = \eta_{\underline{K}} - \mu_F, \quad \text{with} \quad \eta_{\underline{K}} = \hbar^2 \underline{K}^2/2m \tag{21}$$

$$E_{\underline{K}} = (\varepsilon_{\underline{K}}^2 + |\mathbf{\Delta}|^2)^{1/2} \tag{22}$$

For simplicity $|u_{\underline{K}}|^2$, $|v_{\underline{K}}|^2$, and $|\mathbf{\Delta}|^2$ will be denoted by $u_{\underline{K}}^2$, $v_{\underline{K}}^2$, and $\mathbf{\Delta}^2$ in the following. The energy $E_{\underline{K}}$ turns out to be the excitation energy for a quasiparticle excitation of the system. The parameter $\mathbf{\Delta}$ has to be determined selfconsistently and is the energy gap in the quasiparticle excitation spectrum. Also the coefficients $u_{\underline{K}}$ and $v_{\underline{K}}$ find a physical interpretation, because $v_{\underline{K}}^2$ and $u_{\underline{K}}^2 = 1 - v_{\underline{K}}^2$ are the occupation probability of the state $\underline{K}$ with an electron and hole, respectively. It is remarked that the expression for $\varepsilon_{\underline{K}}$ as given in eq.(21) is only valid for the equilibrium case where the chemical potential of the condensate, $\mu_{c,p}$, is equal to μ_F. In general, μ_F has to be replaced by $\mu_{c,p}$. In the presence of an electrostatic potential, Φ, $\varepsilon_{\underline{K}}$ remains unchanged, because $\eta_{\underline{K}}$ has to be replaced by $\eta_{\underline{K}} - e\Phi$ and at the same time μ_F and $\mu_{c,p}$, respectively, have to be replaced by electrochemical potentials. See section 5.3 for a detailed discussion.

In Fig. 8 the Cooper pair formation is illustrated. Although the effective interaction potential is only nonzero in the region $2\delta K$, there are all conduction electrons condensed into the superconducting state at $T = 0\,K$. For finite temperatures quasiparticle excitations of the ground state occur. The

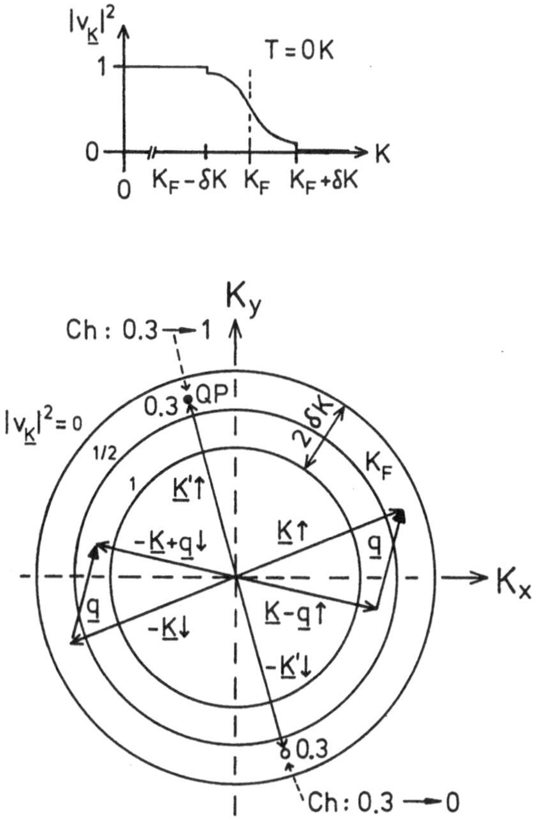

Fig. 8: *Illustration of Cooper pair formation and quasiparticle excitations in a superconductor.*

A plane cut through the $\underline{K}$ *space is drawn. The range* $2\delta K$ *for which the effective interaction potential is nonzero is indicated. The region is very small compared to the Fermi wave number* K_F *and has been magnified in the sketch. The occupation probability of a* $\underline{K}$ *state in a Cooper pair with an electron,* $|v_{\underline{K}}|^2$*, changes across the region* $2\delta K$*. It is nearly zero at the outer border and increases to about one at the inner border. The abrupt changes at the borders are an artifact of the approximation used by BCS for the effective interaction potential.*

The scattering process of Fig. 7 is redrawn into this figure. Furthermore an example for a quasiparticle excitation (QP) at $\underline{K}'\uparrow$ *is given. The value of* $|v_{\underline{K}'}|^2$ *is indicated and also the change of the occupation probability (Ch) due to the excitation of a quasiparticle.*

nature of such an excitation is also illustrated in the figure: A quasiparticle of wave number vector $\underline{K}$ and spin σ is an electron definitely occupying the state $\underline{K}, \sigma$ with its mate $-\underline{K}, -\sigma$ being definitely empty. Here, $\sigma = +1, -1$ denote the spin directions $\uparrow$ and $\downarrow$, respectively.

The excited states are described by 'Bogolubov operators' which are applied to the ground state Ψ_{BCS} and create ($\mathbb{B}^*_{\underline{K},\sigma}$) or annihilate ($\mathbb{B}_{\underline{K},\sigma}$) quasiparticle excitations. The creation operator is for instance given by

$$\mathbb{B}^*_{\underline{K},\sigma} = u^*_{\underline{K}} \, \mathbb{C}^*_{\underline{K},\sigma} - \sigma v_{\underline{K}} \, \mathbb{C}_{-\underline{K},-\sigma} \tag{23}$$

It creates an electron with an amplitude $u^*_{\underline{K}}$ in the state $\underline{K},\sigma$ and annihilates at the same time an electron of amplitude $v_{\underline{K}}$ in the state $-\underline{K},-\sigma$. The Bogolubov operators fulfil the exchange relations for Fermion operators. Thus, the created quasiparticles are Fermions.

The effect of a quasiparticle excitation is to block a pair state from participating in the pairing interaction. It is evident that this disturbance enhances the energy of the system above the energy of the ground state. The result of the BCS theory for the excitation energies as given in eq.(22) and illustrated in Fig.9 shows that there is a temperature dependent energy gap in the excitation spectrum.

Quasiparticle excitations may be generated by tunnel injection (or extraction) of electrons from (or into) a normal conductor. Therefore, the superconductor for instance blocks the penetration of electrons until their energy is above the pair chemical potential (which is equal to μ_F in equilibrium) by at least an amount of the energy gap. This is the reason why these experiments are a method to measure the gap.

The effective charge $Q_{\underline{K}}$ of a quasiparticle depends on the effective number of electrons $\tilde{n}$ added to the system if a quasiparticle with wave number vector $\underline{K}$ is excited. Because $\tilde{n} = u^2_{\underline{K}} - v^2_{\underline{K}}$, it follows [56]

$$Q_{\underline{K}} = (u^2_{\underline{K}} - v^2_{\underline{K}})(-e) \tag{24}$$

Due to its effective charge the character of the quasiparticle continuously changes in the region $K_F \pm \delta K$ from 'electron-like', $Q_{\underline{K}} = -e$, at the upper border to 'hole-like', $Q_{\underline{K}} = e$, at the lower border. At K_F it is $Q_{\underline{K}_F} = 0$. For the quasiparticle at $\underline{K}'$ in Fig. 8 it is $v^2_{\underline{K}'} = 0.3$ and thus $u^2_{\underline{K}'} = 0.7$, so that $Q_{\underline{K}'} = -0.4e$ and the quasiparticle is 'more electron-like'.

The excitation spectrum in Fig.9 has two branches, a 'more electron-like', $\varepsilon_{\underline{K}} > 0$, and a 'more hole-like', $\varepsilon_{\underline{K}} < 0$, with $Q_{\underline{K}} = 0$ at $\varepsilon_{\underline{K}} = 0$. This can easiestly be seen by remembering that $Q_{\underline{K}} = (\varepsilon_{\underline{K}} / E_{\underline{K}})(-e)$.

In equilibrium the probability for the occupation of a quasiparticle state is governed by the Fermi function and the overall charge of the quasiparticles vanishes. Also in a nonequilibrium situation the branches of the excitation spectrum may be equally overpopulated so that there is no charge imbalance and the quasiparticles can simply be characterized by an effective temperature T^* which is larger than the lattice temperature. This is for instance the case for a light irradiated sample [57]. For tunnel injection experiments or in a phase-slip center, however, the two branches of the excitation spectrum are unequally populated and a charge imbalance Q^* is generated [56].

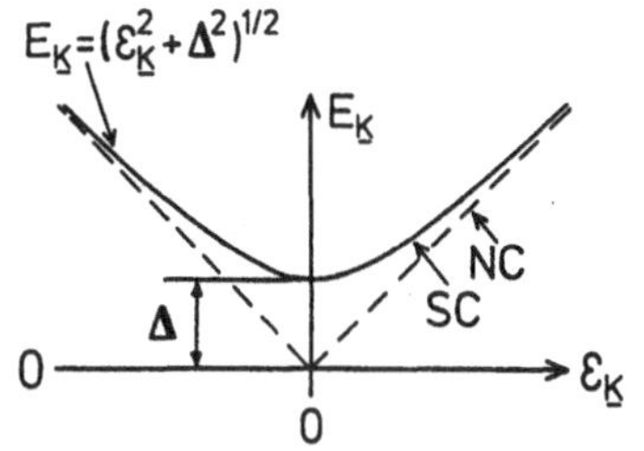

Fig. 9: *Quasiparticle excitation spectrum of a superconductor according to BCS (full line, SC) together with the corresponding spectrum of the normal conductor (dashed line, NC) with $E_K = |\varepsilon_K|$ and $E_K = \varepsilon_K$ for $\varepsilon_K < 0$ and $\varepsilon_K > 0$, respectively. Furthermore , $\Delta(T)$ is the energy gap.*

5.3. Charge Imbalance and Quasiparticle/Pair Electrochemical Potentials

In this section we discuss the electrochemical potentials in a superconductor and their relation with the charge imbalance. In a superconductor two chemical potentials and their related electrochemical potentials are introduced, assigned to the Cooper pairs and quasiparticles, respectively.

First we discuss the physical meaning of the electrochemical pair potential. Then it is shown that in equilibrium the chemical potential of the pairs is equal to the chemical potential in the normal conducting state as well for zero temperature as also for nonzero temperature.

Next, the charge imbalance of the excitation spectrum of a superconductor is defined and its relation with the chemical potential of the pairs is derived. It turns out that the pair chemical potential is different from the chemical potential (of the electrons) in the normal conducting state if there is a charge imbalance in a superconductor.

Then we turn to the problem of the chemical potential of the quasiparticles and first discuss in which situation this potential may be a meaningful quantity. Then a relation is derived between the chemical potentials of the pairs and the quasiparticles in the presence of a charge imbalance. It turns out that both potentials are equal for vanishing charge imbalance, but differ if the charge imbalance is nonzero. In equilibrium the chemical quasiparticle potential is, thus, equal to the chemical potential in the normal state. In nonequilibrium the chemical potential of the pairs, usually, is slightly larger than the normal state chemical potential, except being close to the critical temperature, where both potentials are nearly equal. Finally, we give a physical picture for the concept of a chemical potential of the quasiparticles.

From this knowledge a formula results describing the dependence of the difference between the electrochemical potentials of pairs and quasiparticles on the charge imbalance. This formula is one of the central results of this section.

In the next step it is shown that the electrochemical potential of the pairs usually is spatially constant throughout the superconductor, except in situations where there is a time-dependent order parameter with a time evolution which changes along the superconductor. For a space and time-independent absolute value of the order parameter a relation between the time derivative of the phase of the order parameter and the electrochemical pair potential is derived which is nothing else but the Josephson relation.

At the end of the present section we summarize the basic results and explain how differences in the electrochemical potentials of pairs and quasiparticles can be detected experimentally.

After this survey of the present section we now will discuss all mentioned facts in detail. We start with an introduction of the electrochemical potentials of pairs and quasiparticles.

The chemical potential μ_{chem} indicates the change with particle number of a state function characterizing a system of N particles [26, 58]. If, for instance, the free enthalpy G is a suitable state function of a system with one kind of particles, it is $\mu_{chem} = \partial G/\partial N$. Then μ_{chem} is the free enthalpy which particles can be added at to a system or removed from it without disturbing equilibrium. This implies that a notional reservoir of particles is in equilibrium with the system, if the particles in the reservoir have the (mean) free enthalpy μ_{chem}. To illustrate this interpretation imagine, for instance, a system obeying the Fermi statistic [26]. At the energy $\mu_{chem} = \mu_F$ the Fermi function has the value 1/2. Due to the shape of the Fermi function the particles exchanged between the system and the reservoir have energies distributed around μ_F and their mean energy must be equal to the chemical potential μ_F. In reality the notional reservoir does not exist and μ_F has to be adjusted to give the correct electron density of a metal. In the presence of an electrostatic potential Φ an electrochemical potential $\mu_{el\,chem} = \mu_{chem} - e\Phi$ is defined.

In a superconductor two electrochemical potentials are introduced, the potential of the pairs, μ_p, and the quasiparticle potential, μ, [8, 32, 56, 59 – 67]. Denoting the chemical potential of the pairs and quasiparticles with $\mu_{c,p}$ and $\mu_{c,q}$, respectively, it follows

$$\mu_p = \mu_{c,p} - e\Phi \qquad (25)$$

$$\mu = \mu_{c,q} - e\Phi \qquad (26)$$

The significance of the potential μ_p is [64], that pairs added to the condensate from a notional reservoir enter at the energy $2\mu_p$. To transfer pairs from the reservoir to the condensate, the occupation probability $v_{\underline{K}}^2$ has to be changed. Nevertheless, transferring a small number of pairs does nearly not change the energy of the condensate and all pairs enter the condensate at the same energy $2\mu_p$. The reason is that the energy of the BCS state is minimized with respect to $v_{\underline{K}}^2$.

In the BCS theory $\mu_{c,p}$ is equal to the single-particle energy $\eta_{\underline{K}}$ with $u_{\underline{K}}^2 = v_{\underline{K}}^2 = 1/2$. This can be seen from eqs.(19) - (22). Thus, $\mu_{c,p}$ is a kind of 'mean band energy for the condensate' related to the pair state with the occupation probability 1/2.

As the chemical potential has to be adjusted in a normal metal to give the correct electron density, this has also to be done for the chemical potential of a superconductor. For instance, for the BCS ground state at T = 0 the electron density is [2]

$$N/\Omega = (2/\Omega) \sum_{\underline{K}} v_{\underline{K}}^2 = 2 \int_{-\mu_{c,p}}^{\infty} v_{\underline{K}}^2(\varepsilon_{\underline{K}}) N_n(\varepsilon_{\underline{K}}) d\varepsilon_{\underline{K}} \tag{27}$$

Here, Ω is the volume of the superconductor and $N_n(\varepsilon_{\underline{K}})$ the normal density of states for one spin direction. The factor of 2 arises from the spin. Since $v_{\underline{K}}^2 = 1$ and 0 for energies below $-\hbar\omega_D$ and above $\hbar\omega_D$, respectively [2], we get

$$N/\Omega = 2 \int_{-\mu_{c,p}}^{-\hbar\omega_D} N_n(\varepsilon_{\underline{K}}) d\varepsilon_{\underline{K}} + \int_{-\hbar\omega_D}^{\hbar\omega_D} [1 - \varepsilon_{\underline{K}}/(\varepsilon_{\underline{K}}^2 + \Delta^2)^{1/2}] N_n(\varepsilon_{\underline{K}}) d\varepsilon_{\underline{K}} \tag{28}$$

Assuming $N_n(\varepsilon_{\underline{K}}) \approx N_0$ between $-\hbar\omega_D$ and $\hbar\omega_D$, the second integral yields $2N_0\hbar\omega_D$ so that

$$N/\Omega = 2 \left(\int_{-\mu_{c,p}}^{-\hbar\omega_D} N_n(\varepsilon_{\underline{K}}) d\varepsilon_{\underline{K}} + N_0\hbar\omega_D \right) = 2 \int_{-\mu_{c,p}}^{0} N_n(\varepsilon_{\underline{K}}) d\varepsilon_{\underline{K}} \tag{29}$$

This is just the result which would be obtained for the normal conducting case where the Fermi function is 1 below $\varepsilon_{\underline{K}} = 0$ and is 0 above. Thus, we get $\mu_{c,p} = \mu_F$.

In the general case (for instance for T > 0) the electron density is given by [56, 67]

$$N/\Omega = (2/\Omega) \sum_{\underline{K}} [u_{\underline{K}}^2 \tilde{f}_{\underline{K}} + v_{\underline{K}}^2 (1 - \tilde{f}_{-\underline{K}})] \tag{30}$$

where $\tilde{f}_{\underline{K}}$ is the occupation probability of the state $\underline{K}$ with a quasiparticle excitation.

In thermal equilibrium $\tilde{f}_{\underline{K}}$ is equal to the Fermi function $f_{\underline{K}}$ given by

$$f_{\underline{K}} = 1/[\exp(E_{\underline{K}})/kT) + 1] \tag{31}$$

where $E_{\underline{K}}$ is the energy of a quasiparticle excitation measured relative to the chemical potential of the condensate. The quantity in square brackets in eq.(30) is just the probability that the electron state $\underline{K}$ is occupied: The first term is the probability that the pair state ($\underline{K}, -\underline{K}$) be empty multiplied with the probability that the quasiparticle state $\underline{K}$ be occupied. The second term is the product of the probability that the pair state be filled and the probability that the quasiparticle state $-\underline{K}$ be empty, that means the probability that the $\underline{K}$ state is not held empty by a quasiparticle excitation at $-\underline{K}$. In the case of a Fermi occupancy it is $f_{-\underline{K}}=f_{\underline{K}}$ so that eq.(30) yields

$$N/\Omega = (2/\Omega) \sum_{\underline{K}} [v_{\underline{K}}^2+(u_{\underline{K}}^2-v_{\underline{K}}^2)\, f_{\underline{K}}] \tag{32}$$

In equilibrium $\sum_{\underline{K}}(u_{\underline{K}}^2-v_{\underline{K}}^2)f_{\underline{K}}=0$, because $u_{\underline{K}}^2-v_{\underline{K}}^2=\varepsilon_{\underline{K}}/E_{\underline{K}}$ is odd in $\varepsilon_{\underline{K}}$, whereas $f_{\underline{K}}(E_{\underline{K}})$ is even in $\varepsilon_{\underline{K}}$. Thus, eq.(32) reduces to eq.(27), and also for $T>0$ the chemical potential of the condensate is equal to μ_F.

Since $u_{\underline{K}}^2-v_{\underline{K}}^2=Q_{\underline{K}}/(-e)$, the vanishing of the sum means that the overall charge of the quasiparticle excitation system is zero (i.e. balanced). In nonequilibrium situations where $\tilde{f}_{\underline{K}}$ is no longer equal to the Fermi function there may be a charge imbalance per unit volume, Q^*, given by[56]*1

$$Q^* = (-2e/\Omega) \sum_{\underline{K}} (u_{\underline{K}}^2-v_{\underline{K}}^2)\, \tilde{f}_{\underline{K}} \tag{33}$$

If there is an excess of more electron-like over more hole-like excitations, it is $Q^*<0$, and $Q^*>0$ if there is an excess of more hole-like over more electron-like excitations. A charge imbalance $Q^*<0$ may for instance be generated by adding more electron-like excitations due to tunnel injection of electrons or at a superconductor / normal conductor (SC/NC) interface. Because the injection region is charged, Cooper pairs with equivalent charge emigrate from this region to maintain overall charge neutrality. Thus, the chemical potential of the pairs will change by an amount $\delta\mu_{c,p}=\mu_{c,p}-\mu_F$ to a value below its equilibrium value μ_F. Denoting the charge of the condensate by $Q_p=(-2e/\Omega)\sum_{\underline{K}} v_{\underline{K}}^2$, it follows by charge neutrality

$$Q^* + Q_p|_{\delta\mu_{c,p}\neq 0} = (-2e/\Omega) \sum_{\underline{K}} v_{\underline{K}}^2|_{\delta\mu_{c,p}=0} \tag{34}$$

or

$$\begin{aligned} Q^* &= (-2e/\Omega) \sum_{\underline{K}} (v_{\underline{K}}^2|_{\delta\mu_{c,p}=0} - v_{\underline{K}}^2|_{\delta\mu_{c,p}\neq 0}) \\ &= -2e\int_0^\infty (v_{\underline{K}}^2|_{\delta\mu_{c,p}=0} - v_{\underline{K}}^2|_{\delta\mu_{c,p}\neq 0})\, N_n(\eta_{\underline{K}})\, d\eta_{\underline{K}} \\ &= -2e\, N_0(-\delta\mu_{c,p}) \end{aligned} \tag{35}$$

so that with $\delta\mu_{c,p}=\mu_{c,p}-\mu_F$ it is

$$Q^* = 2\,e\,N_0\,(\mu_{c,p}-\mu_F) \tag{36}$$

As expected $\mu_{c,p}<\mu_F$ for $Q^*<0$ and $\mu_{c,p}>\mu_F$ for $Q^*>0$.

---

*1 In the literature also different definitions of Q^* are used. For a discussion see appendix 1.

Let us now discuss the chemical potential of the quasiparticles: We start from eq.(33) considering that only deviations from equilibrium contribute to Q^*, so that $\tilde{f}_{\underline{K}}$ can be replaced by $\delta\tilde{f}_{\underline{K}}=\tilde{f}_{\underline{K}}-f_{\underline{K}}(E_{\underline{K}})$ where $E_{\underline{K}}=((\eta_{\underline{K}}-\mu_{c,p})^2+\Delta^2)^{1/2}$ and $\mu_{c,p}$ is the actual 'local' value of the chemical potential of the condensate as for instance given in eq.(36). Thus, $\delta\tilde{f}_{\underline{K}}$ describes the deviation from a kind of 'local equilibrium' where the occupation probability of the quasiparticle excitations is given by a Fermi function in which the energies $E(\underline{K})$ are measured relative to the 'local' value of $\mu_{c,p}$. Introducing furthermore $Q_{\underline{K}}$ yields

$$Q^* = (2/\Omega) \sum_{\underline{K}} Q_{\underline{K}}\, \delta\tilde{f}_{\underline{K}} \tag{37}$$

Now we assume that the occupation probability $\tilde{f}_{\underline{K}}$ of a quasiparticle state can be described by a Fermi function, however, with energies $E_{\underline{K}}$ replaced by $E'_{\underline{K}}=((\eta_{\underline{K}}-\mu_{c,q})^2+\Delta^2)^{1/2}$, where $\mu_{c,q}=\mu_{c,p}+\delta\mu_{c,q}$. Thus, it is assumed that the quasiparticles, although not being in local equilibrium with the condensate come into equilibrium with themselves (keeping the total charge associated with the quasiparticles fixed) so that they are characterized by a chemical potential $\mu_{c,q}$ which is shifted by an amount $\delta\mu_{c,q}$ relative to the local value of the chemical potential of the condensate[65, 67]. This assumption is allowed close to the critical temperature[65]. Thus, it is in first order to $\delta\mu_{c,q}$,

$$\delta\tilde{f}_{\underline{K}} = (\partial f_{\underline{K}}/\partial E_{\underline{K}})(\partial E_{\underline{K}}/\partial\mu_{c,p})\,\delta\mu_{c,q} \tag{38}$$

with

$$\partial E_{\underline{K}}/\partial\mu_{c,p} = -\varepsilon_{\underline{K}}/E_{\underline{K}} = -Q_{\underline{K}}/(-e) \tag{39}$$

where

$$\varepsilon_{\underline{K}} = \eta_{\underline{K}} - \mu_{c,p} \tag{40}$$

leading to

$$\delta\mu_{c,q} = Q^*/[(-2/\Omega e) \sum_{\underline{K}} Q_{\underline{K}}^2\,(-\partial f_{\underline{K}}/\partial E_{\underline{K}})] \tag{41}$$

In equilibrium, where $Q^*=0$, it follows $\delta\mu_{c,q}=0$ so that $\mu_{c,q}=\mu_{c,p}$ which is for vanishing charge imbalance equal to μ_F. Thus, it is $\mu_{c,q}=\mu_F$ in equilibrium.

In a nonequilibrium situation with $Q^*\neq 0$ the sum in eq.(41) has to be evaluated (for details see the appendix). For $\Delta \ll kT$ it follows in first order of Δ/kT,

$$\delta\mu_{c,q} = Q^*/[-2eN_0(1-\pi\Delta/4kT)] \tag{42}$$

or

$$Q^* = 2eN_0\,(\mu_{c,p} - \mu_{c,q})(1 - \pi\Delta/4kT) \tag{43}$$

Very close to T_{c0} it is $\pi\Delta/4kT \ll 1$, so that

$$Q^* \approx 2eN_0(\mu_{c,p} - \mu_{c,q}) \tag{44}$$

Thus, $\mu_{c,p} < \mu_{c,q}$ for $Q^* < 0$ and $\mu_{c,p} > \mu_{c,q}$ for $Q^* > 0$.

If we compare the result of eq.(36) with that of eq.(43) it follows $(\mu_{c,p} - \mu_F) = (\mu_{c,p} - \mu_{c,q})(1 - \pi\Delta/4kT)$, that means $\mu_{c,q}$ is slightly larger than μ_F. Very close to T_{c0} eq.(36) has to be compared with eq.(44), yielding that close to the critical temperature $\mu_{c,q} \approx \mu_F$ also in the nonequilibrium case.

The physical picture of the concept of a chemical potential for the quasiparticles is as follows:

If, for instance, electrons are injected into a superconductor so that there is created an excess of more electron-like over more hole-like excitations ($Q^* < 0$), Cooper pairs emigrate from the injection region to maintain electrical neutrality so that $\mu_{c,p}$ becomes smaller. For a given $\underline{K}$ state in the environment of K_F, thus, the occupation probability for the occupation by a Cooper pair becomes smaller. Furthermore the effective charge $Q_{\underline{K}}$ and the probability for the occupation by a thermal quasiparticle, $f_{\underline{K}}$, changes.

The idea is now that the system of injected and thermal quasiparticles rearranges (for instance by the occupation of $\underline{K}$ states which Copper pairs have vanished from) so that the initial Q^* remains unchanged and the system can be described by a Fermi function with a chemical potential $\mu_{c,q}$ which is different from the actual $\mu_{c,p}$. For temperatures close to T_{c0} it turns out that $\mu_{c,q} \approx \mu_F$ so that the occupation probability of a given $\underline{K}$ state with a quasiparticle is nearly the same as in equilibrium. The charge $Q_{\underline{K}}$, however, is different because it has to be calculated with respect to the actual local value of $\mu_{c,p}$. For an illustration see Fig.10. We remark that in a real experiment usually not only a charge imbalance Q^* is created but also a real overpopulation of the excitation spectrum which may lead to an effective temperature T^* of the quasiparticle system which is enhanced over the lattice temperature [57].

From eqs.(43)and(44), respectively, we see that close to T_{c0} the difference of the electrochemical potentials μ of quasiparticles and μ_p of pairs is given by

$$\mu - \mu_p = Q^*/[-2eN_0(1 - \pi\Delta/4kT)] \approx Q^*/(-2eN_0) \tag{45}$$

Thus, $\mu > \mu_p$ for $Q^* < 0$ (excess of more electron-like over more hole-like excitations) and $\mu < \mu_p$ for $Q^* > 0$ (excess of more hole-like over more electron-like excitations).

Now we come to a detailed discussion of the spatial dependence of μ_p, especially for the case of a time-dependent order parameter:

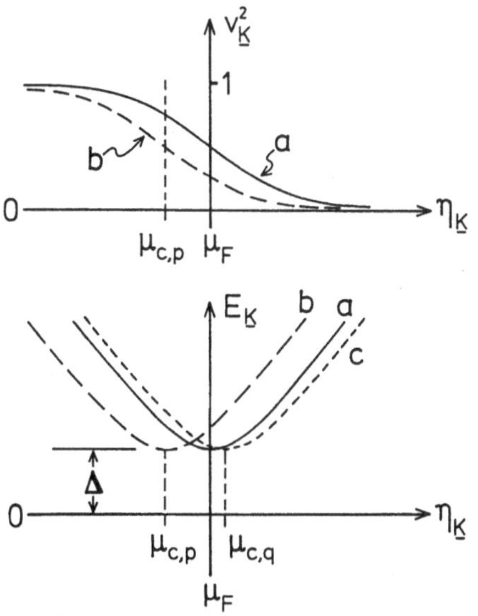

Fig. 10: *Occupation probability for a $\underline{K}$ state with a free electron energy $\eta_{\underline{K}} = \hbar^2 \underline{K}^2/2m$ by a Cooper pair, together with the corresponding excitation spectrum.*
(a) Equilibrium case, $Q^* = 0$, $\mu_{c,p} = \mu_F$, $\mu_{c,q} = \mu_F$, $E_{\underline{K}} = ((\eta_{\underline{K}} - \mu_F)^2 + \Delta^2)^{1/2}$.
(b) Nonequilibrium case, $Q^* < 0$, $\mu_{c,p} < \mu_F$, $\mu_{c,q} > \mu_F$, $E_{\underline{K}}^b = ((\eta_{\underline{K}} - \mu_{c,p})^2 + \Delta^2)^{1/2}$.
(c) Slope $E_{\underline{K}}^c = ((\eta_{\underline{K}} - \mu_{c,q})^2 + \Delta^2)^{1/2}$ *which determines the occupation probability* $\tilde{f}_{\underline{K}}$ *of the nonequilibrium excitation spectrum (case (b)) with quasiparticles, because* $\tilde{f}_{\underline{K}} = f_{\underline{K}}(E_{\underline{K}}^c)$.
Here, Δ, μ_F, $\mu_{c,p}$, *and* $\mu_{c,q}$ *denote the energy gap, the chemical potential in the normal conducting case, the chemical potential of the Cooper pairs, and the chemical potential of the quasiparticles, respectively. The distance between* μ_F *and* $\mu_{c,q}$ *has been enhanced for clarity.*

Usually (except for instance at a phase-slip center) μ_p is spatially constant throughout the superconductor. This can be seen by using the historical introduction of μ_p (Ginsburg, 1943) and its connection with the order parameter of the GL theory [64]:

The London equations are given by [1]

$$\mathrm{rot}\,(\Lambda_L \underline{j_s}) = -\,\underline{B} \tag{46}$$

$$(d/dt)\,(\Lambda_L \underline{j_s}) = \underline{E} \tag{47}$$

where $\underline{B}$ and $\underline{E}$ are the magnetic and electrical field within the superconductor, respectively, and Λ_L is London's constant. Due to the Maxwell equations it is $\mathrm{rot}\,\underline{E} = -\dot{\underline{B}} = -\mathrm{rot}\,\dot{\underline{A}}$ so that $\underline{E} = -\dot{\underline{A}} - \underline{\nabla}\tilde{\Phi}$, where $\underline{A}$ is the vector potential generating $\underline{B}$ and $\tilde{\Phi}$ is a scalar potential which is assumed to be equal to $\mu_p/(-e)$, yielding

$$(d/dt)(\Lambda_L \underline{j_s}) = -\dot{\underline{A}} + (1/e)\,\underline{\nabla}\mu_p \tag{48}$$

If we now take the GL equation for $\underline{j_s}$ from section 5.1 and insert $\psi = |\psi|\exp(i\varphi)$ with $|\psi|$ space and time independent and $\Lambda_L = m'/n_s e'^2 = m'/|\psi|^2 e'^2$ with $e' = -2e$ we get

$$(d/dt)(\Lambda_L \underline{j_s}) = -\dot{\underline{A}} - (\hbar/2e)\underline{\nabla}\dot{\varphi} \tag{49}$$

A comparison between eqs.(48) and (49) yields $\hbar\dot{\varphi}' = -2\mu_p'$ with $\dot{\varphi}' = \dot{\varphi} - c_{\dot{\varphi}}$ and $\mu_p' = \mu_p - c_p$, where $c_{\dot{\varphi}}$ and c_p are constant in space. Setting $c_{\dot{\varphi}} = 0$ leads to

$$\hbar\dot{\varphi} = -2\mu_p' \tag{50}$$

Thus, for $\dot{\varphi} = 0$ (or $\dot{\varphi}$ spatially constant) it is μ_p spatially constant. As will be shown below it is $c_p = \mu_F$ if charge imbalance is the only source for an electrostatic potential in the superconductor for $\dot{\varphi} = 0$. Together with the choice $c_{\dot{\varphi}} = 0$ this leads to the result that $\dot{\varphi} = 0$ in all cases where $\mu_p = \mu_F$ and means nothing else that in the present representation the time evolution of the phase is measured relative to that of the superconductor which the electrochemical pair potential is not changed in, compared to the equilibrium state in the absence of an electrostatic potential.

We remark that eq.(50) can also be obtained from the microscopic theory[64]. Furthermore, the following two equations are equivalent to eq.(50) and, therefore, also used in the literature to introduce the electrochemical potential of the Cooper pairs:

$$i\hbar(\partial\psi/\partial t) = 2\mu_p'\,\psi \tag{51}$$

$$2\mu_p' = (i\hbar/2|\psi|^2)\,(\psi^*(\partial\psi/\partial t) - \psi(\partial\psi^*/\partial t)) \tag{52}$$

A proof of the equivalence can be made by introducing $\psi = |\psi|\exp(i\varphi)$. Here, $|\psi|$ is time independent but not necessarily space independent as must be assumed for a comparison with the London equations[2]. This is similar for the result of the microscopic theory[64], yielding $d\mathbf{\Delta}(\underline{r},t)/dt = -2i\mu_p'\mathbf{\Delta}(\underline{r},t)/\hbar$ and leading to eq.(50) if $\mathbf{\Delta}(\underline{r},t) = |\mathbf{\Delta}(\underline{r})|\exp(i\varphi(\underline{r},t))$.

Certainly eq.(50) is valid in equilibrium. Moreover, the microscopic derivation also holds if it is assumed that, although the quasiparticles being not in equilibrium with the condensate, the condensate itself is in a quasi-equilibrium in the sense that $v_{\underline{K}}^2$ is still given by the BCS expression with the only difference that $\varepsilon_{\underline{K}}$ is now measured relative to $\mu_{e,p}$ instead of relative to μ_F. This is expected to be the case for a constant current carrying NC/SC border where the quasiparticle distribution function does not change in time. For the moment we bypass the question how far it is actually allowed to apply eq.(50) and, thus, the Josephson relation (see below) in situations where the absolute value of the order parameter and μ_p are time dependent, and refer to section 5.9 for a discussion of this problem.

For $\dot{\varphi} = 0$ in eq. (50) it is $\mu_p' = 0$ so that $c_p = \mu_p$ where $\mu_p = \mu_{c,p} - e\,\Phi = \mu_F + Q^*/2eN_0 - e\,\Phi$. Since c_p is spatially constant, also $Q^*/2eN_0 - e\,\Phi$ must be equal to a constant. Setting this constant to zero (leading to $c_p = \mu_F$) means that for $\dot{\varphi} = 0$ charge imbalance is the only source for the electrostatic potential and Φ should be denoted by Φ_{Q^*} in this case. Thus, eq. (50) may be rewritten by

$$\hbar\,\dot{\varphi} = -2(\mu_p - \mu_F) \tag{53}$$

As long as $\dot{\varphi} = 0$ it is thus $\mu_p = \mu_F$. For $\dot{\varphi} \neq 0$ it is, however, $\mu_p - \mu_F = -(1/2)\,\hbar\,\dot{\varphi}$. If again $\mu_p = \mu_F + Q^*/2eN_0 - e\,\Phi$, but now with $\Phi = \Phi_{Q^*} + \Phi_{\dot{\varphi}}$ so that $\mu_p = \mu_F - e\,\Phi_{\dot{\varphi}}$, it follows $\Phi_{\dot{\varphi}} = (1/2e)\,\hbar\,\dot{\varphi}$. Thus, there is an electrostatic potential difference between two places $\underline{r}_1$ and $\underline{r}_2$ with different time evolutions of the phase. In other words, with $\varphi_{12} = \varphi(\underline{r}_1) - \varphi(\underline{r}_2)$ and $V_{12}(t) = \Phi_{\dot{\varphi}}(\underline{r}_1) - \Phi_{\dot{\varphi}}(\underline{r}_2)$, eq. (53) just delivers the Josephson relation [2, 68]

$$d\,\varphi_{12}(t)/dt = 2e\,V_{12}(t)/\hbar \tag{54}$$

The electrostatic potential difference $V_{12}(t)$ due to the different spatial values of $\Phi_{\dot{\varphi}}$ in general has to be measured by detecting the potential μ_p. Detecting μ one would obtain this potential difference only for potential probes placed far away from the nonequilibrium region, where $Q^* = 0$.

In situations where μ_p is constant throughout the superconductor the electrical field $\underline{E}$ within the superconductor is given by (for $\dot{\underline{A}} = 0$)

$$\underline{E} = -\underline{\nabla}\Phi = -(1/e)\,\underline{\nabla}(-\mu_p + \mu_F + Q^*/2eN_0) = -\,\underline{\nabla}Q^*/2e^2N_0 \tag{55}$$

A situation where $\dot{\varphi} = 0$ and, thus, μ_p is constant throughout the superconductor is realized if a normal current penetrates the superconductor at a normal conductor/superconductor (NC/SC) interface. In Fig. 11 a sketch is given for the behaviour of the charge imbalance at an NC/SC interface [56, 64], where Q^* relaxes exponentially with a characteristic length Λ_{Q^*} and the quasiparticle current decays exponentially into supercurrent. Since it is assumed that $\Delta/kT \rightarrow 0$, as for instance for temperatures close to T_{c0}, there is no Adreev reflection [56].

Differences between μ and μ_p can be measured with an NC/SC potential probe pair. The SC probe has the same μ_p as the sample and a charge imbalance decays exponentially into the probe so that $\mu = \mu_p$ at the end of the probe where it is connected with a normal conducting lead of a voltmeter. Thus, the voltmeter measures μ_p although its normal conducting lead adjusts to the quasiparticle potential in the potential probe. An NC probe contains electrons which are in equilibrium with the quasiparticles in the sample and, thus, it measures μ. Using this method measurements were performed at NC/SC boundaries [60, 61, 69], during the injection and extraction through tunnel barriers [62, 63], and in the direct vicinity of phase-slip

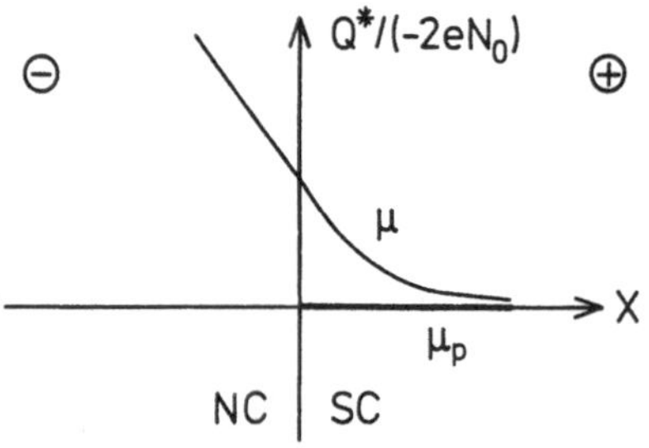

Fig. 11: *Behaviour of the charge imbalance, Q^*, at a normal conductor/superconductor (NC/SC) boundary for $\Delta/kT \rightarrow 0$. For the indicated polarity of the battery there is an excess of more electron-like over more hole-like excitations in the superconductor, so that $Q^* < 0$. The sketch at the same time gives the slope of the electrostatic potential, Φ, the electrochemical potential of the pairs, μ_p, and the electrochemical potential of the quasiparticles, μ, respectively, because $Q^*/(-2eN_0) = -e\Phi$ and $Q^*/(-2eN_0) \approx \mu - \mu_p$.*

centers [8, 70 - 73]. Moreover, nonequilibrium quasiparticles caused by a temperature gradient can be detected by this technique [74 - 77].

We summarize this section: In a superconductor two electrochemical potentials are introduced, $\mu_p = \mu_{c,p} - e\Phi$ for the pairs and $\mu = \mu_{c,q} - e\Phi$ for the quasiparticles. First, $\mu_{c,p}$ is discussed. In equilibrium (for T=0 and T>0) we have $\mu_{c,p} = \mu_F$ and in the nonequilibrium case $Q^* = 2eN_0(\mu_{c,p} - \mu_F)$. Then $\mu_{c,q}$ is discussed. In equilibrium it is $\mu_{c,q} = \mu_F$. In nonequilibrium close to T_{c0} one gets $Q^* = 2eN_0(\mu_{c,p} - \mu_{c,q})(1 - \pi\Delta/4kT) \approx 2eN_0(\mu_{c,p} - \mu_{c,q})$, yielding $\mu_{c,q} \approx \mu_F$, and, furthermore, $\mu - \mu_p = Q^*/[-2eN_0(1 - \pi\Delta/4kT)] \approx Q^*/(-2eN_0)$. Next, μ_p is connected with the phase of the order parameter by $\hbar\dot{\varphi} = -2(\mu_p - \mu_F)$ which is nothing else but the Josephson relation $d\varphi_{12}(t)/dt = 2eV_{12}(t)/\hbar$. Then we show that in situations where μ_p is constant throughout the superconductor the electric field in the superconductor is given by $\underline{E} = -\underline{\nabla} Q^*/2e^2N_0$. Finally the properties of an NC/SC boundary are sketched and methods for the detection of μ and μ_p are discussed.

5.4. Relaxation of Charge Imbalance

In general the creation of charge imbalance involves the creation of quasiparticles carrying an energy above thermal energies. Processes that restore equilibrium usually involve cooling and charge relaxation. In Fig. 12 we summarize several processes that contribute to the relaxation of quasiparticle charge. These are inelastic processes with phonons such as the scattering to lower energy and, thus, lower charge on the same branch or to lower energy and lower charge with opposite sign on the other branch of the excitation

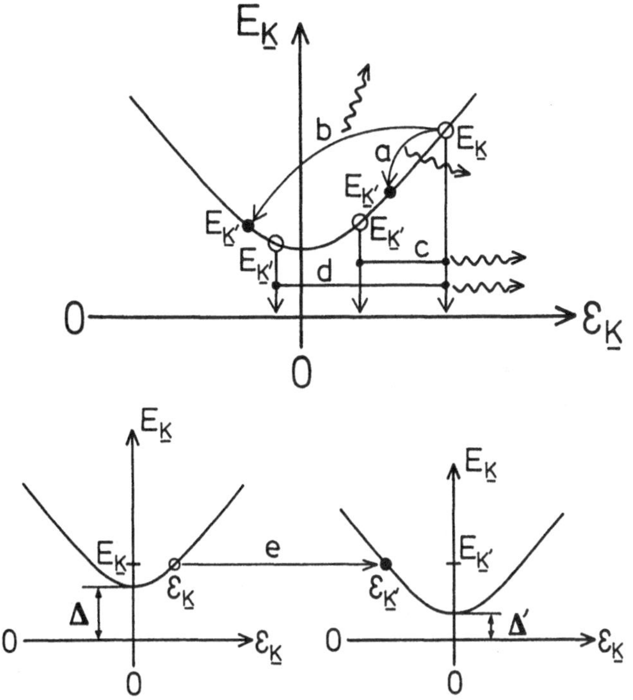

Fig. 12: *Scattering processes that contribute to the relaxation of the charge of a quasiparticle with energy* $E_{\underline{K}}$.

Inelastic processes (with phonon emission): (a) scattering to a state $E_{\underline{K}'}$ *of lower charge on the same branch, (b) scattering to a state* $E_{\underline{K}'}$ *of lower charge with opposite sign on the other branch, (c) recombination with a quasiparticle near the gap edge on the same branch, (d) recombination with a quasiparticle near the gap edge on the other branch.*

Elastic processes in the presence of gap anisotropy: (e) scattering of a quasiparticle at constant energy $E_{\underline{K}} = E_{\underline{K}'}$. *The absolute value and the sign of the energy* $\varepsilon_{\underline{K}}$ *are different from those of* $\varepsilon_{\underline{K}'}$. *The figure shows the excitation spectrum at two different regions of the* $\underline{K}$ *space where the energy gaps are different.*

spectrum. Other inelastic processes are the recombination with a quasiparticle of the same or the other branch. For superconductors with an anisotropic gap also elastic scattering from non-magnetic impurities can contribute to a Q^* relaxation.

The scattering probability of a quasiparticle in a superconductor from a state $\varepsilon_{\underline{K}}$ to a state $\varepsilon_{\underline{K}'}$ and the annihilation probability of two excitations are reduced compared to that of a one electron state in a normal conductor by so called 'coherence factors' [2, 78]. These factors are

$$(u_{\underline{K}} u_{\underline{K}'} - v_{\underline{K}} v_{\underline{K}'})^2 = (1/2)[1 + (\varepsilon_{\underline{K}} \varepsilon_{\underline{K}'} - \Delta^2)/E_{\underline{K}} E_{\underline{K}'}] \tag{56}$$

for scattering, and

$$(v_{\underline{K}} u_{\underline{K}'} + u_{\underline{K}} v_{\underline{K}'})^2 = (1/2)[1 - (\varepsilon_{\underline{K}} \varepsilon_{\underline{K}'} - \Delta^2)/E_{\underline{K}} E_{\underline{K}'}] \tag{57}$$

for recombination (or creation) of quasiparticles.

These coherence factors have to be considered and the fact that the quasiparticle charge is a function of energy. For $\Delta/kT \ll 1$ (that means $T \approx T_{c0}$) then already a very rough qualitative discussion leads to an estimate for the charge imbalance relaxation rate due to inelastic electron-phonon processes [9, 56, 66, 67]: The only inelastic scattering processes for which the coherence factor and the change in quasiparticle charge both are appreciably large are those in which either the initial or the final quasiparticle state has an energy between Δ and 2Δ. Now, the simplifying assumption is made that there are excess quasiparticles on one branch uniformly distributed from Δ to kT. Then, only a fraction of order Δ/kT of all inelastic scattering events leads to a significant relaxation of the charge imbalance, so that the overall charge relaxation rate for phonon scattering $1/\tau_{Q^*in}$ is smaller than the inelastic scattering rate $1/\tau_E$ by a factor of the order Δ/kT. The same result is obtained for the relaxation of Q^* by recombination of quasiparticles. Calculations of the charge relaxation rate due to inelastic electron-phonon scattering verify these qualitative considerations.

In an isotropic superconductor charge imbalance relaxation by elastic scattering from non-magnetic impurities is not possible, because the coherence factor for branch crossing with $E_{\underline{K}} = E_{\underline{K}'}$ and $\varepsilon_{\underline{K}} = -\varepsilon_{\underline{K}'}$ is zero. This, however, would be the only possibility for charge relaxation by elastic processes. In an anisotropic superconductor $|\varepsilon_{\underline{K}}|$ and $|\varepsilon_{\underline{K}'}|$ are different although $E_{\underline{K}} = E_{\underline{K}'}$ so that the coherence factor does not vanish and Q^* may relax by elastic scattering processes.

The first derivation of an appropriate relaxation time was done by Schmid and Schön for dirty ($\ell \ll \xi_0$) superconductors. They obtained a time that close to T_{c0} is equal to the charge imbalance relaxation time τ_{Q^*} (see eq. (50) of ref. 79, eq. (17) of ref. 80, and p. 452 of ref. 81). Close to T_{c0} it is, thus,

$$\tau_{Q^*} = (4kT_{c0}/\pi\Delta)(\tau_E/2\Gamma)^{1/2}(1 + \hbar^2\Gamma/\Delta^2\tau_E)^{1/2} \tag{58}$$

If $\hbar^2\Gamma \ll \Delta^2\tau_E$, it follows

$$\tau_{Q^*} = (4kT_{c0}/\pi\Delta)(\tau_E/2\Gamma)^{1/2} \tag{59}$$

where

$$\Gamma = 1/\tau_s + 1/2\tau_E + (D/2)(4m^2 v_s^2/\hbar^2 - \Delta^{-1}\partial^2\Delta/\partial \underline{r}^2) \tag{60}$$

Here, τ_s is the elastic spin-flip scattering time, $D = \ell v_F/3$ is the diffusion constant and v_s the superfluid velocity. Furthermore, $\tau_E = \tau_{E=0}(T_{c0})$ is the lifetime of electrons in the normal state at the Fermi level μ_F ('E = 0') and at T_{c0} due to electron-phonon collisions [80] ('inelastic collision time' [81]). In general the inelastic collision time is temperature dependent, but non-divergent at the critical temperature, so that in the direct environment of T_{c0} it can be estimated by its constant value at T_{c0}.

The four terms in eq.(60) consider relaxation processes due to elastic spin-flip scattering against magnetic impurities or in the presence of a magnetic field, inelastic collisions with phonons and elastic scattering in the presence of a supercurrent or a spatial dependence of the gap. In the following we will discuss the different relaxation processes considered in eqs.(59) and (60) somewhat in detail:

For the case of relaxation due to inelastic phonon collisions only it is [81], $\Gamma = 1/2\tau_E$ yielding

$$\tau_{Q^*in} = (4kT_{c0}/\pi\Delta)\,\tau_E \tag{61}$$

This result has also been obtained by several other authors, without being restricted to the dirty limit [65-67, 82] and, thus, eq.(61) seems to be valid quite generally. It is remarked that in ref.65, moreover, the role of anisotropy in the inelastic electron-phonon scattering rate $1/\tau_E$ has been discussed.

To consider the effect of inelastic phonon collisions and magnetic impurities [56, 83] or a magnetic field [56, 84] one has to set $\Gamma = 1/\tau_s + 1/2\tau_E$, leading to

$$\tau_{Q^*in,m} = (4kT_{c0}/\pi\Delta)\,\tau_E\,(1+2\tau_E/\tau_s)^{-1/2} \tag{62}$$

In the case of relaxation due to inelastic phonon collisions and a supercurrent [56, 85, 86] it is $\Gamma = 1/2\tau_E + (D/2)(4m^2v_s^2/\hbar^2)$. At the critical current density, j_c, it is v_s equal to the critical velocity [2], $v_c = \hbar/2m\sqrt{3}\,\xi(T)$. In the dirty limit the GL coherence length $\xi(T)$ is given by (see eq.(6) with $\ell \ll \xi_0$), $\xi_D(T) = 0.853\,\xi_0^{1/2}\ell^{1/2}(T_{c0}/(T_{c0}-T))^{1/2}$. Introducing, furthermore, $\xi_0 = 1.781\,\hbar v_F/\pi^2 kT_{c0}$ from eq.(7) and [2, 81, 87] $\tau_{GL} = \pi\hbar/8k(T_{c0}-T)$ it follows $\Gamma = 1/2\tau_E + 1/6\tau_{GL}$ so that at the critical current density

$$\tau_{Q^*in,j_c} = (4kT_{c0}/\pi\Delta)\,\tau_E\,(1+\tau_E/3\tau_{GL})^{-1/2} \tag{63}$$

For $1/2\tau_E \ll 1/6\tau_{GL}$ that means $\tau_E \gg 3\tau_{GL}$, current induced charge imbalance relaxation dominates. At the critical current it is then

$$\tau_{Q^*j_c} = (4kT_{c0}/\pi\Delta)\,(3\tau_E\tau_{GL})^{1/2} \tag{64}$$

Since $\tau_{GL} \sim (T_{c0}-T)^{-1}$, charge imbalance relaxation by a supercurrent will be more important for lower temperatures than for T close to T_{c0}. In ref.86 the importance of current induced relaxation has been discussed for several different materials.

It is remarked that τ_{GL} governs the temporal behaviour of small changes of the magnitude of the order parameter (more precisely of $|\psi|^2$) in the limit where its equilibrium value is zero. This limit is, for instance, realized in experiments dealing with the decay of thermal fluctuations above the critical temperature. Thus, τ_{GL} is a gap relaxation time in the gapless limit (see chapters 7-5 and 8-3 of ref. 2, refs. 81, 87, and 88, and chap. 5.9 of the present work).

The last contribution in Γ depends on the kind of spatial dependence of the gap. Assuming, for instance, the dependence in a region to be given by $\mathbf{\Delta}(\underline{r}) = \mathbf{\Delta}(0)[1-\exp(-r/\xi_D)]$, it follows, if the spatial dependence of the gap is the only relaxation mechanism

$$\tau_{Q^*g} = (4kT_{c0}/\pi\mathbf{\Delta})(\tau_E\tau_{GL})^{1/2} \tag{65}$$

An important case is the consideration of gap anisotropy that means the consideration of a direction dependence of the gap due to anisotropic properties of the crystal lattice [89]. This can, however, not be done in the dirty limit ($\ell \ll \xi_0$), because already for $\ell \approx \xi_0$ most of the anisotropy is eliminated. The reason is that in the presence of impurities the pairing is no longer of the kind ($\underline{K}\uparrow, -\underline{K}\uparrow$), because such a pair is broken up within the distance ξ_0 it would ordinarily occupy. Instead of this, the pairing is of time reversed exact eigenstates of the dirty metal which are a superposition of plane waves with different wave number vectors and, thus, are less able to take advantage of crystal anisotropy [90, 91].

Chi and Clarke [56, 92] and Pethick and Smith [66, 67] obtained an expression for the charge imbalance relaxation time τ_{Q^*el} due to relaxation by elastic scattering in the presence of an anisotropic gap. The result is valid for clean superconductors as well as for superconductors with reduced electron mean free path ℓ due to impurity or surface scattering. Using the relation between the branch crossing rate and the Q^* relaxation rate (see eq. (2.17) of ref. 92), it is the angle-averaged relaxation rate [66, 67]

$$1/\tau_{Q^*el} = 2(1/\tau)(\langle(\delta\mathbf{\Delta})^2\rangle/2\varepsilon_{\underline{K}}^2)(E_{\underline{K}}/|\varepsilon_{\underline{K}}|) \tag{66}$$

Here, $\langle(\delta\mathbf{\Delta})^2\rangle = \langle(\mathbf{\Delta}-\langle\mathbf{\Delta}\rangle)^2\rangle$ is the mean-square deviation of the gap from the average value, where the tapered brackets denote an angular average and $\tau = \ell/v_F$ is the impurity scattering time in the normal metal. Introducing [56, 92] the normalized mean square gap anisotropy [89], $\langle a^2\rangle = \langle(\delta\mathbf{\Delta})^2\rangle/\langle\mathbf{\Delta}\rangle^2$, with $\langle a^2\rangle = \langle a^2\rangle_0/[1+(\hbar/2\tau\langle\mathbf{\Delta}\rangle)^2]$, where $\langle a^2\rangle_0$ is the value for clean bulk material and using $\varepsilon_{\underline{K}} = (E_{\underline{K}}^2-\langle\mathbf{\Delta}\rangle^2)^{1/2}$, it follows as in ref. 92

$$\tau_{Q^*el} = \tau\{[1+(\hbar/2\tau\langle\mathbf{\Delta}\rangle)^2]/\langle a^2\rangle_0\}\{(E_{\underline{K}}^2-\langle\mathbf{\Delta}\rangle^2)^{3/2}/E_{\underline{K}}\langle\mathbf{\Delta}\rangle^2\} \tag{67}$$

For $\hbar/2\tau\langle\mathbf{\Delta}\rangle \approx 1.1(\xi_0/\ell)[T_{c0}/(T_{c0}-T)]^{1/2} \ll 1$ and $E_{\underline{K}} \gg \langle\mathbf{\Delta}\rangle$ it follows

$\tau_{Q^*el} = \tau\, E_K^2/\langle a^2\rangle_0 \langle \Delta\rangle^2$. In the case $\hbar/2\tau\langle \Delta\rangle \gg 1$ and again $E_K \gg \langle \Delta\rangle$ it is $\tau_{Q^*el} = \hbar^2 E_K^2/4\tau\langle a^2\rangle_0 \langle \Delta\rangle^4$.

For temperatures very close to the critical temperature, charge imbalance relaxation due to gap anisotropy can be neglected compared to charge imbalance relaxation by inelastic electron-phonon processes. This can be seen by the computations shown in Figs. 15, 2, and 3 of refs. 56, 92, and 93, respectively. The reason is that $\tau_{Q^*in} \sim \Delta^{-1}$, while $\tau_{Q^*el} \sim \langle \Delta\rangle^{-2}$ (very close to T_{c0} it is even $\tau_{Q^*el} \sim \langle \Delta\rangle^{-4}$) so that τ_{Q^*el} becomes larger than τ_{Q^*in} for small $\langle \Delta\rangle$, that means for temperatures close to T_{c0}.

In the calculations discussed above for the inelastic charge imbalance relaxation, inelastic electron-electron scattering has not been considered. This mechanism also contributes to the charge relaxation [94] with a relaxation time which is close to T_{c0}, as well for clean as also for dirty samples, given by [95]

$$\tau_{Q^*ee} = (4kT_{c0}/\pi\Delta)\,\tau_{ee} \tag{68}$$

where τ_{ee} is the inelastic electron-electron scattering time [96 - 99].

As mentioned in the beginning of this section, the analytical results given here for the charge imbalance relaxation times are only valid close to the critical temperature. The dimensionless factor F^* as introduced in refs. 56, 80, 92, and 100 is, thus, very close to unity so that we set $F^* = 1$. For lower temperatures numerical evaluations have to be performed. They show, for instance, that for lower temperatures the time τ_{Q^*in} differs from the expression given in eq. (61) and depends on the quasiparticle energy [56, 66, 67, 92, 94, 100, 101].

It is remarked that charge imbalance relaxation due to spin-flip scattering or a supercurrent has also been considered using a Boltzmann equation approach [56, 66, 67, 83, 85, 94]. However, as far as a comparison with experiments has been done a better agreement with the theory of Schmid and Schön (as discussed above) has been obtained [56, 83, 85, 94].

In earlier work the relaxation time τ_Q of the quasiparticle number branch imbalance Q has been calculated [63, 102, 103] and considered in connection with the relaxation of differences of the electrochemical potentials of quasiparticles and pairs [16, 17, 63, 70, 104]. A truer view is that the relaxation of charge imbalance Q^* governs the relaxation of the electrochemical potentials [65, 66, 80, 105].

The relation between τ_Q and τ_{Q^*} is discussed in ref. 80 for the case of low quasiparticle energy, yielding that at least for $T \to T_{c0}$ it is $\tau_Q = \tau_{Q^*}$. In ref. 92 it is shown that for the case of inelastic electron-phonon scattering $\tau_{Q\,in} = \tau_{Q^*in}$ for T close to the critical temperature. Thus, close to T_{c0}, $\tau_{Q\,in}$ is also given by eq. (61).

The early calculation of $\tau_{Q\,in}$ by Tinkham [102] yields a value which has to be multiplied by a factor $2/\pi$ to get the result of eq. (61). [See eq.(37)

of ref. 102 and consider remark 27 of ref. 16 which is based on remark 10 of ref. 80. If comparing the present discussion with a similar one in chap. 4.2.1. of ref. 40, note that the numerical factor in eq. (5) of ref. 16 has been approximated.]

Tinkham also derived an expression for the branch imbalance relaxation time τ_{Qel} due to gap anisotropy which has to be multiplied by a factor $(E_{\underline{K}}^2 - \langle \mathbf{\Delta} \rangle^2)^{1/2} / E_{\underline{K}} = N_0 / N_q$ to get the result of eq.(67) of the present work [see eqs. (46) and (48) of ref. 102]. Here, N_q is the quasiparticle density of states [2]. For $E_{\underline{K}} \gg \langle \mathbf{\Delta} \rangle$ it is $N_0/N_q \approx 1$ and the factor becomes unimportant.

To get a more explicit expression for τ_{Q^*in} from eq. (61), we insert the expression for the energy gap of a weak-coupling superconductor close to T_{c0}, given by [2]

$$\mathbf{\Delta}(T) \approx \mathbf{\Delta}(0)\, 1.74\, (1 - T/T_{c0})^{1/2} \tag{69}$$

with

$$\mathbf{\Delta}(0) = 1.76\, k\, T_{c0} \tag{70}$$

We thus get

$$\tau_{Q^*in} = \tau_{Q^*in}(0)\, (T_{c0}/\Delta T)^{1/2} \tag{71}$$

where $\Delta T = T_{c0} - T$ and $\tau_{Q^*in}(0) = 0.42\, \tau_E$.

It is remarked that for deriving eq. (70) a reduction of the energy gap due to a transport current has not been considered: At the critical current of a thin wire it is $|\psi| / |\psi_0| = (2/3)^{1/2}$. Since $\psi(\underline{r}) \sim \mathbf{\Delta}(\underline{r})$, where $\mathbf{\Delta}(\underline{r})$ is the order parameter of the microscopic theory [5], also $|\mathbf{\Delta}(\underline{r})|$ is suppressed by a factor of $(2/3)^{1/2}$ below its value $|\mathbf{\Delta}_0(\underline{r})|$ in the absence of currents and magnetic fields. Interpreting $|\mathbf{\Delta}(\underline{r})|$ as the energy gap would lead to the statement that the gap $\mathbf{\Delta}$ at the critical current is suppressed by a factor of $(2/3)^{1/2}$ beyond its value given in eq. (69). Although this identification cannot generally be made (see pp. 105, 119, and 261 - 262 of ref. 2) it seems to be reasonable to assume a depression of the energy gap by a factor of $(2/3)^{1/2}$ for a homogeneous filament near T_{c0}. This can be seen from the theory [106, 107] but also from direct measurements of the energy gap at the critical current [73, 108]. Introducing the factor $(2/3)^{1/2}$ into eq.(69) would lead to $\tau_{Q^*in}(0) = 0.51\, \tau_E$.

From eq.(71) we see that τ_{Q^*in} is proportional to $\Delta T^{-1/2}$, diverging for $T \rightarrow T_{c0}$. Introducing an expression for $\langle \mathbf{\Delta} \rangle$ into eq.(67) would lead to the result that τ_{Q^*el} is at least proportional to ΔT^{-1}. Very close to T_{c0} it is even $\tau_{Q^*el} \sim \Delta T^{-2}$.

The result for τ_{Q^*in} depends on the inelastic collision time τ_E. Since τ_E is a very important quantity, a detailed discussion of this time will be given in the following.

A value for τ_E has been calculated by Tinkham [see eq. (28) of ref. 102], namely

$$\tau_E = (\tau_\Theta / 8.4)(\Theta / T_{c0})^3 \tag{72}$$

where [16, 32, 102]

$$\tau_\Theta = (\rho_\Theta \ell_\Theta / \rho_{298K} v_F)(298\,K / \Theta) \tag{73}$$

Here, Θ is the Debye temperature, and $\rho_\Theta \ell_\Theta = \rho_n \ell$ as introduced in section 2.3. Furthermore, ρ_{298K} is the phonon-induced temperature dependent part of the resistivity at room temperature. For pure metals, ρ_{298K} can be replaced with sufficient accuracy by the measured total resistivity of pure bulk material which has a very small residual resistance ratio $\rho^* = R_n / (R_{298K} - R_n) = \rho_n / \rho_{298K}$. In general (for instance for alloys) the measured total resistivity at room temperature (sometimes called ρ_{RT}) is also for bulk material $\rho_{298K,tot} = \rho_{298K} + \rho_n$ so that $\rho_{298K} = \rho_{298K,tot} / (1 + \rho^*)$, if Mathiessen's rule [26] is assumed to be valid.

Kaplan et al. [103] calculated quasiparticle lifetimes for weak and strong-coupling dirty superconductors. To plot their results in a uniform, material independent form a unit time τ_0 has been introduced. This universal form is only valid for the case where the electron-phonon coupling function $\tilde{\alpha}^2 F(\omega)$ of the Eliashberg theory is approximately proportional to ω^2, where ω is the frequency of the phonons. This should be a good approximation for weak-coupling superconductors. We explicitly remark that the time τ_0 is not a quasiparticle relaxation time, but is related to τ_E by [80]

$$\tau_E = \tau_0 / 8.4 \tag{74}$$

Similar to Tinkham's result for τ_E, the expression for τ_0 also contains a factor T_{c0}^{-3}. Kaplan et al. summarized values of τ_0 for several pure materials and some Pb based alloys (see Tab. 1 of ref. 103).

There are several methods to measure τ_E: The 'classical' arrangement is an NC1/I/SC/I/NC2 double tunnel junction. Here, 'I' denotes the insulator. A charge imbalance is generated in the superconductor, for instance by the injection of electrons from the first normal conductor (NC1) and detected as a voltage appearing between the superconductor (SC) and the second normal conductor (NC2) [56, 62, 92, 94, 105, 109]. Another tunneling injection method uses an SC1/I/SC2/I/SC3 double tunnel junction consisting of three superconductors, SC1, SC2, and SC3 which are separated by two insulators. Quasiparticles are injected from SC1 into SC2 generating a structure in the SC2/I/SC3 detector characteristic [110 - 112]. A further method using a double tunnel junction is the investigation of the enhancement of the energy gap due to quasiparticle redistribution caused by injected quasiparticles or by quasiparticle extraction [110, 113 - 115]. Very recently a novel tunneling technique has been developed to determine τ_E from the low-voltage resistance of an SC/I/NC tunnel junction [116 - 117].

Also the order parameter relaxation time, τ_Δ, which characterizes the relaxation of the Cooper pair density in the presence of a nonzero gap, depends on τ_E. There are several methods to measure the order parameter relaxation time. One method is to illuminate a superconductor which is part of a tunnel junction of two superconductors with a short laser pulse and to measure the real time response of the junction voltage to the perturbation [110, 118, 119]. Another possibility is to determine the time delay of the voltage response to a supercritical current pulse [9, 120 - 122]. Moreover, measurements of the critical direct current with a small superimposed alternating current have been performed [9, 123, 124]. In other experiments the conductivity of a superconductor is measured applying a large, chopped dc and a small continuous ac current [125 - 128].

Information about τ_E is also obtained from quasiparticle recombination time measurements [92, 103, 104, 129 - 135]. A serious problem is that the measured quasiparticle recombination times are usually enhanced over the intrinsic value by phonon trapping effects [104, 131, 132]. Furthermore, measured times are in most cases decay times of an excess quasiparticle distribution. Calculated results for the intrinsic recombination time usually represent the time τ_r for the recombination of a given quasiparticle. Since two quasiparticles vanish in each recombination event, most measured times have to be multiplied with a factor of two before comparing them with theory [103, 131] which gives the relationship between τ_r and τ_0 so that $\tau_E = \tau_0/8.4$ can be calculated. For sufficiently low temperatures and quasiparticles at the gap edge there is an analytic approximation for $\tau_0 \tau_r^{-1}$, while, generally, numerical results have to be used (see eq. (14) of ref. 103 or Fig. 2 of ref. 103).

Furthermore, experimental values for τ_E can be obtained from the differential resistance of a phase-slip center [20, 33, 40, 84, 136], spatially resolved measurements of the electrochemical quasiparticle potential near a phase-slip center [70, 72, 73], and the time evolution of the voltage at a phase-slip center [137], the boundary resistance of superconducting/normal interfaces [61, 93, 138, 139], the flux flow behaviour [140], the 'foot structures' in the characteristics of short weak links [87, 141 - 143], and the low frequency border of microwave radiation-stimulated superconductivity [144 - 146].

A summary of experimental values for τ_E is given in the appendix of the present work.

Some authors mention several experimental results and give a critical discussion [56, 92, 94, 140, 147, 148]. It turns out that for most materials the experiments deliver reasonable values for an inelastic electron-phonon scattering time. In Al properties are more complicated. The temperature dependence of τ_E is much stronger than the predicted T_{c0}^{-3} law yields and the dependence of τ_E on the electron mean free path (or residual resistivity) is not in agreement with a theory considering electron-phonon and electron-electron collisions in three-dimensional samples at zero temperature [92, 96]. A strong dependence of the measured inelastic scattering time on film thickness and sheet resistance is observed [140, 146, 148]. A probable explanation is that experiments with dirty Al films measure the inelastic electron-electron scattering time τ_{ee} as calculated for a two-dimensional sample instead of τ_ε [94, 98, 99, 140, 146, 148, 149].

In the following we will give some information about electron-electron scattering times:

In a pure three-dimensional metal the electron-electron collision rate is given by [26]

$$1/\tau_{ee} \approx v_F\, n\, \sigma_0\, (k\,T/\eta_F)^2 \tag{75}$$

Here, n is the electron density, η_F the Fermi energy and σ_0 the effective cross-sectional area for screened Coulomb interaction, that means $\sigma_0 \approx 10^{-19}\,m^2$.

The electron-electron scattering rate for an electron with energy $\varepsilon_{\underline{K}}$ in a three-dimensional metal with finite mean free path ℓ at $T = 0\,K$ is given by [96]

$$1/\tau_{ee} = (\pi\,\varepsilon_{\underline{K}}^2/8\,\hbar\,\eta_F)\,[1 + (4\sqrt{3}/\pi)\,(\hbar/p_F\ell)^{3/2}\,(\eta_F/\varepsilon_{\underline{K}})^{1/2}] \tag{76}$$

Here, $p_F = m\,v_F$ is the Fermi momentum.

For a disordered dirty two-dimensional metal it is [97, 150, 151]

$$1/\tau_{ee} = (k\,T/2\,\eta_F\,\tau)\,\ln(T_1/T) \tag{77}$$

with $T_1 = [2\,m\,e^4/k\,(4\,\pi\,\varepsilon_0)^2\,\hbar^2]\,(k_F\ell)^3$ yielding $T_1 = 6.3 \cdot 10^5\,(k_F\ell)^3\,K$. Here, it is $\tau = \ell/v_F$ and k_F the Fermi wave number. Furthermore, ε_0 denotes the influence constant. This result is obtained from eq. (3.8) of ref. 97 by inserting all quantities for the two-dimensional case [97, 150]. It is remarked that here 'dirty' means $\ell < \hbar\,v_F/k\,T$.

For thicker two-dimensional systems with a film thickness d for which still $d^2 < \hbar\,v_F\,\ell/3\,k\,T$ but already $k_F\,d > 1$ the inelastic electron-electron scattering rate is suppressed by a factor $\pi/k_F\,d$, so that [97]

$$1/\tau_{ee} = (\pi/k_F\,d)\,(k\,T/2\,\eta_F\,\tau)\,\ln(T_1/T) \tag{78}$$

with $T_1 = (32/27)\,[m\,e^4/(4\,\pi\,\varepsilon_0)^2\,2\,\hbar^2]\,(k_F\ell)^3$ yielding $T_1 = 1.9 \cdot 10^5\,(k_F\ell)^3\,K$. This value of T_1 is different from the strictly two-dimensional case because we used the expression for the diffusion constant in three dimensions [151], $D = \ell\,v_F/3$ for its calculation.

In the case of thin wires, two factors of $1/k_F\,d$ will suppress the mechanism further and $1/\tau_{ee} \sim T^{1/2}$ is predicted. This, however, seems to be in disagreement with some experiments [97].

For comparison with experiment it is of advantage to introduce the sheet resistance $R_\square$ of a square film with length L and width L and thickness d which is given by $R_\square = \rho L / d L = \rho / d$. The resistivity is given by [26], $\rho = m / n e^2 \tau$, where n is the electron density that means the number of electrons N per volume Ω.

In two dimensions it is $N = 2\pi k_F^2 / (2\pi / L)^2$, where the factor of 2 arises from the two spin directions. Furthermore, it is $\Omega = L^2 d$. Inserting all quantities leads to $R_\square = 2\pi \hbar^2 / e^2 \ell m v_F$. Since $\eta_F = m v_F^2 / 2$ it follows $1/2\eta_F \tau = e^2 R_\square / 2\pi\hbar^2$ and, for instance, eq. (78) yields

$$1/\tau_{ee} = (\pi / k_F d)(e^2 R_\square / 2\pi\hbar^2) k T \ln(T_1/T) \tag{79}$$

It is remarked that there exists another calculation of $1/\tau_{ee}$ confirming the $kT\ln(T_1/T)$ temperature dependence but yielding a prefactor which is two times smaller and a temperature T_1 which is a factor of 4 larger [152].

Finally, we return to the inelastic electron-phonon scattering time τ_E. A brief discussion of the dependence of τ_E on the electron mean free path ℓ will be given in the following [96, 151, 153, 155]:

According to Keck and Schmid [154] the electron-phonon collision rate for electrons at the Fermi level can be calculated from the electron-phonon coupling function $\tilde{\alpha}^2 F(\hbar\omega)$, where ω is the phonon frequency. The results depend on the temperature T. We replaced T by T_{c0} to obtain an expression for the electron-phonon collision rate $1/\tau_E$ at the Fermi level and at T_{c0}. We thus get [154]

$$1/\tau_E = (4\pi/\hbar) \int [\tilde{\alpha}^2 F(\hbar\omega) / \sinh(\hbar\omega / k T_{c0})] \, d(\hbar\omega) \tag{80}$$

The integral runs over all phonon energies. The integrand is very small for large phonon energies so that the borders of the integral can be chosen to be 0 and ∞. For arbitrary electron mean free path it is [154]

$$\tilde{\alpha}^2 F(\hbar\omega) = C(\hbar\omega)^2 [\phi_L(\omega\ell/c_L) + (c_L/c_T)^4 2\phi_T(\omega\ell/c_T)] \tag{81}$$

where

$$C = g_L^2 N_0 (\hbar q_D / 2 p_F k \Theta_L)^2 \tag{82}$$

$$\phi_L(\tilde{x}) = (2/\pi)[(\tilde{x} \operatorname{arctg} \tilde{x})/(\tilde{x} - \operatorname{arctg} \tilde{x}) - 3/\tilde{x}] \tag{83}$$

$$\phi_T(\tilde{x}) = (3/\pi)[\tilde{x}^{-4}(2\tilde{x}^3 + 3\tilde{x} - 3(\tilde{x}^2+1)\operatorname{arctg} \tilde{x})] \tag{84}$$

Here, c_L and c_T are the longitudinal and transverse sound velocity, respectively. Furthermore [153, 154, 156, 157], $g_L^2 = p_F^4 / 9 m^2 \rho c_L^2$ where ρ is the density and $p_F = m v_F$ is the Fermi momentum, $N_0 = m^2 v_F / 2\pi^2 \hbar^3$ is the number of electronic states per volume and energy interval at the Fermi energy for one spin direction, $k\Theta_L = \hbar q_D c_L$ with q_D the Debeye wave number. Inserting all quantities yields

$$C = (v_F^3 m^2 / 72 c_L^4 \pi^2 \hbar^3 \rho) \qquad (85)$$

For pure materials with $\ell \to \infty$ (where ℓ is the electron mean free path for impurity scattering as for instance calculated from the residual resistance) it is $\phi_L = 1$ and $\phi_T = 0$, so that $\tilde{\alpha}^2 F(\hbar\omega) = C(\hbar\omega)^2$. Then

$$1/\tau_{E\infty} = (4\pi/\hbar)\, C\, (kT_{c0})^3 \int_0^\infty [(\hbar\omega/kT_{c0})^2/\sinh(\hbar\omega/kT_{c0})]\, d(\hbar\omega/kT_{c0}) \qquad (86)$$

The numerical evaluation of the integral yields a value of 4.28, so that

$$1/\tau_{E\infty} = 4.28\, (v_F^3 m^2 / 18 c_L^4 \pi \hbar^4 \rho)\, (k T_{c0})^3 \qquad (87)$$

For materials with finite mean free path ℓ it is

$$\frac{1}{\tau_E}(\ell) = \frac{1}{4.28\,\tau_{E\infty}} \int_0^\infty \frac{\left(\frac{\hbar\omega}{kT_{c0}}\right)^2 \left[\phi_L\left(\frac{\omega\ell}{c_L}\right) + \left(\frac{c_L}{c_T}\right)^4 2\,\phi_T\left(\frac{\omega\ell}{c_T}\right)\right]}{\sinh\left(\frac{\hbar\omega}{kT_{c0}}\right)}\, d\left(\frac{\hbar\omega}{kT_{c0}}\right) \qquad (88)$$

The value of the integral depends on ℓ and has to be calculated with a computer. We evaluated $1/\tau_E$ from eq. (88) for several materials as a function of ℓ. A plot of $(\tau_E/\tau_{E\infty})(\ell)$ for Zn, Al, In, Sn, and Pb is given in the appendix, where we also tabulated $\tau_{E\infty}$ for these materials. In all cases the qualitative behaviour is similar. First $\tau_E/\tau_{E\infty}$ decreases significantly with decreasing ℓ, then it increases again for very short mean free paths. For Zn and Al this increase is very strong and occurs already for ℓ between 10 nm and 1 nm. For Sn, In, and Pb, the increase becomes weaker and happens at mean free paths well below 1 nm.

As far as known to the author there is no experimental test for the mean free path dependence of τ_E as discussed in this section. The reason may be that for Al, where a lot of measurements with samples of different electron mean free paths have been performed, electron-electron scattering is expected to dominate the electron-phonon mechanism, and no further attention was drawn to the problem of the mean free path dependence of τ_E. Therefore, in the present work we compare experimental data for τ_E as collected in Tab. A1 with the theoretical prediction for $(\tau_E/\tau_{E\infty})(\ell)$. For details see the appendix. Since the calculated values for $\tau_{E\infty}$ are much too large, these values have been determined from experiments. It turns out that evidence for the predicted behaviour can be obtained from several measurements. Nevertheless, the situation is unclear, because other experiments seem to contradict the prediction.

The inelastic electron-phonon collision time, τ_E, has been discussed somewhat in detail in this section because it is a very important parameter for the problem of charge imbalance relaxation: It appears in the expression for τ_{Q^*} which is the charge imbalance relaxation time for steady-state experiments where $\mu_{c,p}$ is time independent, so that $\dot{Q}^* = (2/\Omega) \sum_{\underline{K}} Q_{\underline{K}} \dot{\tilde{f}}_{\underline{K}}$. For non-steady state experiments with $\dot{\mu}_{c,p} \neq 0$, so that $\dot{Q}_{\underline{K}} \neq 0$, it is

$\dot{Q}^* = (2/\Omega)\sum_{\underline{K}}(Q_{\underline{K}}\dot{\tilde{f}}_{\underline{K}} + \dot{Q}_{\underline{K}}\tilde{f}_{\underline{K}})$. As the time variation of the pair chemical potential leads to a change of the effective charge for all quasiparticles there is a large effect on the charge imbalance relaxation. It is theoretically predicted [158, 159]*1 that the dynamic charge-imbalance relaxation time is τ_E rather than τ_{Q^*}. A time-resolved observation of charge imbalance decay following pulse injection of quasiparticles, for instance, should yield τ_E as relaxation time and not τ_{Q^*}.

5.5. Energy Mode

An arbitrary quasiparticle disequilibrium contains two components for which $\delta\tilde{f}_{\underline{K}}(\varepsilon_{\underline{K}})$ is odd and even, respectively [8, 9, 79]. The odd type of disequilibrium, the 'charge imbalance' or 'transverse mode', can only be excited by charged perturbations, such as particle injection. It has been discussed in detail in the previous two sections. In general these perturbations at the same time excite the even type of disequilibrium, also called 'longitudinal', 'energy', or 'temperature mode'. In the pure form the even disequilibrium is created by neutral perturbations, such as photons or phonons.

In the even type of disequilibrium the charge of the quasiparticle system is balanced. Nevertheless, the occupation probability of a given quasiparticle state deviates from equilibrium. This may be due to a redistribution of quasiparticles or by a quasiparticle overpopulation.

While a pure charge imbalance mode does not influence the magnitude of the energy gap, the energy mode changes the size of the gap from its equilibrium value. The reason is that in the charge imbalance mode the actual distribution function $\tilde{f}_{\underline{K}}$, entering the selfconsistent BCS gap equation instead of the equilibrium function $f_{\underline{K}}$, is only different from $f_{\underline{K}}$ by terms odd in $\varepsilon_{\underline{K}}$. In the energy mode there are, however, even terms giving rise to a nonvanishing contribution to the gap equation [9]. The nonequilibrium gap may be larger than in the equilibrium situation or smaller.

A stimulation of superconductivity occurs for instance if quasiparticle excitations are removed from the gap edge to higher energies by electromagnetic high-frequency radiation or phonons [9, 107, 144 - 146, 160 - 190]. The reason is that $\underline{K}$-states near K_F are more important for Cooper pair formation than states further away from K_F. If a $\underline{K}$-state is occupied by a quasiparticle excitation, it is blocked for the Cooper pair formation, which is less serious far away from K_F. The effect of this redistribution of quasiparticles is a strengthening of Cooper pair formation leading to an enhancement of the energy gap although the total number of excitations has not changed. This mechanism has been proposed by Eliashberg [191, 192].

*1 See also section 5.7

Moreover, Chang and Scalapino showed that there is an additional stimulation mechanism present. The recombination rate of quasiparticles depends on their energy, being higher for higher energies than for lower energies. This leads to a reduction of the quasiparticle density in relation to the equilibrium situation [193 - 195]. Gap enhancement can not only be obtained by high-frequency radiation but also by the injection of quasiparticles, leading again to a redistribution of excitations, and by the extraction of quasiparticle excitations [110, 113 - 115, 196, 197].

A stimulation of superconductivity by an external perturbation is the intuitively unexpected result. One would rather expect a depression of the energy gap or the critical current below their equilibrium values. A depression of the energy gap or critical current for intense quasiparticle injection has indeed been observed for tunnel junctions [198 - 206]. Another most puzzling result is, however, that the perturbed film may switch into a state with two or more different gaps [207 - 214]. There have been proposed different explanations, namely a switching between the states with different gaps [214, 215] or a spatially inhomogeneous state with simultaneously existing gaps [216 - 221]. The spatially inhomogeneous state with coexisting gaps has been detected experimentally [222, 223]. Furthermore, phonon injection from a normal conducting heater leads to a reduction of the critical current of a superconducting film which cannot be understood by simple heating [224 - 230]. Finally, illumination is a very prominent method to weaken the superconducting properties of a sample [110, 118, 119, 231 - 237].

The reason for the weakening of the superconducting properties is an overpopulation of the excitation spectrum caused by the perturbation. Thus, $\underline{K}$ states are blocked for Cooper pair formation leading to a decrease of the energy gap.

Two different quasi-thermal models have been proposed for the explanation of the properties of a superconductor in which an external source creates an overpopulation of quasiparticle excitations. These are the 'μ^* model' of Owen and Scalapino [104, 217, 219, 238, 239] and the 'T^* model' of Parker [104, 217, 219, 240].

In the μ^* model of Owen and Scalapino [238] it is assumed that the quasiparticles are in equilibrium among themselves in a Fermi-Dirac distribution characterized by a chemical potential μ^* and by a temperature T equal to the lattice or phonon temperature and also to the bath temperature if there are no simple heating effects. The chemical potential μ^* is assumed to be different from the pair chemical potential $\mu_{c,p}$. More explicitly, it is assumed that the nonequilibrium distribution function is given by $\tilde{f}_{\underline{K}} = [\exp((E_{\underline{K}} + \mu_{c,p}) - \mu^*)/kT) + 1]^{-1}$, where $E_{\underline{K}}$ is the energy of a quasiparticle excitation measured relative to the chemical potential of the condensate. In equilibrium it is $\mu^* = \mu_{c,p}$ and $\tilde{f}_{\underline{K}} = f_{\underline{K}}$ as given in eq. (31). In the literature $\mu_{c,p}$ is usually set to zero and μ^* then directly denotes the deviation of the chemical potential of the quasiparticles from $\mu_{c,p}$. The effect of a $\mu^* > \mu_{c,p}$ is to enhance the occupation probability of a given state $E_{\underline{K}}$, independent of the sign of the corresponding $\varepsilon_{\underline{K}}$. The energies $E_{\underline{K}}$ remain unchanged. Thus, μ^* is quite different from $\mu_{c,q}$ as introduced in chap. 5.3 and μ^* cannot be detected by a measurement of electrochemical potentials of pairs and quasiparticles.

The model treats the quasiparticle excitations as a Fermi gas at temperature T, but more numerous than in the complete equilibrium situation. Therefore, it must be assumed that the recombination time for quasiparticles to form pairs is much greater than the characteristic time for their thermalization to the lattice temperature T. Both times were calculated by Kaplan et al. [103] as a function of temperature for several values of the quasiparticle energy. Comparing the results of Figs. 1 and 2 of that work, one expects this assumption to be only valid for higher quasiparticle energies and temperatures far below T_{c0}.

This conclusion has also been drawn by Parker, thus developing the T^* model [240]. In this model it is again assumed that the quasiparticles are characterized by a Fermi-Dirac distribution, but now determined by a chemical potential which is equal to $\mu_{c,p}$ and by an effective temperature T^* which is greater than the bath temperature T, so that $\tilde{f}_{\underline{K}} = [\exp(E_{\underline{K}}/kT^*) + 1]^{-1}$. It is assumed that the perturbation (for instance optical radiation) increases the number of phonons with energy $\hbar\omega$ greater than 2Δ but leaves the number of phonons with energy less than 2Δ unchanged. The high energy phonons are assumed to be characterized by an effective temperature T^*, while the phonons of less energy are assumed to remain characterized by the ambient temperature T. The properties of the nonequilibrium superconductor are assumed to be the thermal equilibrium properties of an ordinary superconductor at the temperature T^*. Thus, in this model the quasiparticles and high energy phonons are in equilibrium at the temperature T^*. The assumptions of the model are valid if the time for the quasiparticles to thermalize with respect to the low-energy phonons is long compared to the intrinsic recombination time τ_r. Furthermore, there must be a phonon-trapping, so that a recombination phonon is far more likely to be reabsorbed by the superconductor with a creation of two quasiparticles than to escape from the superconductor.

Both models predict a decrease of the energy gap with increasing excess quasiparticle density. While in the μ^* model the gap shows a step-like transition to zero at high quasiparticle densities, the T^* model predicts a continuously decreasing gap [238, 240]. There are several experiments showing an agreement with the T^* model which is as excellent as the agreement with the μ^* model [240]. Furthermore, there are some experiments which agree satisfactorily with the μ^* model, and others which agree with the T^* model [9].

Chang, Lai, and Scalapino suggested an improvement of the T^* model [82, 241]. They numerically solved the coupled set of nonlinear kinetic Boltzmann equations governing the distributions of quasiparticles and phonons in a driven superconductor. For the case of tunnel injection as external drive and a weak thermal coupling between the superconductor and the temperature bath they got numerical results that can be reasonably approximated by a straightforward extension of Parker's model. They found that the T^* Fermi distribution is indeed an excellent one-parameter fit to the quasiparticle distribution. Assuming the quasiparticles Fermi distributed they solved the kinetic equations for the phonon distribution leading to a distribution function B_{T,T^*} which is a weighted average of Bose distributions $b_{T^*}(\hbar\omega, kT^*)$ and $b_T(\hbar\omega, kT)$. For $\hbar\omega > 2\Delta$ the function B_{T,T^*} is reasonably approximated by b_{T^*} as assumed in Parker's model. For $\hbar\omega < 2\Delta$ the distribution B_{T,T^*} leads only to Parker's result, b_T, if the phonon escape time, τ_{es}, is much larger than the scattering time, τ_s^{ph}, for a phonon with a quasiparticle. In general, also b_{T^*} will have a significant weight in B_{T,T^*}. A measure of the thermal coupling is the ratio of the time τ_{es} and the zero temperature pair breaking lifetime τ_0^{ph} denoted by τ_B for finite temperatures. Chang et al. assumed weak thermal coupling to be characterized by $\tau_{es} \gg \tau_0^{ph}$, implying $\tau_{es} \gg \tau_B$. For a detailed discussion of these characteristic phonon times see refs. 103, 195, and 241.

In Parker's model the effective temperature T^* is implicitly given by

$$(N/N_T)^2 \approx (T^*/T)^3 [\int_{X_G^*}^{\infty}(x^2 dx/(e^x-1))/\int_{X_G}^{\infty}(x^2 dx/(e^x-1))] \qquad (89)$$

where $X_G = 2\Delta(T)/kT$ and $X_G^* = 2\Delta(T^*)/kT^*$.

This equation connects the actual quasiparticle number per volume, N, with the effective quasiparticle temperature T^*. Here, N_T denotes the thermal equilibrium quasiparticle number per volume at the bath temperature T.

In the direct vicinity of the critical temperature (that means for T only a few millikelvin below T_{c0}) the quasiparticle number can be explicitly connected with T^*. If the sample is still in the superconducting state also T^* can only have values between T and T_{c0}. Thus, in a good approximation it is $T \approx T^*$ leading to $X_G \approx X_G^*$. Then the quotient of the integrals in eq. (89) can be set to unity so that we get simply $(N/N_T) \approx (T^*/T)^{3/2}$. Since $T^* = T + \delta T^*$ with $\delta T^* \ll T$ it is $T^*/T = 1 + \delta T^*/T$ with $\delta T^*/T \ll 1$ and, thus, $(T^*/T)^{3/2} \approx (1 + 3\delta T^*/T)^{1/2} \approx 1 + (3/2)(\delta T^*/T)$, if we neglect contributions proportional to $(\delta T^*/T)^2$ and $(\delta T^*/T)^3$, and in the last step expanded the square root. This yields, with $T \approx T_{c0}$,

$$\delta T^* \approx (2/3)\, T_{c0}\, (N/N_T - 1) \qquad (90)$$

which is a very useful approximation close to the critical temperature.

A third model was suggested by Elesin [219, 242 - 244]. It is based on the analysis and solution of kinetic equations for the quasiparticle distribution function $\tilde{f}_{\underline{K}}$ and for the phonon distribution function. The equations were solved on a computer and $\tilde{f}_{\underline{K}}$ is obtained as a function of $\varepsilon_{\underline{K}}/\Delta(0)$. The numerical results can be approximated within 5 - 10% by a Fermi distribution with μ^* and T^* as fitting parameters so that $\tilde{f}_{\underline{K}} \approx [\exp((E_{\underline{K}} - \mu^*)/kT^*) + 1]^{-1}$ if $\mu_{c,p}$ is set to zero. Plots for both parameters as a function of the strength of the external quasiparticle source ('source power') are given in Figs. 4 and 3 of refs. 219 and 243, respectively. Different curves are obtained for different sets of values for τ_{es}/τ_B and T. It is always $\mu^* < 0$ and T^* in most cases larger than T_{c0} at a finite value of the gap, indicating that in this representation T^* cannot be considered as a temperature. Experiments with double tunnel junctions confirm these results [245]. For high injection energies of the quasiparticles it is found that $\mu^* < 0$ and $T^* > T_{c0}$. Concerning the gap, it is found that $\Delta/\Delta(0)$ decreases with increasing source power. There are regions of source powers in which the gap is also pedicted to increase with increasing source power. In these regions $\Delta/\Delta(0)$ is a double valued function.

For the case of a quasiparticle distribution function with 'overheating' character (that means $\tilde{f}_{\underline{K}}(\varepsilon_{\underline{K}}=0) < f_{\underline{K}}(\varepsilon_{\underline{K}}=0)$) it is $\tilde{f}_{\underline{K}} = \tilde{f}_{\underline{K},0} + \tilde{f}_{\underline{K},1}$, where $\tilde{f}_{\underline{K},0}$ satisfies the kinetic equation for the quasiparticle distribution function with the coherence factors (see chap. 5.4) set to zero and $\tilde{f}_{\underline{K},1}$ is the 'coherence correction'. For a superconductor with small $\Delta/\Delta(0)$ it is $\tilde{f}_{\underline{K},1} = -\Delta^2\tilde{\varphi}/\pi E_{\underline{K}}$. Here, $\tilde{\varphi}$ is a function that measures the 'nonequilibrium of the system'. As T and τ_{es}/τ_B increase, $\tilde{\varphi}$ decreases, tending to zero but always remaining a positive quantity. For small Δ the coherence correction can be related to μ^* by $\mu^*/4kT_{c0} = \tilde{f}_{\underline{K},1}(\varepsilon_{\underline{K}}=0)$. For $T \to T_{c0}$ (that means $\Delta \to 0$) and $\tau_{es}/\tau_B \gg 1$, the coherence correction $\tilde{f}_{\underline{K},1}$ vanishes with the consequence that also μ^* approaches zero. In this case it is $\tilde{f}_{\underline{K},0} = \tilde{f}_{\underline{K}}$, where $\tilde{f}_{\underline{K}} \approx [\exp(E_{\underline{K}}/kT^*)+1]^{-1}$ as in the T^* model. Thus, for strong phonon trapping and temperatures close to the critical temperature the T^* model becomes a good approximation.

Finally, some remarks on the relaxation of the energy mode. When the external disturbance which brings the superconductor out of equilibrium is switched off, the superconductor will relax to the thermal equilibrium state with the thermal equilibrium value of the gap. The relaxation time for this process is the gap or order parameter relaxation time τ_Δ, given by [9, 79, 81, 87]

$$\tau_\Delta = (\pi^3/7\zeta(3))(kT_{c0}/\Delta)\tau_E \tag{91}$$

where $\zeta(3) = 1.202...$.

The result is obtained for a dirty superconductor and $\Delta \ll kT_{c0}$, that means for temperatures close to the critical temperature. The first derivation has been given by Schmid and Schön from the microscopic theory [79]. The same result can be derived without being restricted to the dirty limit within the framework of a Boltzmann equation supplemented by the BCS gap equation [65]. In that work also the result for τ_Δ in the presence of an anisotropic electron-phonon scattering rate is discussed.

Inserting the gap for a weak-coupling superconductor close to T_{c0} from eq. (69) yields

$$\tau_\Delta = 1.2\,\tau_E(1-T/T_{c0})^{-1/2} \tag{92}$$

The result is valid for a relaxation by inelastic electron-phonon collisions only. A more general result, valid for dirty superconductors close to the critical temperature has been obtained by Eckern and Schön [81, 246]

$$\tau_{\Delta} = (\pi^3/7\zeta(3))(kT_{c0}/\Delta)\tau_E[(1+\hbar^2\Gamma^2/\Delta^2)^{1/2} - \hbar\Gamma/\Delta + \hbar/2\Delta\tau_E] \qquad (93)$$

where Γ is given by eq. (60) of the present work.

5.6. Collective Excitations

Besides the charge-imbalance mode and the energy mode also collective excitation of the whole Cooper pair and quasiparticle system may occur [65 - 67, 81, 247 - 257]. While there are several sound-like propagating collective oscillations well established in neutral superfluids [248, 250, 258], none of these modes has been observed in superconductors *1.

First and fourth sound are associated with density oscillations and their frequencies are shifted up to the plasma frequency in a charged system with long-range Coulomb interaction such as a superconductor. The plasma frequency is little modified by the superconducting transition [250], so that these oscillations are just plasma oscillations which do not differ from those in the normal state. Second sound-like excitations are restricted to temperatures extremely close to T_{c0} (the range is only of order $10^{-5}\,T_{c0}$) [248, 250].

For the case of a (hypothetical) neutral superconductor oscillations of the phase of the order parameter with an acoustic dispersion relation have been predicted ('Anderson-Bogolubov mode'). Due to the periodical variation in time and space of the phase, supercurrents are flowing, leading to a spatial variation of the superfluid density. Thus, for a charged superconductor this mode will also lead to plasma oscillations.

Moreover, oscillations of the magnitude of the order parameter at a frequency $\omega_{OP} = 2\Delta/\hbar$ have been predicted by Schmid [250, 259]. Also exciton-like modes corresponding to excited pair states [2, 52, 53] have been discussed [247]. However, it does not exist any conclusive experimental evidence for these modes [250].

The only collective excitation which is well established by theory [65 - 67, 81, 250, 254 - 256] and experiment [249, 251 - 253] is the mode with an acoustic dispersion relation found by Carlson and Goldman. Described in

---

*1 In superfluid ^{4}He there are different sound-like excitations. First sound is a density wave in which the superfluid and the normal fluid component move in phase with the same velocity. In the second sound mode both components move in the opposite direction so that the total current vanishes and, thus, the density remains constant in space. Third sound is a surface wave. Fourth sound is a motion of the superfluid component only, due to a strong damping of the normal component.

terms of a two fluid model it is a propagating wave in which the density of the superfluid part (Cooper pairs) oscillates. The resulting space charge induces a counterflow of the normal part (quasiparticles). To preserve overall charge neutrality, from regions of enhanced Cooper pair density more electron-like quasiparticles are removed into regions of lowered Cooper pair density and more hole-like excitations are flowing vice versa. The resulting total space charges are so small that they can be neglected in the continuity equation [250]. The motion of the normal part is responsible for the damping of the wave. This damping is small close to T_{c0}, where the superfluid density is small and where the superfluid oscillations can, thus, be compensated by small oscillations of the normal part.

Although in the Carlsson-Goldman mode normal current and supercurrent are flowing in opposite directions, the mode is no second sound phenomenon. In the second sound mode, the matter is locally in thermodynamic equilibrium and the basic restoring force is due to temperature changes [248, 258]. In the Carlsson-Goldman mode there are departures from local equilibrium, because the chemical (and electrochemical) potentials of pairs and quasiparticles differ at a given locus. The reason is that the charge of the quasiparticle system is out of balance. In a region of enhanced Cooper pair density there is an excess of more hole-like over more electron-like excitations so that $Q^* > 0$. In a region of lowered Cooper pair density, more electron-like excitations dominate and it is, thus, $Q^* < 0$. The restoring force in the Carlsson-Goldman mode is given by the gradient of the Cooper pair chemical potential [65, 250] or, in other words (if T is close to T_{c0} and $\mu_{c,q}$ is a meaningful quantity), by differences between the chemical potentials of quasiparticles and Cooper pairs [67].

To give more quantitative results, close to T_{c0} the Carlsson-Goldman mode can be described as a longitudinal oscillation, where the space and time dependent displacement amplitude $u_s(\underline{r}, t)$ of the superfluid part is given by [250, 254]

$$u_s(\underline{r}, t) = u_s(0,0)\exp(i\,\underline{q}_{CG}\cdot\underline{r} - i\,\omega_{CG}\,t) \tag{94}$$

where

$$\omega_{CG} = -iF \pm (G - F^2)^{1/2} \tag{95}$$

For arbitrary electron mean free path ℓ it is [250]

$$F = (1/2)(n_{se}/n)(v_F/\ell) \tag{96}$$

and

$$G = c_{CG}^2\, q_{CG}^2 \tag{97}$$

with

$$c_{CG}^2 = (n_{se}/n)(4kT/\pi\Delta)(v_F^2/3) \tag{98}$$

Here c_{CG} is the phase velocity of the collective mode in the limit of a bare wave with vanishing damping ($F \to 0$) and q_{CG} is the absolute value of the wave number vector $\underline{q}_{CG}$ of the collective excitation. In the limit of a bare wave also the group velocity is given by c_{CG}. The ratio of the superfluid density $n_{se} = 2n_s$ (the number of electrons bound to Cooper pairs per volume, 'super-electron density') and the total electron density, n, is given by [250]

$$n_{se}/n = (7\zeta(3)/4\pi^2)(\Delta/kT)^2\chi \tag{99}$$

where χ is given by eq.(8) of the present work. The last equation can be obtained from the GL theory (see eq.(4) in section 5.1.) for the case of a space independent order parameter which is identified with the energy gap. Then $n_{se} = 2|\psi|^2$ and $n = m^3 v_F^3/3\pi^2\hbar^3$ can be inserted from the free electron model [26].

In the dirty limit ($\ell \ll \xi_0$) it is $\chi = (0.752\,\xi_0/\ell)^{-1}$, leading to $F = \pi\Delta^2/4\hbar kT$ and $c_{CG} = (2D\Delta/\hbar)^{1/2}$ as derived by Schmid and Schön [254], where $D = \ell v_F/3$. In the clean limit ($\ell \gg \xi_0$) it is $\chi = 1$, leading to $F = (7\zeta(3)/8\pi^2)(\Delta/kT)^2(v_F/\ell)$ $= (1 - T/T_{c0})(v_F/\ell)$ and $c_{CG} = (7\zeta(3)\,v_F^2\Delta/3\pi^3 kT)^{1/2}$ as given by Pethick and Smith [65]*1, where c_{CG} is in agreement with the result of Artemenko and Volkov [255].

The parameter F describes the damping of the mode. As can be seen from eq.(96) the damping is small for temperatures close to T_{c0}, where $n_s/n \ll 1$. Furthermore, the damping is predicted to be smaller in clean samples than in dirty specimens.

Due to the damping, the mode only propagates short distances. An estimate for the decay length λ_d can be made, considering that for small damping as well the phase velocity of the mode, $(G - F^2)^{1/2}/q_{CG}$, as also its group velocity, $(d/dq_{CG})(G - F^2)^{1/2}$, are approximately given by c_{CG}. Thus, the mode propagates with almost the bare wave velocity c_{CG}, while the absolute value of $u_s(\underline{r},t)$ decreases by a factor $1/e$ in the time $\tau_d = 1/F$, leading to the estimate $\lambda_d \approx c_{CG}/F$ as given in ref. 250. Then it follows from eqs. (96), (98), and (99) for arbitrary electron mean free path ℓ

$$\lambda_d \approx [(64\pi/7\zeta(3))(D\ell/v_F)(kT/\Delta)^3(1/\chi)]^{1/2} \tag{100}$$

In the dirty limit it is $\lambda_d = [32\hbar D(kT)^2/\pi^2\Delta^3]^{1/2}$ and in the clean limit it is $\lambda_d = [(64\pi/7\zeta(3))(D\ell/v_F)(kT/\Delta)^3]^{1/2}$. To give an impression of its magnitude, the length λ_d may be compared with the charge imbalance relaxation length due to inelastic electron-phonon collisions $\Lambda_{Q^*in} = (D\tau_{Q^*in})^{1/2}$. With τ_{Q^*in} from eq.(61) and λ_d from eq.(100) it follows

$$\lambda_d/\Lambda_{Q^*in} = [(16\pi^2/7\zeta(3))(kT_{c0}/\Delta)^2(\ell/v_F)(1/\chi\tau_E)]^{1/2} \tag{101}$$

---

*1 For $T \to T_{c0}$, the parameter β of that work becomes zero. Then F is obtained from eq.(5.67) and c_{CG} is given by eq.(5.68) (or p. 161) of that publication.

For samples in the dirty limit λ_d is usually smaller than Λ_{Q^*in} although the temperature may be only a few millikelvins below T_{c0}. In the clean limit, however, λ_d may exceed Λ_{Q^*} by several times, if the temperature is some millikelvins below T_{c0}.

In the results given above the only damping mechanism is the motion of the quasiparticles which is dissipative due to scattering from impurities. Other damping mechanisms such as, for instance, a diffusion of quasiparticles are neglected. A more general calculation (in the dirty limit) leads to the following results, valid close to the critical temperature [250],

$$F = (1/2)\{(n_{se}/n)(v_F/\ell)+(Dq_{CG}^2/\pi)[2\ln(8\Delta/\hbar\omega_{CG})-2]+\tau_E^{-1}+\Gamma_0\} \qquad (102)$$

$$c_{CG}^2 = (2D\Delta/\hbar)(1-2Dm^2v_s^2/\Delta) \qquad (103)$$

with $\Gamma_0=\Gamma-1/2\tau_E$, where Γ is given by eq.(60). The second term in eq.(102) reflects damping by quasiparticle diffusion. Furthermore, inelastic electron-phonon collisions and other pair breaking mechanisms (as summarized in Γ_0) further increase the damping. The velocity c_{CG} is reduced if a supercurrent is flowing through the sample.

5.7. Dynamics of Charge Imbalance

In the previous sections several ingredients of a nonequilibrium state in a superconductor have been discussed: There may be a charge imbalance in the quasiparticle system, leading to differences between the chemical (and electrochemical) potentials of pairs and quasiparticles. In steady-state situations, where $\mu_{c,p}$ is independent of time, the charge imbalance relaxes with a typical time τ_{Q^*}. If $\mu_{c,p}$ is time dependent, τ_E rather than τ_{Q^*} governs the relaxation. Furthermore, the energy mode may be excited, leading to a change of the magnitude of the energy gap. Next, there is the possibility of a collective excitation (Carlsson-Goldman mode) which is a propagating fluctuation of the Cooper pair density, damped by a counterflow of quasiparticle charge. The frequency of this mode is assumed to be so large that charge imbalance relaxation within a cycle may be neglected. In addition there may be simple heating.

In an arbitrary nonequilibrium all phenomena may be excited together. Nobody has considered this case. However, there are descriptions containing charge imbalance relaxation and collective excitations in situations where $\mu_{c,p}$ is not necessarily independent of time [66, 67, 158, 250, 260, 261]. A very compact formulation within a two-fluid picture was given by Kadin, Smith, and Skocpol (KSS) [158], resulting in a single differential equation in space and time for the quasiparticle charge imbalance Q^*. This equation can be derived as follows:

As mentioned at the end of section 5.4., for the nonstationary case with $\mu_{c,p} \neq 0$ it follows from eq. (33), inserting $Q_{\underline{K}}$ from eq. (24),

$$\dot{Q}^* = (2/\Omega)\sum_{\underline{K}}(Q_{\underline{K}}\dot{\tilde{f}}_{\underline{K}} + \dot{Q}_{\underline{K}}\tilde{f}_{\underline{K}}) \tag{104}$$

Since $Q_{\underline{K}} = (\varepsilon_{\underline{K}}/E_{\underline{K}})(-e)$, with $E_{\underline{K}} = (\varepsilon_{\underline{K}}^2 + \Delta^2)^{1/2}$ and $\varepsilon_{\underline{K}} = \eta_{\underline{K}} - \mu_{c,p}$, it is $\dot{Q}_{\underline{K}} = (-e)\dot{\mu}_{c,p}(\partial/\partial\mu_{c,p})(\varepsilon_{\underline{K}}/E_{\underline{K}}) = e\dot{\mu}_{c,p}\Delta^2/E_{\underline{K}}^3$. Here, $\mu_{c,p}$ is the actual 'local' value of the chemical potential of the condensate. Furthermore, eq. (36) yields $\dot{\mu}_{c,p} = \dot{Q}^*/2eN_0$. Thus,

$$\dot{Q}^* = (2/\Omega)(\sum_{\underline{K}} Q_{\underline{K}}\dot{\tilde{f}}_{\underline{K}})/[1-(1/\Omega N_0)(\sum_{\underline{K}}\tilde{f}_{\underline{K}}\Delta^2/E_{\underline{K}}^3)] \tag{105}$$

The evaluation of the denominator for temperatures close to T_{c0} yields $\pi\Delta/4kT$. For details see section A.4 of the appendix. Thus,

$$\dot{Q}^* = (4kT/\pi\Delta)[(2/\Omega)\sum_{\underline{K}} Q_{\underline{K}}\dot{\tilde{f}}_{\underline{K}}] \tag{106}$$

In general, the factor in square brackets is equal to the sum of all mechanisms leading to a change of $\tilde{f}_{\underline{K}}$. For a relaxation of charge imbalance by inelastic scattering, a relaxation time approximation, $\dot{\tilde{f}}_{\underline{K}} = -\delta\tilde{f}_{\underline{K}}/\tau_{Q^*}$, may be used. Considering eq. (37) this results in

$$\dot{Q}^* = (4kT/\pi\Delta)(-Q^*/\tau_{Q^*}) \tag{107}$$

Since $\delta\tilde{f}_{\underline{K}} = \tilde{f}_{\underline{K}} - f_{\underline{K}}(E_{\underline{K}})$, this equation describes how $\tilde{f}_{\underline{K}}$ relaxes to the local equilibrium distribution characterized by $f_{\underline{K}}(E_{\underline{K}})$. For the case of relaxation due to inelastic electron-phonon scattering only, it is $\tau_{Q^*} = \tau_{Q^*in}$ given by eq. (61), yielding $\dot{Q}^* = -Q^*/\tau_E$, as already discussed at the end of section 5.4. As the distribution function is also changed if there is an outflow of quasiparticles due to a 'normal' current density, $\underline{j_n}$, a term $\underline{\nabla}\cdot\underline{j_n}$ enters eq. (107). Thus, it is

$$\dot{Q}^* = (4kT/\pi\Delta)(-Q^*/\tau_{Q^*} - \underline{\nabla}\cdot\underline{j_n}) \tag{108}$$

This is the first of the basic equations needed for the derivation of the desired differential equation.

The second basic equation needed relates $\underline{j_n}$ with $\underline{\nabla}Q^*$ and $(d/dt)\underline{j_s}$. It can be derived as follows: We start from eq. (49) and insert $\dot{\varphi}$ from eq. (53). Considering that μ_F is space independent and multiplying the whole equation with the normal conductivity [26], $\sigma = ne^2\tau/m$, where $\tau = \ell/v_F$, yields

$$\tau_{0R}\, d\underline{j_s}/dt = -\sigma\dot{\underline{A}} + (\sigma/e)\underline{\nabla}\mu_p \tag{109}$$

where $\tau_{0R} = \sigma\Lambda_L = (n/n_{se})\tau$.

From eq. (45) it follows that $\mu_p \approx Q^*/2eN_0 + \mu$ for temperatures close to T_{c0}. With $\sigma/2e^2N_0 = \ell v_F/3 = D$ we get

$$\tau_{0R}\, d\underline{j_s}/dt = -\sigma\dot{\underline{A}} + (\sigma/e)\underline{\nabla}\mu + D\underline{\nabla}Q^* \tag{110}$$

From section 5.3 we know that $\mu = \mu_{c,q} - e\Phi$ with $\mu_{c,q} \approx \mu_F$. Thus, it is $\underline{\nabla}\mu \approx -e\underline{\nabla}\Phi$. Using furthermore Maxwell's equation (see again section 5.3) leads to $-\sigma\dot{\underline{A}} + (\sigma/e)\underline{\nabla}\mu \approx \sigma\underline{E}$. For temperatures close to T_{c0}, where Δ is small, most of the quasiparticles are hole-like or electron-like so that the identification $\sigma\underline{E} = \underline{j_n}$ is allowed. It then follows

$$\tau_{OR}\, d\underline{j_s}/dt - \underline{j_n} = D\underline{\nabla}Q^* \tag{111}$$

as the second basic equation.

This equation shows that τ_{OR} is the equilibrium response time of the supercurrent to a change in the total current: In the equilibrium case where $Q^* = 0$ a sudden change in the total constant current, $\underline{j} = \underline{j_s} + \underline{j_n}$, will initially be carried as a normal current. The reason is that quasiparticles and Cooper pairs are accelerated and there are much more quasiparticles than pairs present if T is close to T_{c0}. The normal current will then decay into supercurrent due to eq. (111), exponentially with the decay time τ_{OR}.

With n_{se}/n as given in eq. (99) the expression for τ_{OR} yields

$$\tau_{OR} = (4\pi^2/7\zeta(3))\,(kT/\Delta)^2\chi^{-1}\tau \tag{112}$$

In the clean limit, where $\chi = 1$, it is thus $\tau_{OR} = \tau/2(1-T/T_{c0})$. Here, we inserted the expression for Δ from eq. (69). In the dirty limit, where $\chi = (0.752\,\xi_0/\ell)^{-1}$, it follows $\tau_{OR} = 2kT_{c0}\hbar/\pi\Delta^2$. It is remarked that $\tau_{OR} = 1/2F$, where F is the damping parameter of the Carlsson-Goldman mode discussed in section 5.6.

It may be comfortable to insert Δ from eq. (69) into eq. (112), yielding $\tau_{OR} = (\ell/2v_F\chi)(T_{c0}/(T_{c0}-T))$ as the expression for the supercurrent response time for arbitrary electron mean free path.

Now, we calculate $\underline{\nabla}\cdot(d\underline{j_n}/dt)$ from the time derivative of eq. (108) and $\underline{\nabla}\cdot(d\underline{j_s}/dt)$ from the divergence of eq. (111). Then we use $\underline{\nabla}\cdot\underline{j} = 0$ due to electroneutrality, leading to $\underline{\nabla}\cdot(d\underline{j_n}/dt) + \underline{\nabla}\cdot(d\underline{j_s}/dt) = 0$. Into the resulting equation we insert $\underline{\nabla}\cdot\underline{j_n}$ from eq. (108), getting the differential equation desired,

$$D\tau_{Q^*}\underline{\nabla}^2 Q^* = \tau_{OR}\tau_\alpha\ddot{Q}^* + (\tau_{OR} + \tau_\alpha)\dot{Q}^* + Q^* \tag{113}$$

with $\tau_\alpha = (\pi\Delta/4kT)\tau_{Q^*}$, where $T \approx T_{c0}$. For the case of charge imbalance relaxation due to inelastic electron-phonon collisions only it is $\tau_{Q^*} = \tau_{Q^*in}$, as given in eq. (61), leading to $\tau_\alpha = \tau_E$.

Thus, we have got a charge imbalance wave equation, describing propagating damped dispersive waves of charge imbalance, given by

$$Q^*(\underline{r}, t) = Q^*(0,0)\exp[(\pm)(i\omega_{KSS}t - i\underline{q}_{KSS}\cdot\underline{r})] \tag{114}$$

with the dispersion relation

$$-\Lambda_{Q*}^{2}\,\underline{q}_{KSS}^{2} = (1\,(\pm)\,i\,\omega_{KSS}\,\tau_{OR})\,(1\,(\pm)\,i\,\omega_{KSS}\,\tau_{\alpha}) \tag{115}$$

where $\Lambda_{Q*} = (D\,\tau_{Q*})^{1/2}$, leading to

$$\omega_{KSS} = (\pm)\,i\,F_{KSS} \pm (G_{KSS} - F_{KSS}^{2})^{1/2} \tag{116}$$

where

$$F_{KSS} = (\tau_{OR} + \tau_{\alpha})/2\,\tau_{OR}\,\tau_{\alpha} \tag{117}$$

and

$$G_{KSS} = 1/\tau_{OR}\,\tau_{\alpha} + \Lambda_{Q*}^{2}\,q_{KSS}^{2}/\tau_{OR}\,\tau_{\alpha} \tag{118}$$

Here, q_{KSS} is the absolute value of the wave number vector, $\underline{q}_{KSS}$, of the charge imbalance wave.

In the exponent of eq. (114) we introduced the two signs, $(\pm)$. The positive sign belongs to the solution given in the work of KSS, while the negative sign leads to a solution as used in section 5.6 for the description of the Carlsson-Goldman mode which we will compare our results with: For $\tau_{\alpha} \gg \tau_{OR}$, it is $F_{KSS} \approx 1/2\,\tau_{OR} = F$. If, furthermore, $\Lambda_{Q*}^{2}\,q_{KSS}^{2} \gg 1$, it follows $G_{KSS} \approx \Lambda_{Q*}^{2}\,q_{KSS}^{2}/\tau_{OR}\,\tau_{\alpha}$. Setting $G_{KSS} = c_{KSS}^{2}\,q_{KSS}^{2}$, we get $c_{KSS}^{2} \approx \Lambda_{Q*}^{2}/\tau_{OR}\,\tau_{\alpha} = c_{CG}^{2}$. The last identity follows by inserting the definition of τ_{α} and using $\tau_{OR} = (n/n_{se})\,\tau$.

Thus, the charge imbalance wave equation leads to a collective mode with a damping parameter F_{KSS} and a bare wave velocity c_{KSS} which are equal to the corresponding quantities of the Carlsson-Goldman mode in the case that the only damping mechanism is the motion of quasiparticles which are scattered at impurities. For the case of charge imbalance relaxation due to electron-phonon collisions only, it is $\tau_{\alpha} = \tau_{E}$ and the condition $\tau_{\alpha} \gg \tau_{OR}$ means that there are no inelastic electron-phonon collisions within the supercurrent response time. As can be seen from eq. (115), the condition $\Lambda_{Q*}^{2}\,q_{KSS}^{2} \gg 1$ is fulfilled if $\omega_{KSS}\,\tau_{OR} \gg 1$ and $\omega_{KSS}\,\tau_{\alpha} \gg 1$, that means in the 'high frequency limit', $\omega_{KSS} \gg \tau_{OR}^{-1}, \tau_{\alpha}^{-1}$.

In the 'low frequency limit', $\omega_{KSS} \ll \tau_{OR}^{-1}, \tau_{\alpha}^{-1}$, it follows from eq. (115) that $q_{KSS}^{2} = -1/\Lambda_{Q*}^{2}$, leading to $q_{KSS} = \pm i\,\Lambda_{Q*}^{-1}$. The same result is obtained from eq. (116) for $\omega_{KSS} \to 0$. Thus, there is a solution for $Q^{*}(\underline{r}, t)$ describing exponential charge imbalance decay with the characteristic decay length Λ_{Q*}. For $\omega_{KSS} \to 0$ the solution becomes completely independent of time and is for $\underline{q}_{KSS}$ parallel to $\underline{r}$ given by $Q^{*}(\underline{r}) = Q^{*}(0)\exp(-r/\Lambda_{Q*})$.

The conclusions drawn for the limiting cases can also directly be seen by discussing the limits of the charge imbalance wave equation: Due to eq. (114) the term $\sim \ddot{Q}^{*}$ is a contribution of order $\tau_{OR}\,\tau_{\alpha}\,\omega_{KSS}^{2}$, while the terms $\sim \dot{Q}^{*}$ and $\sim Q^{*}$ are contributions of order $(\tau_{OR} + \tau_{\alpha})\,\omega_{KSS}$ and unity, respectively. Thus, in the high frequency limit the term $\sim \ddot{Q}^{*}$ dominates leading to $D\,\tau_{Q*}\,\underline{\nabla}^{2}\,Q^{*} \approx \tau_{OR}\,\tau_{\alpha}\,\ddot{Q}^{*}$. This equation is solved by

$Q^*_h(\underline{r},t)=Q^*(0,0)\exp(i c_{KSS,h} q_{KSS} t - i \underline{q}_{KSS}\cdot\underline{r})$, with $c_{KSS,h}=\Lambda_{Q*}/(\tau_{OR}\tau_\alpha)^{1/2}$. In the low frequency limit the term $\sim Q^*$ in the charge imbalance wave equation dominates, leading to $D\tau_{Q*}\underline{\nabla}^2 Q^* \approx Q^*$. This equation is solved by $Q^*_l(\underline{r},t)= Q^*(0,0)\exp(-i\underline{q}_{KSS,l}\cdot\underline{r})$ with $q_{KSS,l}=\pm i\Lambda^{-1}_{Q*}$.

Next, some remarks are made concerning the work of Kulik[262], who derived a system of equations governing the dynamics of superconductivity in the electric field and considered the interaction of electrons with the Carlsson-Goldman collective oscillations of the order parameter. Kulik quantizes the oscillator so that the collective oscillations become quantized 'particles' which he calls 'bogolons', and studies the influence of electron-bogolon scattering on the Carlsson-Goldman mode. The system of equations contains two unknown variables, the phase φ of the order parameter and the scalar potential $\Phi(\underline{r},t)$ of the electromagnetic field. Solving for φ, the equation system leads to an equation for $\Phi(\underline{r},t)$, given by

$$\underline{\nabla}^2\Phi(\underline{r},t)-\Lambda^{-2}_{Ku}(\omega_{Ku})\Phi(\underline{r},t) = 0 \qquad (119)$$

where ω_{Ku} is the frequency of the collective mode and $\Lambda_{Ku}(\omega_{Ku})$ is the decay length of the scalar potential (or the penetration depth of the electric field into the superconductor) given by

$$\Lambda^2_{Ku}(0)/\Lambda^2_{Ku}(\omega_{Ku}) = (1-i\omega_{Ku}\tau_{eb})(1-i\omega_{Ku}\tau_{OR}) \qquad (120)$$

where

$$\Lambda^2_{Ku}(0) = (n/m N_s \tau^{-1}\tau^{-1}_{eb}) \qquad (121)$$

with $N_s=(\pi\Delta/2kT)N_0$ and $\tau=\ell/v_F$. Furthermore, τ^{-1}_{eb} is the electron-bogolon scattering rate given by $\tau^{-1}_{eb}=\tau^{-1}_{ebO}(kT/\Delta)$ where $\tau^{-1}_{ebO}=8\pi k^2T^2/\eta_F p_F \ell$ for a three-dimensional metal and $\tau^{-1}_{ebO}=8\pi(k^3T^3/\eta_F)^{1/2}(p_F d)^{-1}(\hbar/p_F\ell)^{1/2}$ for a film with thickness $d \ll \xi_0$. It is remarked that $\Lambda^2_{Ku}(0)=c^2_{CG}\tau_{OR}\tau_{eb}$ follows from eq. (121).

With

$$\Phi(\underline{r},t) = \Phi(0,0)\exp(i\underline{q}_{Ku}\cdot\underline{r}-i\omega_{Ku}t) \qquad (122)$$

we get from eq. (119) $q^2_{Ku}=-\Lambda^{-2}_{Ku}(\omega_{Ku})$ and with eq. (120)

$$-q^2_{Ku}\Lambda^2_{Ku}(0) = (1-i\omega_{Ku}\tau_{eb})(1-i\omega_{Ku}\tau_{OR}) \qquad (123)$$

in complete analogy to eq. (115).

Thus, we get

$$\omega_{Ku} = -iF_{Ku}\pm[G_{Ku}-F^2_{Ku}]^{1/2} \qquad (124)$$

where

$$F_{Ku} = (\tau_{OR}+\tau_{eb})/2\tau_{OR}\tau_{eb} \qquad (125)$$

and

$$G_{Ku} = 1/\tau_{0R}\tau_{eb} + \Lambda^2_{Ku}(0)\, q^2_{Ku}/\tau_{0R}\tau_{eb} \qquad (126)$$

It should be noted that for $\tau_{eb} \gg \tau_{0R}$ we have $F_{Ku} = 1/2\tau_{0R} = F$ and for $\Lambda^2_{Ku}(0)\, q^2_{Ku} \gg 1$ we get $G_{Ku} \approx \Lambda^2_{Ku}(0)\, q^2_{Ku}/\tau_{0R}\tau_{eb}$. Setting $G_{Ku} = c^2_{Ku} q^2_{Ku}$ it follows that $c^2_{Ku} = \Lambda^2_{Ku}(0)/\tau_{0R}\tau_{eb} = c^2_{CG}$.

The results of Kulik are very similar to those of KSS. In Kulik's results the mode is damped by the motion of quasiparticles scattered at impurities and by electron-bogolon scattering, characterized by τ_{0R} and τ_{eb}, respectively. In the KSS formulation τ_{0R} and τ_α determine the damping. Kulik characterizes the charge imbalance by $\Phi(\underline{r}, t)$, while KSS use $Q^*(\underline{r}, t)$. Both quantities are closely related, because $\Phi(\underline{r}, t) = Q^*/2e^2 N_0 - (1/e)(\mu_p - \mu_F)$, according to eqs. (25) and (36). As long as $\dot{\varphi}$ is zero or constant, the difference $\mu_p - \mu_F$ is zero or constant and $\Phi(\underline{r}, t)$ is proportional to $Q^*(\underline{r}, t)$.

In the high frequency limit, $\omega_{Ku} \gg \tau_{0R}^{-1}, \tau_{eb}^{-1}$, the relation $\Lambda^2_{Ku}(0)\, q^2_{Ku} \gg 1$ is fulfilled, leading to a damped propagating mode. In the low frequency limit, $\omega_{Ku} \ll \tau_{0R}^{-1}, \tau_{eb}^{-1}$, it is $q^2_{Ku} = -1/\Lambda^2_{Ku}(0)$, describing an exponential decay of $\Phi(\underline{r}, t)$, with the characteristic decay length $\Lambda_{Ku}(0)$.

It is remarked that the decay length $\Lambda_{Ku}(0)$ inserting τ_{eb} for the three-dimensional case can be expressed as $\Lambda_{Ku}(0) = (3^{1/2} D m/\hbar)\, \xi_0(T_{c0})$. Thus, it is $\Lambda_{Ku}(0) \sim \ell$ and does not diverge for $T \to T_{c0}$, whereas $\Lambda_{Q^*} = (D\tau_{Q^*})^{1/2}$ is proportional to $\ell^{1/2}$ and diverges like $(1 - T/T_{c0})^{-1/4}$ for T approaching T_{c0}. The ratio of both lengths is given by $\Lambda_{Ku}(0)/\Lambda_{Q^*} = 1.55\,(m/\hbar)(\ell v_F/\tau_E)^{1/2}\, \xi_0(T_{c0})(1 - T/T_{c0})^{1/4}$, if only inelastic electron-phonon scattering is considered in τ_{Q^*}. We calculated the ratio, using bulk material values, assuming $\ell = 1\,\mu m$ and $T_{c0} - T = 5\,mK$, and estimating τ_E after Tinkham. It is found that $\Lambda_{Ku}(0)/\Lambda_{Q^*} = 74, 25, 35, 24$, and 10 for Pb, Sn, In, Al, and Zn, respectively. Thus, charge imbalance relaxation due to inelastic electron-phonon scattering seems to be dominant in these materials. For a further discussion of Kulik's results see section 7.2.

Finally, the conclusion of this section is that propagating modes and exponential charge imbalance decay are two limiting cases of the charge imbalance dynamics in a superconductor.

5.8. Models for a Phase-Slip Center

5.8.1. The Phase-Slip Phenomenon

The first step to come to an understanding of the nature of the dissipative state in a quasi-one-dimensional superconductor is to imagine what happens to the superconducting state in a superconductor with a potential difference between its ends.

As has been discussed in detail in section 5.3., the time evolution of the phase of the superconducting order parameter $\psi(\underline{r}) = |\psi(\underline{r})| \exp(i\varphi(\underline{r}))$ is different at any two places $\underline{r}_1$ and $\underline{r}_2$ with an electrostatic potential difference $V_{12}(t)$, and the time development of the phase difference $\varphi_{12} = \varphi(\underline{r}_1) - \varphi(\underline{r}_2)$ is given by the Josephson relation $\dot{\varphi}_{12} = 2eV_{12}(t)/\hbar$, leading to

$$\varphi_{12}(t) = \varphi_{12}(0) + (2e/\hbar)\int_0^t V(t')\,dt' \tag{127}$$

Here and in the following we dropped the index '1 2' at the voltage.

If the time average of the voltage is nonzero, but for instance $\langle V(t)\rangle > 0$, the phase difference increases as a function of time. This leads to an increase of $\underline{\nabla}\varphi(\underline{r})$ and thus to an increase of the superfluid velocity $\underline{v}_s = (\hbar/m')\underline{\nabla}\varphi(\underline{r})$. The consequence is an increase of the supercurrent density which is given by $\underline{j}_s = -(2e/m')|\psi(\underline{r})|^2\hbar\underline{\nabla}\varphi(\underline{r})$, as obtained from eq. (2) by neglecting the vector potential. However, with increasing $\underline{v}_s$ the absolute value of the order parameter decreases[2]. In the homogeneous case, where $|\psi(\underline{r})| = |\psi|$ is space independent, the critical current density will be reached as soon as $|\psi|^2/|\psi_0|^2 = 2/3$ and the superconductor will enter the normal conducting state.

To avoid this, a phase-loss mechanism is needed to reduce the phase difference. For this purpose 'phase-slip processes' are assumed to occur in the sample, each reducing the phase difference $\varphi_{12}(t)$ by 2π. A stationary state can be reached if in time average the phase difference generated by the voltage is destroyed by phase-slip processes. The period τ_{PSC} of the phase-slip process is given by

$$2\pi = \varphi_{12}(\tau_{PSC}) - \varphi_{12}(0) = (2e/\hbar)\int_0^{\tau_{PSC}} V(t')\,dt' = (2e/\hbar)\langle V(t)\rangle\,\tau_{PSC} \tag{128}$$

The repetition frequency of the phase-slip process is then

$$\omega_{PSC} = 2\pi/\tau_{PSC} = (2e/\hbar)\langle V(t)\rangle \tag{129}$$

which is just the Josephson frequency [68, 263].

Already Langer and Ambegaokar invoke phase-slip processes for a description of the fluctuation governed phase transition from the superconducting to the normal state in a quasi-one-dimensional superconductor[264]. They pointed out that the order parameter before and after the phase-slip is a solution of the GL equations, but belonging to different local minima of the GL free enthalpy. Langer and Ambegaokar calculated the free energy barrier which the system has to overcome before it can change to another local minimum. For vanishing transport current this barrier is given by $\delta F_0 = (8\sqrt{2}/3)(B_{cth}^2/2\mu_0)A\xi(T)$, where A is the cross sectional area of the sample. Here, $B_{cth}^2/2\mu_0$ is the difference of the free enthalpy of the normal and the superconducting state per volume, and $A\xi(T)$ is a volume of length $\xi(T)$ of the superconductor.

The result for δF_0 leads to the interpretation that the fluctuating region of reduced order parameter extends over some $\xi(T)$. In its center the absolute value of the superconducting order parameter becomes very small in a region of length $\sim\xi(T)$ and, for a small constant current $\underline{j}_s$, thus, $\underline{\nabla}\varphi(\underline{r})$

becomes very large. In a three-dimensional graphical representation (with axes Im ψ, Re ψ, and space coordinate, x), the windings of the complex order parameter would become very close in this region[2]. Now a certain distance within the region of small order parameter becomes completely normal for a moment, thus decoupling the phases to the left and to the right. Then the superconducting state over the 'normal distance' reastablishes, but with an order parameter which has lost one winding that means with a phase difference which has changed by 2π [2, 263]. After the fluctuation has vanished, the order parameter has thus changed its phase by 2π.

In this description the phase-loss occurs somewhere in the fluctuating region. Only phase changes of $\pm 2\pi$ (or multiples $\pm \tilde{n}\, 2\pi$) are possible, because the reestablished order parameter has to join the order parameter on both sides of the formerly normal conducting distance with correct magnitude and phase. It is not clear how large the normal distance really is. It may extend over $\sim \xi(T)$ in a real fluctuating situation. On the other hand, the space needed for one winding of the order parameter would be enough. This space may shrink to one point in an ideal situation where $|\psi(\underline{r})|$ approaches zero without any disturbance. Phase-slips of $\pm \tilde{n}\, 2\pi$ are very improbable due to energy arguments[265] and we concentrate on phase changes of $\pm 2\pi$. Finally, if there is a transport current, the transitions with a phase change of -2π are more probable than those of $+2\pi$, because the energy barrier is smaller in the first case. The reason is that the system performs a transition into a state of lower free enthalpy [264].

Additional insight into the nature of the phase-slip process is obtained by a description in a 'particle picture':

Imagine a quasi-one-dimensional superconductor with an impressed transport current some millikelvin below the critical temperature. There may be a region of somewhat weakened superconductivity ('weak link') in the superconductor. If the transport current (which is carried by the few electrons still being condensed to Cooper pairs close to T_{c0}) is now slowly enhanced the superconducting state first breaks down in the weak link. The few Cooper pairs which carried the current with a high drift velocity are broken to quasiparticle excitations the drift velocity of which is reduced by inelastic scattering. For a moment there is normal conductivity in the weak-link region and all electrons carry the transport current. The superconductor now 'recognizes' that although being below T_{c0}, there are electrons with a drift velocity smaller than the critical velocity of the condensate. Thus, there is no reason against condensating Cooper pairs from these electrons. The pairs are accelerated to carry the current as a supercurrent - up to the critical drift velocity - and the process repeats. This picture may be rough but it shows the nature of phase slip most clearly, namely to dissipate the energy supplied to the system as heat rather than to convert it into kinetic energy of the condensate.

5.8.2. The RSM Model

Rieger, Scalapino, and Mercereau (RSM) [265] used time dependent GL equations and invoke the phase-slip concept to calculate the time behaviour of the supercurrent density in a region of weakened superconductivity, localized between two pieces of always strongly superconducting material (weak link) in a quasi-one-dimensional superconductor which is driven by an overcritical current.

The idea is that the attempt of applying an overcritical current to the equilibrium superconductor leads to an acceleration of the Cooper pairs associated with an increase of the supercurrent and a decrease of the order parameter. When the critical current of the weakest point is reached the order parameter there begins to decay toward zero within a region of about one coherence length. Soon the order parameter becomes very small and can be destroyed by fluctuations so that the phase-slip process can happen.

RSM assume that the superconductor remains in a phase-coherent state as long as the amplitude of the order parameter is nonzero and that the supercurrent in this case can be calculated by the time dependent GL theory. The phase slippage is grafted onto the problem by assuming that a phase slip of -2π occurs instantaneously if the free energy of the region $\xi(T)$ becomes larger than the free energy related to an order parameter with a phase difference reduced by 2π.

By numerical calculations RSM obtain the time dependent supercurrent density $\overline{j}_s(t)$, spatially averaged over the entire region of weakened superconductivity. Their astonishing result is that $\overline{j}_s(t)$ does not decay toward zero after the critical current density j_c of the weak link is reached, but oscillates at a frequency which increases with increasing applied total current density j. The reason is that their numerical calculations deliver an order parameter which indeed grows again after the phase-slip event. The oscillation frequency of the supercurrent is found to be equal to the Josephson frequency as given in eq. (129).

RSM pointed out that for impressed total current density, j, the part $j - j_s(t)$ has to be carried as 'normal' current so that the time averaged voltage across the weak superconductor is given by

$$\langle V(t)\rangle (I) = R(I - \langle \overline{I}_s(t)\rangle) \tag{130}$$

Here, $I = Aj$ and $\overline{I}_s(t) = A\overline{j}_s(t)$ are the total current and the supercurrent, respectively. RSM assumed R to be equal to the normal residual resistance of the weakly superconducting region.

Furthermore, an analytical approximation for the current-phase relation of the supercurrent is given [265 - 267], namely

$$\overline{I}_s(t) = (1/2)\, I_c\, (1 + \cos\varphi_{12}(t)) \tag{131}$$

where $\varphi_{12}(t)$ as given by eq. (127) with $V(t) = R(I - \overline{I}_s(t))$ is the phase difference across the weak link.

For $I \gg I_c$ the current-phase relation predicts a nearly sinusoidal time dependence of the supercurrent. [As $|\overline{I}_s(t)| \le I_c$ it is then $V(t) \approx RI$ yielding $\varphi_{12}(t) \approx \varphi_{12}(0) + (2e/\hbar)RIt$.] For $I \approx I_c$ the time dependence of the supercurrent is nonsinusoidal. This behaviour is in agreement with the numerical calculations.

It is remarked that the knowledge of the current-phase relation is essential for the description of the superconducting properties. Thus, $I_s(\varphi_{12})$ has been measured for several weak-link structures [268 - 271]. These measurements show that $I_s(\varphi_{12})$ is nearly sinusoidal for point contacts and short microbridges but may strongly deviate from a sinusoidal dependence for long microbridges. A change from a sinusoidal to a nonsinusoidal behaviour of $I_s(\varphi_{12})$ with increasing length of the weak link is also expected from theory [272].

An interesting property is the shape of the $\langle V(t)\rangle(I)$ curve. For $I \gg I_c$ the numerical calculations of RSM lead to a straight line with a zero voltage intercept of $0.5\,I_c$. In the environment of the critical current the curve is rounded and bends down to zero voltage at the critical current.

5.8.3. The SBT Model

The RSM model does not take into account the relaxation of nonequilibrium quasiparticle excitations produced during the phase-slip cycle. The consequence is that simply the normal resistance of the weak section enters the calculation. A description including the behaviour of nonequilibrium quasiparticles is given in the model for a phase-slip center proposed by Skocpol, Beasley, and Tinkham (SBT) [32]. The model involves a relaxation oscillation of the order parameter at Josephson frequency as the RSM model does but with a length scale for the flow of 'normal' currents set by the 'quasiparticle diffusion length', Λ, rather than by $\xi(T)$.

The characteristic features of the SBT model [2,8,32,273] are sketched in Fig. 13. Similar to the RSM model (or the relaxation oscillator model of Notarys and Mercereau [274]) phase slip is assumed to happen in a localized region of length of order $\xi(T)$. Thus, one may imagine a 'core region' of $\sim 2\xi(T)$ where strong variations of the superconducting order parameter occur. The absolute value of the superconducting order parameter is assumed to be driven to zero only in the 'core' itself, situated at the point X_{PSC} in the center of the core region.

The behaviour of the supercurrent is calculated using the time independent GL equations (1) and (2) together with the Josephson relation as given in eq.(53) and a suitable assumption about the behaviour of μ_p: In a quasi-one-dimensional homogeneous filament the absolute value of the supercurrent density is given by $\underline{j_s} = -(2e/m')|\psi|^2 \hbar \tilde{q}\, \hat{\underline{x}}$, where $|\psi|^2 = |\psi_0|^2(1-\xi^2(T)\tilde{q}^2)$ with $\tilde{q} = d\varphi/dx$ and $\hat{\underline{x}}$ the unit vector along the axis of the filament. [The last relation (for $|\psi|^2$) is obtained from the GL equation for the order parameter by neglecting the vector potential, introducing the phase $\varphi(x)$ of the order parameter by $\psi(x) = |\psi|\exp(i\varphi(x))$,

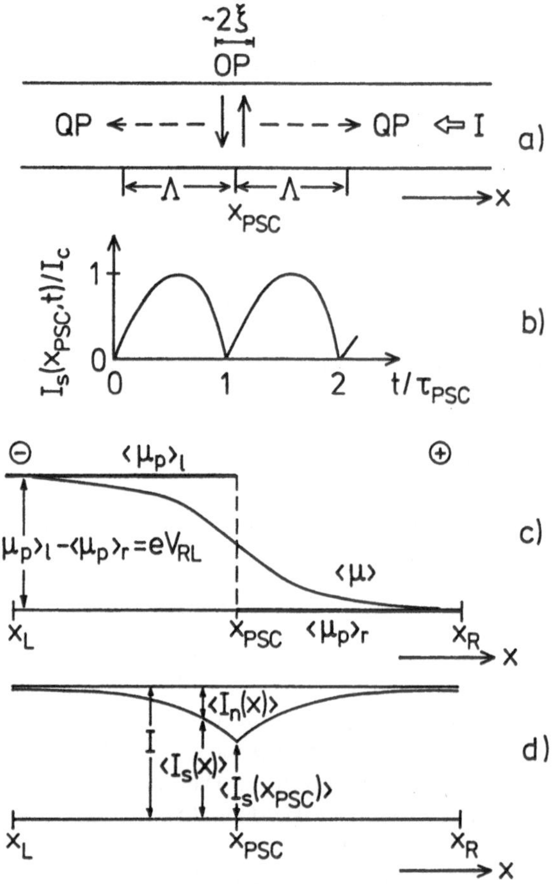

Fig. 13: *Characteristic features of the SBT model of a phase-slip center.*

a) Sketch of a quasi-one-dimensional superconductor. For total currents I above I_c a phase-slip center appears at X_{PSC}, leading to a relaxation oscillation of the order parameter, 'OP', on a length scale $\xi(T)$. The phase-slip center emitts nonequilibrium quasiparticles which diffuse into the bordering superconductor on a length scale of the quasiparticle diffusion length, Λ. Here, x denotes the space coordinate.

b) Time behaviour of the supercurrent $I_s(X_{PSC},t)$ normalized by I_c, where $\tau_{PSC} = \hbar\pi/eV_{RL}$ is the Josephson frequency.

c) Time averaged electrochemical potentials of the Cooper pairs, $\langle\mu_p\rangle$, and the quasiparticles, $\langle\mu\rangle$. Here, the indices 'l, L' and 'r, R' denote the left and the right side of the phase-slip center, respectively. The use of the indices at the voltage V_{RL} is in analogy to the index convention used in eq. (54) of chap. 5.3. Furthermore, x_L and x_R are two points far away from the phase-slip center, where $\mu = \mu_p$. Here, ⊖ and ⊕ indicate the polarity of the battery.

d) Behaviour of the time averaged supercurrent and 'normal' current at different points.

and furthermore using the definitions for α, β, and $\xi^2(T)$.] The space derivative of the Josephson relation yields $\hbar \partial\tilde{q}/\partial t = -2\partial\mu_p/\partial x$, so that $\tilde{q}(x,t) = \tilde{q}(x,0) - (2/\hbar)\int_0^t(\partial\mu_p/\partial x)\,dt$.

Thus, the time evolution of j_s depends on the assumption about the behaviour of μ_p during the phase-slip cycle. The time average $\langle\mu_p\rangle$ is assumed to be constant in space with a step-like change at X_{PSC}. In the core region the instantaneous values $\mu_p(x)$ are assumed to oscillate around $\langle\mu_p\rangle$, while $\mu_p(x) = \langle\mu_p\rangle$ further away from the core. [Thus, the voltage V_{RL} between x_R and x_L is actually not time dependent. It represents the time averaged voltage between any two points x_r and x_l situated right and left of X_{PSC} in the core region.] During the phase-slip cycle the Cooper pairs are first accelerated and then decelerated. In the deceleration part of the cycle the phase-slip event with the 'snap back' of the phase difference takes place. SBT assumed that the variation of μ_p during the acceleration part can be approximated by $\mu_{p,a}(x) = (1/2)\Delta\mu_p(1-\tanh[(x-X_{PSC})/\tilde{a}\xi]) + \langle\mu_p\rangle_r$, where $\Delta\mu_p = \langle\mu_p\rangle_l - \langle\mu_p\rangle_r$. For the deceleration part SBT assumed that $\mu_{p,d}(x)$ remains equal to $\langle\mu_p\rangle$ further away from the core region, but has a gradient everywhere which is $-\gamma$ times that of $\mu_{p,a}(x)$ and is discontinuous at X_{PSC}. Assuming, furthermore, that the deceleration part persists for $1/\gamma$ times as long as the acceleration part leads to a constant time average $\langle\mu_p\rangle$ on both sides of the core. Taking $\gamma \gg 1$ means modeling an instantaneous snap back of the phase.

Thus, it is $\tilde{q}(x,t)\xi = \tilde{q}(x,0)\xi + \{\Delta\mu_p/\hbar\tilde{a}\cosh^2[(x-X_{PSC})/\tilde{a}\xi]\}t$ during the acceleration part. Setting $\tilde{q}(X_{PSC},0) = 0$ means that after the snap back the phase-slip cycle starts with a superfluid velocity which is zero in the core. Then the Cooper pairs are accelerated and j_s grows. At $\tilde{q}(X_{PSC},t)\xi = 1/\sqrt{3}$ it is $|\psi|^2/|\psi_0|^2 = 2/3$ in the core (that means $j = j_c$) and the superconducting state becomes unstable, leading to a collaps of the order parameter and a decrease of j_s. SBT assume that the remaining Cooper pairs are still accelerated, leading to a further increase of $\tilde{q}$. Although the order parameter collapses, j_s and $|\psi|^2$ are calculated using the expressions given above, until it is $\tilde{q}(X_{PSC},t)\xi = 1$, yielding $|\psi|^2 = 0$ in the core, and the snap-back event can take place.

The constant $\tilde{a}$ in the expression for $\mu_{p,a}(x)$ has been set to π. Then it is $\tilde{q}(X_{PSC},t)\xi = 1$ for $t_{sb} = \hbar\pi/\Delta\mu_p$ and the phase-slip cycle repeats at the Josephson frequency. Furthermore, the enhancement of the phase difference along the filament during the phase-slip cycle as given by $\int_{-\infty}^{+\infty}[\tilde{q}(x,t_{sb}) - \tilde{q}(x,0)]\,dx$ yields exactly a value of 2π.

The time-averaged supercurrent density in the core is obtained by averaging $j_s(\tilde{q}(X_{PSC}))$ over $\tilde{q}$ from zero to $1/\xi$, yielding $\langle j_s(X_{PSC})\rangle/j_c = 3\sqrt{3}/8 = 0.65$. Since the model assumes an instantaneous reastablishment of the order parameter after the snap back, SBT argued that the time averaged supercurrent density may be somewhat overestimated.

Now, the idea is that the quasiparticle excitations generated during the phase-slip cycle in the core region diffuse into the surrounding superconductor. The more electron-like ones are assumed to migrate toward the positive pole of the battery while the more hole-like ones migrate to the negative pole. The charge imbalance Q^* thus created causes the electrochemical quasiparticle potential to be different from the electrochemical potential of the pairs. As has been discussed in detail in section 5.3 it is $\mu > \mu_p$ in a region where there is an excess of more electron-like over more hole-like excitations, whereas $\mu < \mu_p$ for an excess of more hole-like over more electron-like excitations. While the quasiparticles are diffusing, the charge imbalance relaxes and the difference between μ and μ_p becomes smaller.

SBT give a description of the time averaged behaviour of $\mu - \mu_p$ only. They apply a differential equation, derived by Pippard, Shepherd, and Tindall [275] for the description of the spatial variation of the hole-electron imbalance at a current carrying NC/SC boundary. Thus, SBT equate, for each region of constant $\langle \mu_p \rangle$,

$$\langle \mu \rangle - \langle \mu_p \rangle = \Lambda^2 \frac{d^2}{dx^2} (\langle \mu \rangle - \langle \mu_p \rangle) \tag{132}$$

where

$$\Lambda = ((1/3)\, \ell\, v_F \tau_2)^{1/2} \tag{133}$$

is the 'quasiparticle diffusion length'. More exactly, Λ is the decay length which a quasiparticle travels by random walk until a scattering process occurs which leads to a reduction of the difference between the electrochemical potentials of pairs and quasiparticles. The time τ_2 is the relaxation time for such a scattering process.

According to eq. (45) it is $\mu - \mu_p \sim Q^*$, that means only scattering processes which reduce the charge imbalance can reduce differences between μ and μ_p. Therefore, τ_2 has to be identified with τ_{Q^*} and Λ is equal to the charge imbalance decay length $\Lambda_{Q^*} = (D \tau_{Q^*})^{1/2}$. Then eq. (132) is just the differential equation for Q^* as derived in section 5.7 from the charge imbalance wave equation in the low frequency limit, describing a time independent exponential spatial charge imbalance decay.

From eq. (132) and the boundary condition that $\mu = \mu_p$ at x_L and x_R it follows

$$\langle \mu \rangle_r - \langle \mu_p \rangle_r = U_r \sinh[(x_R - x)/\Lambda] \tag{134}$$

for x ranging from X_{PSC} to x_R. A similar expression, namely $\langle \mu \rangle_l - \langle \mu_p \rangle_l = U_l \sinh[(x - x_L)/\Lambda]$, is obtained for the left side of X_{PSC}. Here, U_r and U_l are arbitrary constants.

In the following, the expressions will be given for example for the right side of X_{PSC}: Due to the symmetry of the problem it is $(\langle \mu \rangle_r - \langle \mu_p \rangle_r)_{x = X_{PSC}} = e V_{RL}/2$ leading to $U_r = e V_{RL}/2 \sinh[(x_R - X_{PSC})/\Lambda]$. Thus, the time averaged electrochemical quasiparticle potential is given by

$$\langle \mu \rangle_r = \langle \mu_p \rangle_r + \frac{e V_{RL}}{2} \frac{\sinh[(x_R - x)/\Lambda]}{\sinh[(x_R - X_{PSC})/\Lambda]} \tag{135}$$

To calculate the 'normal' that means quasiparticle current density, $\underline{j_n}$, it is used that the part of the total current density, $\underline{j}$, which is not carried as supercurrent, is flowing as normal current. This condition, $\underline{j} - \underline{j_s} = \underline{j_n}$ is the consequence of assuming negligible departures from electrical neutrality. If one furthermore takes the normal current density to be given by $\underline{j_n} = (1/e\rho_n)(d\mu/dx)\hat{\underline{x}}$ where ρ_n is the residual resistivity, it follows, for the absolute values of the current densities, that $j - \langle j_s(x)\rangle = \langle j_n(x)\rangle = (U_r/e\rho_n\Lambda)\cosh[(x_R - x)/\Lambda]$. The value of U_r is fixed because there is a phase-slip center at $x = X_{PSC}$ with $\langle j_s(X_{PSC})\rangle = 0.65\, j_c$. Multiplying the current densities with the cross sectional area A, we get the behaviour of the normal current,

$$\langle I_n(x)\rangle = \left(\cosh \frac{x_R - x}{\Lambda}\right)\left(\cosh \frac{x_R - X_{PSC}}{\Lambda}\right)^{-1}(I - \langle I_s(X_{PSC})\rangle) \qquad (136)$$

The voltage V_{RL} developed by the phase-slip center is given by the sum of $(\langle\mu\rangle_r - \langle\mu_p\rangle_r)_{x=X_{PSC}}$ and $(\langle\mu_p\rangle_l - \langle\mu\rangle_l)_{x=X_{PSC}}$, devided by e, leading to

$$V = \frac{\rho_n\Lambda}{A}\left(\tanh\frac{x_R - X_{PSC}}{\Lambda} + \tanh\frac{X_{PSC} - x_L}{\Lambda}\right)(I - \langle I_s(X_{PSC})\rangle) \qquad (137)$$

If $x_R - X_{PSC}$ and $X_{PSC} - x_L$ are both much larger than Λ, it is

$$V = (2\rho_n\Lambda/A)(I - \langle I_s(X_{PSC})\rangle) \qquad (138)$$

In the last two equations we dropped the index 'RL' at the voltage.

The differential resistance developed by a phase-slip center is given by $dV/dI = 2\rho_n\Lambda/A$. Since in the early experiments on tin microbridges and tin whiskers [19, 32] the differential resistance was found to be independent of temperature and not to diverge like $(T - T_{c0})^{-1/4}$, SBT in their original work concluded that a temperature-independent relaxation time τ_E rather than τ_{Q*} should be identified with τ_2 [2, 8, 32, 273]. Concerning this problem we refer to chapter 7 for a more detailed discussion.

The regular step-like structure experimentally observed is explained by the successive appearance of phase-slip centers with increasing total current. For an ideal homogeneous filament the first phase-slip center should appear in the middle between the superconducting contacts. In a real sample the critical current may somewhat vary in space due to inhomogeneities. Then the first phase-slip center occurs at the locus of the smallest local critical current $I_c(x)$. The next phase-slip center then appears at some other point of the wire where the supercurrent becomes equal the local critical current. In the environment of a phase-slip center the supercurrent is smaller than the total current due to the flow of quasiparticle currents. Therefore, additional phase-slip centers will avoid places close to active phase-slip centers. As long as the phase-slip centers are well separated each contributes to the total differential resistance with the same amount.

Tinkham formulated this idea more quantitatively for an ideal homogeneous filament [276], taking as criterion for the appearance of an additional phase-slip center that the time averaged supercurrent $\langle I_s(x)\rangle = I - \langle I_n(x)\rangle$ somewhere exceeds the critical current I_c. It is assumed that the number of active phase-slip centers in the sample is enhanced by one even if the criterium is fulfilled at several places in the sample. The reason may be that once a phase-slip center becomes active at one of the places, a rearrangement of the centers hinders the appearance of phase-slip centers at the other places.

The voltage $V_{\tilde{m}}$ developed by $\tilde{m}$ phase-slip centers in a filament of length L is then obtained from eq. (137) by replacing $x_R - X_{PSC}$ and $X_{PSC} - x_L$ both by $L/2\tilde{m}$ and by multiplying the resulting voltage by the number of centers, yielding

$$V_{\tilde{m}} = \frac{\rho_n L}{A} \left(\frac{2\tilde{m}\Lambda}{L} \tanh \frac{L}{2\tilde{m}\Lambda} \right) (I - \beta_{SBT} I_c) \tag{139}$$

with $\beta_{SBT} = \langle I_s(X_{PSC})\rangle / I_c$, where $\langle I_s(X_{PSC})\rangle$ is still the time averaged supercurrent of an isolated phase-slip center.

To calculate the maximum total current $I_{max,\tilde{m}}$ which can be carried with $\tilde{m}$ phase-slip centers being active in the sample, we remember that for a single phase-slip center the normal current $\langle I_n(x)\rangle$ has its smallest value at x_L and x_R where $\langle\mu\rangle = \langle\mu_p\rangle$. Assuming that also for $\tilde{m}$ phase-slip centers the electrochemical potentials are equal midway between the centers, the smallest normal current is obtained from eq. (136) by setting $x = x_R$ and $x_R - X_{PSC} = L/2\tilde{m}$. At the locus of the smallest normal current the supercurrent first reaches the critical current if I is enhanced. Thus, concerning the existence of $\tilde{m}$ phase-slip centers, the criterion $\langle I_s(x)\rangle \le I_c$ leads to the condition $I - (I - \beta_{SBT} I_c)(\cosh[L/2\tilde{m}\Lambda])^{-1} \le I_c$, yielding

$$\frac{I_{max}}{I_c} = \frac{\cosh(L/2\tilde{m}\Lambda) - \beta_{SBT}}{\cosh(L/2\tilde{m}\Lambda) - 1} \tag{140}$$

Plotting these results for a given value of $L/2\Lambda$ leads to a step-like $V-I$ characteristic starting at I_c [276]. Between the voltage jumps the characteristic is straight. All straight parts of the characteristic show the same extrapolated zero voltage intercept $\beta_{SBT} I_c$. The distance in current between the voltage jumps increases with the number of voltage steps. For the first few steps it is rather small.

We have discussed the SBT model somewhat in detail, because it describes the basic mechanism of a phase-slip center in a very transparent way. To remain transparent, several simplifications are necessary. One of them is to give only time averaged information about the behaviour of the charge imbalance in the quasiparticle system. Concerning the quasiparticles, the phase-slip center is described as if just behaving like a constant-current driven SC/NC/SC system with the NC part shrinked to a quasiparticle source of vanishing size. This description is nothing else but the low frequency limit of the charge imbalance dynamics as governed by the charge imbalance wave equation. The more general case of a time dependent behaviour of the charge imbalance in a phase-slip center is rather envolved. It will be discussed in the following section.

5.8.4. The KSS Model

The Kadin-Smith-Skocpol (KSS) model for a phase-slip center[158] in a quasi-one-dimensional superconductor close to T_{c0} is based on the charge imbalance wave equation (eq. (113)) as derived in section 5.7. KSS only consider charge imbalance relaxation by inelastic electron-phonon scattering so that $\tau_{Q*} = \tau_{Q*in}$, $\Lambda_{Q*} = \Lambda_{Q*in}$, and $\tau_{\alpha} = \tau_{E}$. Since the charge imbalance wave equation is based on a representation of the quasiparticle excitation spectrum which does not take into account pair breaking effects of the supercurrent[277], KSS neglect these effects in all quantities needed for the description of the charge imbalance.

As in the SBT model properties of the core region and of the charge imbalance are separately described. In the KSS model the core region is represented by an ideal Josephson element of zero resistance and negligible extent characterized by a current-phase relation being a free parameter of the model. This Josephson element is assumed to excite charge imbalance waves with a temporal and spatial evolution as given by the charge imbalance wave equation. The excitation of collective oscillations occurs due to the periodic emission of quasiparticle charge from the Josephson element into the surrounding superconductor. Depending on the sign of the quasiparticle charge, Cooper pairs escape from (or migrate into) the emission region to maintain overall electric neutrality. Since the phase-slip cycle repeats at the Josephson frequency also the frequency of the Carlsson-Goldman mode thus excited is expected to be the Josephson frequency. Thus, the KSS model uses a less detailed representation of the core region than the SBT model does, but it generalizes the description of the charge imbalance to ac effects.

To develop their model in a quantitative form, KSS pointed out that the charge-imbalance wave equation has full formal analogy to the well-known telegraph equation. Charge imbalance waves behave like electrical signals propagating along a simple transmission line. The properties of the phase-slip center are obtained by coupling the Josephson oscillator to a transmission line equivalent circuit.

In more detail, in the one-dimensional case the telegraph equation for the absolute value $E = E(x,t)$ of the electric field is given by

$$(\partial^2/\partial x^2)E = \tilde{L}\tilde{C}\ddot{E} + (\tilde{R}\tilde{C} + \tilde{G}\tilde{L})\dot{E} + \tilde{R}\tilde{G}E \qquad (141)$$

Here, $\tilde{L}$ is a series inductance per unit length, $\tilde{R}$ is a series resistance per unit length, $\tilde{C}$ a leakage capacitance, and $\tilde{G}$ a leakage conductance.

This equation yields the charge imbalance wave equation, eq. (113), if we replace E by Q^* and make the identifications

$$\tilde{R} = 1/\sigma A; \quad \tilde{G} = 2N_0 e^2 A/\tau_{Q^*in} = \sigma A/\Lambda^2_{Q^*in}$$
$$\tilde{L} = \tau_{OR}/\sigma A; \quad \tilde{C} = 2N_0 e^2 A(\pi\Delta/4kT) = (\sigma A/\Lambda^2_{Q^*in})\tau_E \tag{142}$$

It is remarked that the expressions given in eq.(142) are determined from the two basic equations (108) and (111) which lead to the charge imbalance wave equation. These equations can be rewritten by introducing the nonequilibrium potential $\Phi_{Q^*} = \Phi - \Phi_{\dot{\varphi}}$ as defined in section 5.3, where $\Phi_{\dot{\varphi}} = (1/2e)\hbar\dot{\varphi}$. Since $\Phi_{Q^*} = Q^*/2e^2N_0$, eqs. (108) and (111) yield $\partial I_n/\partial x + \tilde{G}\Phi_{Q^*} + \tilde{C}\dot{\Phi}_{Q^*} = 0$ and $\tilde{L}\dot{I}_s - \tilde{R}I_n - \partial\Phi_{Q^*}/\partial x = 0$, respectively. On the other hand, these equations follow from the transmission line equivalent of Fig. 14a using Kirchhoff's laws.

In this transmission line equivalent the resistive line and the inductive line represent the channels for the 'normal' current I_n carried by the quasiparticles and for the supercurrent I_s carried by the condensate, respectively. The corresponding electrochemical potentials along the lines are

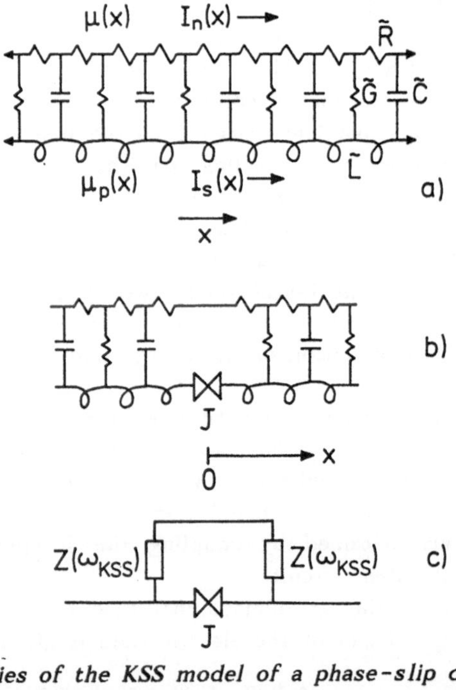

Fig. 14: *Basic properties of the KSS model of a phase-slip center [158].*
a) Transmission line equivalent of the charge imbalance wave equation.
b) Equivalent circuit of a phase-slip center with the ideal Josephson element (J) in the inductive (condensate) line at x = 0.
c) Equivalent circuit valid for currents and potentials at the origin.

$\mu(x)$ and $\mu_p(x)$. According to eq. (45) the voltage between the lines is given by $(\mu(x)-\mu_p(x))/(-e)=\Phi_{Q^*}(x)$. The voltages along the lines are given by $(\mu(x)-\mu_F)/(-e)=\Phi_{\dot\varphi}+\Phi_{Q^*}$ and $(\mu_p(x)-\mu_F)/(-e)=\Phi_{\dot\varphi}$ for the quasiparticle and pair channel, respectively. Here, we used results from section 5.3, namely $\mu=\mu_p-Q^*/2eN_0$ and $\mu_p=\mu_F-e\Phi_{\dot\varphi}$.

The constants of eq. (142) have a transparent physical interpretation, too: The relaxation of Φ_{Q^*} is governed by $\tilde C/\tilde G=\tau_E$, the ratio $\tilde L/\tilde R=\tau_{OR}$ is the supercurrent response time if $\Phi_{Q^*}=0$, $\tilde R$ is the normal resistance per unit length, and $\tilde L=\Lambda_L/A$ is the kinetic inductance of the supercurrent in the London equations. Furthermore, it is $\tilde C=Q_nA/\Phi_{Q^*}$, where $Q_n=(\pi\Delta/4kT)Q^*$ is the 'free charge density' [67, 158]. Thus, $\tilde C$ differs from Q^*A/Φ_{Q^*} by the factor $\pi\Delta/4kT$ which may be interpreted as effective 'dielectric constant'. It considers the fact that a free charge leads to a much larger Q^*, because $\mu_{c,p}$ is readjusted to maintain overall charge neutrality. Finally, $\tilde G=AQ^*/\Phi_{Q^*}\tau_{Q^*in}$, so that, considering eq. (107), it follows $\dot Q_n=-(\tilde G/A)\Phi_{Q^*}$ and, thus, $\tilde G$ connects the relaxation rate of the free charge density with the nonequilibrium potential Φ_{Q^*}. Here, A is the cross-sectional area.

Important properties which characterize the behaviour of charge imbalance waves are their phase velocity and their decay length. These quantities are obtained from the dispersion relation given in eq. (115). KSS take ω_{KSS} to be a real quantity and q_{KSS} to be complex, instead of q_{KSS} to be real and ω_{KSS} to be complex as done in section 5.7. The advantage of this representation is that it delivers directly the spatial behaviour of the wave and makes it easy to identify ω_{KSS} with the frequency of the Josephson oscillator modeling the core region. The phase velocity is then given by $v_{KSS}=\omega_{KSS}/\mathrm{Re}(q_{KSS})$ and the decay length by $\lambda_{d,KSS}=1/|\mathrm{Im}(q_{KSS})|$. Here, $\mathrm{Re}(q_{KSS})$ and $\mathrm{Im}(q_{KSS})$ denote the real and imaginary part of q_{KSS}, respectively. In the special case where $\tau_{OR}=\tau_E$ (i.e. $\tilde L/\tilde R=\tilde C/\tilde G$) there is no dispersion. The phase velocity and the decay length are then given by $v_{KSS}=\Lambda_{Q^*in}/\tau_{OR}$ and $\lambda_{d,KSS}=\Lambda_{Q^*in}$ and do not depend on ω_{KSS}. Furthermore, in the low-frequency limit where $\omega_{KSS}\ll\tau_{OR}^{-1}$, τ_E^{-1}, it follows $v_{KSS}\approx\Lambda_{Q^*in}/[\frac{1}{2}(\tau_E+\tau_{OR})]$ and $\lambda_{d,KSS}\approx\Lambda_{Q^*in}$. In the high-frequency limit where $\omega_{KSS}\gg\tau_{OR}^{-1},\tau_E^{-1}$, it is $v_{KSS}\approx\Lambda_{Q^*in}/(\tau_{OR}\tau_E)^{1/2}$ and $\lambda_{d,KSS}\approx\Lambda_{Q^*in}(\tau_{OR}\tau_E)^{1/2}/[\frac{1}{2}(\tau_E+\tau_{OR})]$. It is remarked that the result for the decay length in the high frequency limit leads to the decay length of the Carlsson-Goldman mode as given in eq. (100) if it is assumed that $\tau_E\gg\tau_{OR}$ so that $\lambda_{d,KSS}\approx 2\Lambda_{Q^*in}(\tau_{OR}/\tau_E)^{1/2}$.

An important tool for the description of a phase-slip center in the transmission-line picture is the impedance of the transmission line, given by

$$Z(\omega_{KSS})=\left(\frac{\tilde R+i\omega_{KSS}\tilde L}{\tilde G+i\omega_{KSS}\tilde C}\right)^{1/2}=\frac{\Lambda_{Q^*in}}{\sigma A}\left(\frac{1+i\omega_{KSS}\tau_{OR}}{1+i\omega_{KSS}\tau_E}\right)^{1/2} \tag{143}$$

For $\tau_{OR}=\tau_E$ and in the low frequency limit it is $Z(\omega_{KSS})=R_{eff}/2$, where $R_{eff}=2\Lambda_{Q^*in}/\sigma A$ is the normal resistance of a length $2\Lambda_{Q^*in}$ of the line. In the high-frequency limit it is $Z(\omega_{KSS})=R_{ac}/2$, where $R_{ac}=R_{eff}(\tau_{OR}/\tau_E)^{1/2}$.

Now a dc current ($\dot{I} = 0$) is assumed to flow through the quasi-one-dimensional superconductor and to generate a phase-slip center. This situation is modeled by placing at $x=0$ an ideal Josephson oscillator into the condensate line (Fig. 14 b). As far as only properties at $x=0$ are considered, the transmission line acts as if the Josephson element was shunted by $Z(\omega_{KSS})$ as sketched in Fig. 14 c. Since far away from the core region of the phase-slip center the total current flows as a supercurrent, the external current source is connected with the condensate line. The supercurrent in the core region is then assumed to be given by $I_s(x=0,t)=I_c f(\varphi_{+-}(t))$, where $\varphi_{+-}(t)=\varphi(x=0^+)-\varphi(x=0^-)$ is the difference between the phases of the order parameter on both sides of the Josephson oscillator. Here, f is a function with amplitude 1 and period 2π, describing the supercurrent-phase relation.

Now we summarize the results of this model and refer to the original KSS paper for more details:

For $I \gg I_c$ most of the current at $x=0$ has to flow as normal current I_n through the shunt impedance. Since $I = I_s + I_n$, the normal current will oscillate, but the oscillations are only of order I_c and, thus, very small compared to the absolute value of the normal current. Therefore, the voltage at the Josephson element is nearly constant ('high voltage dc behaviour') as in the SBT model. The consequence is that $\varphi_{+-}(t)$ increases linearly with time (see for instance eq. (127)) so that the time average of the supercurrent in the core can be replaced by the average over φ_{+-}. In the high-voltage dc limit the excitation of charge imbalance waves is only a small disturbance of the dc behaviour. Therefore, the properties of the phase-slip center can be calculated by setting $\omega_{KSS}=0$, so that the shunt impedance is given by $Z(0)$. It then follows (from eq. (53) of ref. 158) for the voltage developed by the phase-slip center

$$V = 2Z(0)[I-\langle I_s(0,t)\rangle] = R_{eff}[I-I_c\langle f\rangle_{\varphi_{+-}}] \qquad (144)$$

where $\langle f\rangle_{\varphi_{+-}} = (1/2\pi)\int_0^{2\pi} f(\varphi_{+-})\,d\varphi_{+-}$.

In the high voltage dc limit the KSS model, thus, reproduces the SBT result for the V-I dependence. Furthermore, the KSS model predicts that $\langle\mu_p\rangle$ is spatially constant and that in the high voltage dc limit $\langle\mu\rangle$ and $\langle I_n\rangle$ decay exponentially with the decay length Λ_{Q^*in}. This is very similar to the SBT result.

In more detail, the following relation between the electrochemical potentials holds quite generally (i. e. without being restricted to the dc limit), namely

$$\mu_p(x,t)-\mu_p(\infty,t) = -\tau_{QR}[\dot{\mu}(x,t)-\dot{\mu}(\infty,t)] \qquad (145)$$

where $\mu_p(\infty,t) = \mu_F - e\,\Phi_{\dot{\varphi}}(\infty,t)$. The time average (over τ_{PSC}) yields

$\langle\mu_p(x,t)\rangle - \langle\mu_p(\infty,t)\rangle = 0$, because $\mu(x,t)-\mu(\infty,t)$ is periodic in time so that $\mu(x,\tau_{PSC})-\mu(\infty,\tau_{PSC}) = \mu(x,0)-\mu(\infty,0)$. Therefore, the time average of μ_p is space independent. This result seems to be valid quite generally.

It is remarked that $\mu_p(\infty,t)$ and $\dot{\mu}(\infty,t)$, appearing in eq.(145), and $\mu(\infty,t)$ are set to zero by KSS. In the dc limit there is no nonequilibrium far away from the phase-slip center, so that $\mu(\infty,t)$ and $\mu_p(\infty,t)$ are equal and actually independent of space and time and may both be set to zero. (In the present work this will not be done.) Indeed $\dot{\mu}(\infty,t)=0$ is the natural value in the dc limit.

In the dc limit, where $\omega_{KSS}=0$, $Q^*(x,t)$ decays exponentially with the decay length Λ_{Q^*in}, and $\mu_p(\infty,t)=\mu_p(\infty)$ is independent of space and time, it is found that

$$I - \langle I_s(x)\rangle = \langle I_n(x)\rangle = \langle I_n(x=0)\rangle \exp(-x/\Lambda_{Q^*in}) \tag{146}$$

$$\langle\mu(x)\rangle - \mu_p(\infty) = -e\langle\Phi_{Q^*}(x)\rangle = -e(\Lambda_{Q^*in}/\sigma A)\langle I_n(x=0)\rangle \exp(-x/\Lambda_{Q^*in}) \tag{147}$$

where $\langle I_n(x=0)\rangle = \sigma A\langle Q^*(0)\rangle/2e^2 N_0 \Lambda_{Q^*in}$.

To set a scale for the change from a complicated space and time dependence of the phase-slip center to the high voltage dc limit, KSS introduce the voltage $V_c = R_{eff}\, I_c$ which would be obtained from eq.(144) for $I = I_c(1+\langle f\rangle_{\varphi_{+-}})$ and which sets a frequency scale by $\omega_c = (2e/\hbar)V_c$. For voltages $\langle V\rangle$ much lower than V_c the Josephson oscillations become more pulse-like and are dominated by higher harmonics of $(2e/\hbar)\langle V\rangle$. In this regime the oscillations of the supercurrent excite charge imbalance waves which have a strong influence on the behaviour of the phase-slip center. For $\langle V\rangle \gg V_c$ we have the high voltage dc limit in which the oscillations of the supercurrent and, thus, the charge imbalance waves excited become negligible small perturbations of a dc behaviour, where the properties of the phase-slip center are governed by the low frequency limit solution of the charge imbalance wave equation with its exponential charge imbalance decay.

KSS show that $\omega_c\tau_{OR} = (4/3\sqrt{3})\,\Lambda_{Q^*in}/\xi(T)$ in the case where the critical current which the phase-slip center appears at is equal to the Ginsburg-Landau result for a quasi-one-dimensional superconductor. Thus, as long as $\Lambda_{Q^*in} \gg \xi(T)$, it is $\omega_c\tau_{OR} \gg 1$. If, furthermore, $\tau_E > \tau_{OR}$, it is $\omega_c\tau_E \gg 1$, too, and the filament is already in the 'high frequency limit' if $\langle V\rangle = V_c$. The temperature range where this is the case depends on the material of the filament and can be obtained from Fig.5 of ref.158 for several superconductors.

In the 'general case' charge imbalance waves govern the behaviour of the phase-slip center. In a first step KSS calculate the normal current in the origin, looking for solutions for $I_n(x=0,t)$ which are periodic with the Josephson frequency $\omega_J = (\mp)(2e/\hbar)\langle V_{(\mp\pm)}\rangle$. Here, the voltage V_{-+} is measured between the left side and the right side of the Josephson oscillator in the origin and may be expressed as $V_{-+}(t) = (-1/e)[\mu_p(x=0^-,t) - \mu_p(x=0^+, t)]$. It is remarked that in the present work e is a positive quantity, while it is negative in the work of KSS. Let us assume that $\langle V_{-+}\rangle < 0$, so that $\omega_J > 0$. It may be comfortable to have the results also in terms of $V_{+-} = -V_{-+}$ which then is positive quantity. The corresponding sign changes are given in parentheses. The formal advantage is that a conventional current flows from $x=0^+$ to $x=0^-$ for $V_{+-} > 0$ and is also positive due to the standard convention.

The normal current in the origin is given by the following system of equations (assuming that $\varphi_{+-}(0)=0$):

$$I_n(x=0,t) = \sum_{\tilde{m}=-\infty}^{+\infty} c_{\tilde{m}}(\omega_J)\, e^{i\tilde{m}\omega_J t} \tag{148}$$

$$c_0 = \langle I_n(x=0,t)\rangle = \langle V_{\substack{-+\\(+-)}}\rangle / R_{eff} \tag{149}$$

$$c_{\tilde{m}}(\omega_J) = -\frac{\omega_J I_c}{2\pi} \int_0^{2\pi/\omega_J} dt\, e^{i\tilde{m}\omega_J t} f(\varphi_{+-}(t)), \qquad \text{for } \tilde{m} \neq 0 \tag{150}$$

$$\varphi_{+-}(t) = \omega_J t \underset{(+)}{-} \frac{4e}{\hbar} - \sum_{\tilde{m}\neq 0} Z(\tilde{m}\omega_J) \frac{c_{\tilde{m}}(\omega_J)}{i\tilde{m}\omega_J} (e^{i\tilde{m}\omega_J t} - 1) \tag{151}$$

$$Z(\tilde{m}\omega_J) = \frac{\Lambda_{Q^*in}}{\sigma A} \left(\frac{1 + i(\tilde{m}\omega_J)\tau_{QR}}{1 + i(\tilde{m}\omega_J)\tau_E} \right)^{1/2} \tag{152}$$

For a given function $f(\varphi_{+-}(t))$ these equations have to be solved by iteration in general. The problem is selfconsistent, because $c_{\tilde{m}}(\omega_J)$ appears in the expression for $\varphi_{+-}(t)$. For the special case of a linear current phase-relation where $f(\varphi_{+-}) = \varphi_{+-}(t)/2\pi$ for $0 \le \varphi_{+-} \le 2\pi$, eq.(150) together with eq.(151) can be explicitly evaluated, yielding

$$c_{\tilde{m}}(\omega_J) = \frac{I_c}{2\pi i \tilde{m}} \left(1 + \frac{I_c}{2\pi i \tilde{m}} \frac{2 Z(\tilde{m}\omega_J)}{\langle V_{\substack{-+\\(+-)}}\rangle} \right)^{-1} \tag{153}$$

Furthermore, the space and time dependent quantities are given by (for $x > 0$)

$$I - I_s(x,t) = I_n(x,t) = \sum_{\tilde{m}} c_{\tilde{m}}\, e^{i(\tilde{m}\omega_J t - q_{\tilde{m}} x)} \tag{154}$$

$$\Phi_{Q^*}(x,t) = \sum_{\tilde{m}} Z(\tilde{m}\omega_J)\, c_{\tilde{m}}\, e^{i(\tilde{m}\omega_J t - q_{\tilde{m}} x)} \tag{155}$$

$$\mu(x,t) - \mu(\infty,t) = \frac{e}{\sigma A} \sum_{\tilde{m}} \frac{c_{\tilde{m}}}{i q_{\tilde{m}}}\, e^{i(\tilde{m}\omega_J t - q_{\tilde{m}} x)} \tag{156}$$

$$\mu_p(x,t) - \mu_p(\infty,t) = \frac{e}{\sigma A} \sum_{\tilde{m}} \frac{-i\tilde{m}\omega_J \tau_{QR}}{i q_{\tilde{m}}}\, c_{\tilde{m}}\, e^{i(\tilde{m}\omega_J t - q_{\tilde{m}} x)} \tag{157}$$

where $q_{\tilde{m}} = q_{KSS}(\tilde{m}\omega_J)$ and $\mu(\infty,t) = \mu_p(\infty,t)$.

Moreover, the voltage measured by probes far away from the phase-slip center is given by $V_{\mp\infty} = (-1/e)[\mu(-\infty,t) - \mu(\infty,t)] = (-1/e)\,2[\mu(0,t) - \mu(\infty,t)]$, where the last equality is due to the symmetry of the problem. This results in

$$V_{\mp\infty} = \langle V_{-+}\rangle + \frac{2\Lambda_{Q^*in}}{\sigma A} \sum_{\tilde{m}\neq 0} \frac{c_{\tilde{m}}\, e^{i\tilde{m}\omega_J t}}{[(1+i\tilde{m}\omega_J\tau_{QR})(1+i\tilde{m}\omega_J\tau_E)]^{1/2}} \tag{158}$$

The important point is that this voltage contains an ac contribution. This contribution vanishes in time average so that $\langle V_{\mp\infty}\rangle = \langle V_{-+}\rangle$.

In time average it is $\langle\mu_p(x,t)\rangle - \langle\mu_p(\infty,t)\rangle = 0$, $I - \langle I_s(x,t)\rangle = \langle I_n(x,t)\rangle = \langle I_n(x=0)\rangle \exp(-x/\Lambda_{Q^*in})$, $\langle\mu(x,t)\rangle - \langle\mu(\infty,t)\rangle = -(e\langle V_{-+}\rangle/2)\exp(-x/\Lambda_{Q^*in})$, and $\langle\Phi_{Q^*}\rangle = (\langle V_{-+}\rangle/2)\exp(-x/\Lambda_{Q^*in})$. Note that $\langle I_n(x=0)\rangle = \langle V_{-+}\rangle/R_{eff} = (\sigma A/\Lambda_{Q^*in})(\langle Q^*(0)\rangle/2e^2N_0)$, if we use eq. (149) and $-e\langle V_{-+}\rangle/2 = (\langle\mu\rangle - \langle\mu_p\rangle)_{x=0} = \langle Q^*(0)\rangle/(-2eN_0)$ according to eq. (45). Setting $x = 0$ in the expression for the time averaged normal current we get $\langle V_{-+}\rangle = R_{eff}[I - \langle I_s(0,t)\rangle]$, where $\langle V_{-+}\rangle = \langle V_{\mp\infty}\rangle$.

Thus, the results of the high voltage dc limit (i.e. also of the SBT model) seem to be recovered in time average. This is true concerning the exponential spatial dependence of the currents and potentials. There is, however, a very important difference in the result for the time averaged supercurrent in the core: In the high voltage dc limit the time average $\langle I_s(0,t)\rangle$ could be replaced by its angle average $\langle I_s(0,\varphi_{+-})\rangle_{\varphi_{+-}} = I_c\langle f\rangle_{\varphi_{+-}}$, because $\varphi_{+-}(t)$ increases linearly with time due to the nearly constant voltage at the Josephson element. This is not allowed in the general case, where the voltage across the core region is a complicated function of time due to the excitation of charge imbalance waves by the Josephson oscillator. Therefore in general $\langle I_s(0,t)\rangle$ is not independent of the voltage $\langle V_{-+}\rangle$ but changes along the voltage-current characteristics. Only for high voltages the SBT behaviour is recovered. For low voltages there is a significant deviation from a straight line as will be discussed below.

In the original KSS paper I_n and Φ_{Q^*} were numerically calculated for the general case [158]. The behaviour in the origin as well as the spatial dependence for a fixed time are plotted in that article. Moreover, the spatial (fixed time) behaviour of the electrochemical potentials are given. In Fig. 15 the result for the normal current and the electrochemical potentials using the linear current-phase relation are redrawn from ref. [158]. The quantities show a distinct structure due to the excitation of charge-imbalance waves.

Finally, we will discuss the resulting V-I characteristics in the general case. These arise from the condition that $I = I_s(x,t) + I_n(x,t)$, which has to be fulfilled for all (x,t), especially for $(x=0, t=0)$, yielding

$$I = I_c f(\varphi_{+-}(t=0)) + \sum_{\tilde{m}} c_{\tilde{m}}(\omega_J) \tag{159}$$

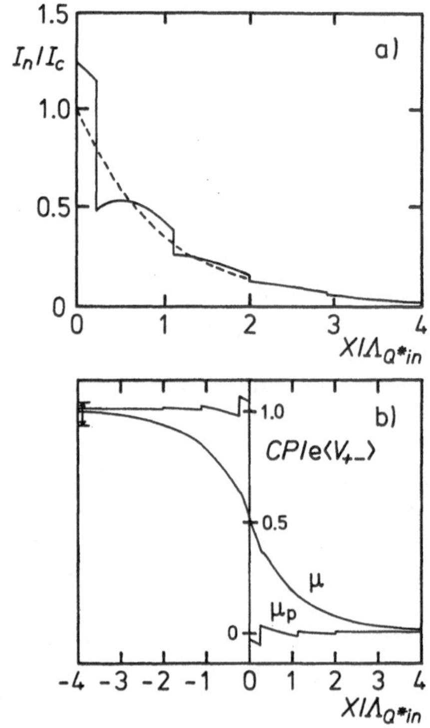

Fig. 15: *Numerical results of the KSS model [158] in the general case. The parameters are chosen to be $\tau_E/\tau_{QR} = 11$, $\omega_c\tau_{QR} = 2.2$, $\langle V\rangle = V_c = I_c R_{eff}$, $I = 1.5\,I_c$, as appropriate for a Sn thin film filament in the dirty limit at $T/T_{c0} = 0.99$. The core of the phase-slip center is localized at $x = 0$. The vertical jumps in the curves correspond to phase-slip events and are discontinuous as an artifact of using the linear current-phase relation.*

a) Spatial dependence of the real part of the normal current $I_n(x,t)$, normalized by I_c, for $x > 0$ and $t = 0.25\,\tau_{PSC}$. The dashed curve is the high voltage dc time average.

b) Spatial dependence of the real part of the electrochemical potentials, 'CP', of Cooper pairs, $\mu_p(x,t)$, and quasiparticle excitations, $\mu(x,t)$, for $t = 0.25\,\tau_{PSC}$. Here, $\mu(\infty,t) = \mu_p(\infty,t)$ has been set to zero and the electrochemical potentials have been normalized by $e\langle V_{+-}\rangle = \langle[\mu_p(x=0^-,t) - \mu_p(x=0^+,t)]\rangle$. The arrow indicates the range of variation of the chemical potentials for $|x| \to \infty$, leading to the ac contribution in the voltage as given in eq. (158).

Since $\omega_J = (2e/\hbar)\langle V_{+-}\rangle$ and $\langle V_{+-}\rangle = \langle V_{\pm\infty}\rangle$ this equation is an explicit expression for I as a function of the voltage $\langle V\rangle$ which may be for instance measured with superconducting contacts far away from the phase-slip center. The current-phase relation $f(\varphi_{+-})$ is open for a free choice. To handle current-phase relations which are discontinuous at $t = 0$, KSS replace the first term in eq. (159) by $(1/2)I_c[f(\varphi_{+-}(t=0^+)) + f(\varphi_{+-}(t=0^-))]$.

The corresponding V - I characteristics have then to be concluded from the I - ⟨V⟩ characteristics. KSS calculated the I - ⟨V⟩ characteristics taking the parameters needed from the set given in Fig. 15 and assuming a linear current phase relation. Coming from currents above the critical current, the curve extends to values with $I < I_c$ and then begins to bend back to the critical current. However, the current range of negative differential resistance is very small and the curve ends abruptly. The reason is that further solutions in the negative differential resistance region do not exist and the curve does not bend back to the critical current.

For qualitative discussions or rough estimations, KSS give an approximate expression for the I - ⟨V⟩ characteristics which makes it unnecessary to solve for the coefficients $c_{\tilde{m}}(\omega_j)$,

$$I = I_G(\langle V\rangle, R_{ac}) + \langle V\rangle (1/R_{eff} - 1/R_{ac}) \tag{160}$$

Here I_G has to be inserted from a generalized resistivity shunted junction (GRSJ) model [272, 278 - 280] where $\langle V\rangle_G$ and the shunt resistance, R_s , are replaced by $\langle V\rangle$ and R_{ac} respectively.

For the GRSJ model it is [158]

$$\langle V\rangle_G = I_c R_s \left\{ \frac{1}{2\pi} \int_0^{2\pi} \frac{1}{\frac{I_G}{I_c} - f(\varphi_{+-})} d\varphi_{+-} \right\}^{-1} \tag{161}$$

leading to

$$\langle V\rangle_G = R_s (I_G^2 - I_c^2)^{1/2} \qquad \text{for} \qquad f(\varphi_{+-}) = \sin(\varphi_{+-}) \tag{162}$$

$$\langle V\rangle_G = I_c R_s / \ln[I_G/(I_G - I_c)] \qquad \text{for} \qquad f(\varphi_{+-}) = \varphi_{+-}/2\pi \tag{163}$$

The I - ⟨V⟩ curves resulting from the approximation of eq. (160) are shown in Fig. 16 for several values of τ_E/τ_{OR}, using the linear current-phase relation. This plot is of high practical use for qualitative discussions, but one should keep in mind that the exact solutions do not completely bend back to the critical current and the KSS model actually does not predict those large regions of negative differential resistance shown in the figure.

The V - I characteristics are then easily obtained from Fig. 16: As long as $\tau_E/\tau_{OR} \leq 1$, close to T_{c0}, the characteristics simply follow the curves given in the figure. As soon as $\tau_E/\tau_{OR} > 1$, at lower temperatures, there will be a voltage jump into the resistive state at I_c for increasing current and a jump back into the superconducting state for decreasing current at the point I_{min} where the curves begin to bend back.

Thus, the model predicts a nonthermal, intrinsic hysteresis. The hysteresis grows with decreasing temperature (that means increasing τ_E/τ_{OR}) but is limited by the condition, $(I_{min}/I_c) = \langle f\rangle_{\varphi_{+-}}$ for $\tau_E/\tau_{OR} \to \infty$.

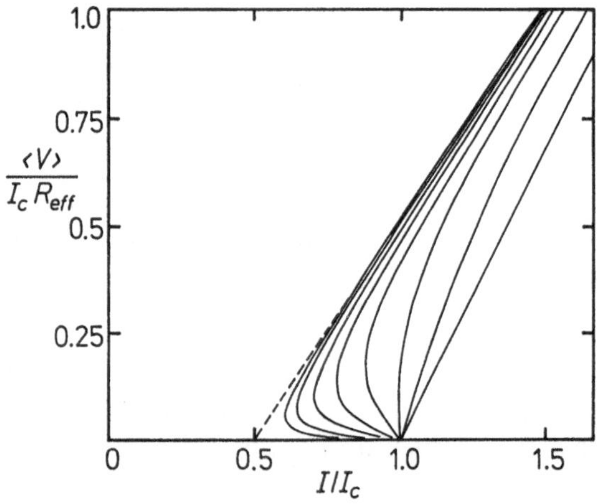

Fig. 16: *I-⟨V⟩ characteristics calculated from the approximate solution, eq.(160), of the KSS model [158] for the linear current-phase relation. The curves belong to different values of τ_E/τ_{0R} which are 1024, 256, 64, 16, 4, 1, 1/4, and 1/16 from left to right. For a detailed discussion see the text.*

The physical reason for the hysteresis is the following: For $\tau_{0R}=\tau_E$ it is $2Z(\omega_{KSS})=R_{eff}$ for all frequencies so that the Josephson element is simply shunted with a frequency independent resistance and the approximation, reducing to the GRSJ result, is equal to the exact solution of the model. This solution starts with infinite slope and then bends to higher currents. For $\tau_{0R}<\tau_E$, the impedance is smaller than R_{eff} and, thus, the amplitude of the voltage oscillations at the borders of the Josephson oscillator has decreased. Thus, the voltage driving the oscillator is closer to being a constant voltage and the curve has to be closer to the high voltage dc limit (dashed line in Fig. 16).

Qualitatively the hysteresis may also be understood by the following rough picture: Due to the excitation of charge imbalance waves each phase-slip cycle generates a region of enhanced quasiparticle charge around the core region with different polarity on the left and on the right side. The backflow of the quasiparticle charge into the core region is hindered by the next emission of quasiparticles. If now the total current is reduced to the critical current, the phase-slip center tries to switch off, because $I_s(0,t)$ does not exceed the critical current any longer. This stops the production of quasiparticles in the core region and leads to a backflow of quasiparticle charge into this region causing a quasiparticle current flowing in opposite direction of the supercurrent $I_s(0,t)$. Then, the constant total current I can only be drawn through the core region, if $I_s(0,t)$ is increased to a value above I_c and the switch-off of the phase-slip center actually does not occur. The times τ_{0R} and τ_E play a role in this process, because τ_{0R} governs the time development of $I_s(0,t)$ and τ_E is expected to be the charge imbalance relaxation time after a pulse-like quasiparticle injection. The switch-off of the phase-slip center is only cancelled if the quasiparticle charge does not relax much faster than the supercurrent will increase and hysteresis will only occur for $\tau_{QR} < \tau_E$.

Finally, it is remarked that the KSS model got confirmation by solving the time-dependent Ginsburg-Landau (TDGL) equations of Kramer and Watts-Tobin (see the next section for a discussion of the TDGL theory) which has been modified to contain the 'capacitive term' proportional to $\dot{\Phi}_{Q*}$ of KSS [281]. Oscillatory phase-slip solutions are then obtained which generate damped waves of charge imbalance, propagating along the quasi-one-dimensional superconductor. Moreover, a description of charge imbalance fluctuations in the KSS model with thermal noise included [282] leads to the same results as found by independent methods [159]. There is also experimental evidence for charge imbalance waves in the vicinity of phase-slip centers [283, 284].

5.9. Time-Dependent Ginsburg-Landau (TDGL) Theory

In 1958 Gorkov pointed out [285] that in his version of the BCS theory of superconductivity the 'anomalous' Green's function*1 varies in time proportional to $\exp(-2i\mu_p t/\hbar)$. The order parameter $\mathbf{\Delta}(\underline{r})$ of the microscopic theory is proportional to this Green's function [46]. Therefore, also $\mathbf{\Delta}(\underline{r})$ and the order parameter of the GL theory, $\psi(\underline{r}) \sim \mathbf{\Delta}(\underline{r})$, should show this time dependence. Gorkov's prediction led Josephson [286] to expect a periodically varying current at a frequency $2eV/\hbar$ between a two-superconductors system with a potential difference, V, between the superconducting regions, because nothing changes physically if the phase difference is changed by a multiple of 2π. Furthermore, it follows eq. (53) of the present work if we introduce the convention of section 5.3 to measure the evolution of the phase relative to the case where $\mu_p = \mu_F$, so that the above expression changes to $\exp(-2i(\mu_p - \mu_F)t/\hbar)$.

In one of the early approaches toward a time dependent GL theory, Anderson, Werthamer, and Luttinger used Gorkov's result for a phenomenological derivation of an 'extra Ginsburg-Landau equation' which connects the charge, the electrical potential and the time dependence of the order parameter [287], thus confirming the results of Stephen and Suhl directly obtained from Green's function theory [288].

In the following we briefly sketch the development of the TDGL theory from these first attempts to the actual version of the TDGL equations. The early stages before 1967 have been summarized by Werthamer [47]. Furthermore, Cyrot [289] reviewed the development up to 1972. There is,

*1 denoted F^+ by Gorkov [285] but F by Lüders and Usadel [46]

however, no summarizing representation available including the generalization of the theory derived very recently.

In 1966 Schmid[290] and Abrahams and Tsuneto[291] proposed a time dependent modification of the Ginsburg-Landau equation. Their work has been discussed in refs. 47 and 265. If, for instance, following Schmid, the TDGL equation for the order parameter is given by[290]

$$\gamma_R\left(\hbar\frac{\partial}{\partial t}-2ie\tilde{\Phi}\right)\psi=-\frac{1}{2m'}\left(-i\hbar\underline{\nabla}-e'\underline{A}\right)^2\psi-\alpha\psi-\beta|\psi|^2\psi \tag{164}$$

The supercurrent density, $\underline{j_s}$, is again given by eq.(2), but now with $\psi=\psi(\underline{r},t)$, while the normal current is $\underline{j_n}=-\sigma(\dot{\underline{A}}+\underline{\nabla}\tilde{\Phi})$. The constants α and β are the same as in the time independent GL theory and have been defined in section 5.1. Note that α in the present work and in Schmid's paper have been defined with a reversed sign and that we formulated the results in terms of m' instead of m.

The potential $\tilde{\Phi}=(\Phi-\mu_{c,q}/e)+\mu_F/e$, so that $-e\tilde{\Phi}=\mu-\mu_F$ and the left hand side of eq.(164) is equal to $\gamma_R(\hbar\partial/\partial t+2i(\mu-\mu_F))\psi$. Furthemore, it is $\gamma_R=(2|\alpha|/\hbar)\tau_R$, where τ_R has the physical interpretation that a deviation of ψ from its equilibrium value $\sqrt{-\alpha/\beta}$ vanishes proportional to $\exp(-t/\tau_R)$. It is shown that $\tau_R=\pi\hbar/16k(T_{c0}-T)$, so that the relaxation time for $|\psi|^2$ is $\tau_{GL}=2\tau_R$ as introduced in section 5.4. Compared to ref. 290 a term μ_F/e has been added in $\tilde{\Phi}$, because μ_F has not been set to zero in the present work. If ψ is independent of time and $\mu=\mu_F$, the left hand side of the TDGL equation yields zero as must be in the equilibrium case.

The potential $\tilde{\Phi}$ in general differs from the electrostatic potential, Φ, because $\mu_{c,q}$ is somewhat different from μ_F, as discussed in section 5.3. Schmid, furthermore, considers that already the chemical potential of electrons in the normal state may differ from μ_F due to screening effects of local deviations of charge neutrality [26, 290]. However, if we neglect the latter and then set $\mu_{c,q}\approx\mu_F$ for temperatures close to T_{c0}, as shown in section 5.3, it follows $\tilde{\Phi}\approx\Phi=\Phi_{Q*}+\Phi_{\dot{\varphi}}$. Naturally, this interpretation of $\tilde{\Phi}$ is a today's view. At that time charge imbalance effects, which lead to a difference of $\mu_{c,q}$ and μ_F and to a contribution Φ_{Q*} in Φ, were not considered.

The right hand side of the TDGL equation is just the negative one of the right hand side of the time independent GL equation for the order parameter, given in eq.(1), and it is therefore equal to $-\delta G_s/\delta\psi^*$ being zero in equilibrium. Here, G_s denotes the Ginsburg-Landau free enthalpy of the superconductor. Thus, the essential idea of the TDGL equation is to have $\dot{\psi}\sim(1/\tau_R)(-\delta G_s/\delta\psi^*)$. This means that the time variation of the order parameter is proportional to the deviation of the system from equilibrium. The additional term containing the electrochemical potential enters the equation to have a gauge invariant formulation.

By proposing the relation for $\tilde{\Phi}$ it has been assumed that the quasiparticles come into equilibrium among themselves so that they can be characterized by a chemical potential. Since the TDGL theory will be applied to problems where there may be strong and rapid variations of the phase and the magnitude of the order parameter and where μ_p may be time-dependent, moreover, the question of the validity of the Josephson relation under such conditions arises and thus the question of what is $\Phi_{\dot{\varphi}}$: If $\hbar\dot{\varphi}=-2(\mu_p-\mu_F)$ holds, which is nothing else than the Josephson relation, it is $\Phi_{\dot{\varphi}}=(1/2e)\hbar\dot{\varphi}$. On the other hand, Gorkov's prediction given in the beginning of this section yields $\varphi=-2(\mu_p-\mu_F)t/\hbar$ so that $\hbar\dot{\varphi}=-2(\mu_p-\mu_F)-2\dot{\mu}_p t$. According to section 5.3 it is $\mu_p=\mu_F-e\Phi_{\dot{\varphi}}$ and thus $\dot{\mu}_p=-e\dot{\Phi}_{\dot{\varphi}}$ from which we get $\Phi_{\dot{\varphi}}=(1/2e)\hbar\dot{\varphi}-\dot{\Phi}_{\dot{\varphi}}t$ or $-\hbar\ddot{\varphi}/2+2e\dot{\Phi}_{\dot{\varphi}}+e\ddot{\Phi}_{\dot{\varphi}}t=0$ if we take the time derivate. We will neglect the time derivatives of $\Phi_{\dot{\varphi}}$ by assuming that they are small. This assumption implies that the part of the electric potential Φ in the superconductor which is related to the evolution of the phase is not too strongly time dependent. With this assumption it is $\Phi_{\dot{\varphi}}=(1/2e)\hbar\dot{\varphi}$ or $\tilde{\Phi}-(1/2e)\hbar\dot{\varphi}=\Phi_{Q^*}$.

There remains the question how long Gorkov's result holds. This should be the case as long as properties do not change too fast so that an instantaneous chemical potential $\mu_{c,p}$ of the pairs (and thus an instantaneous μ_p) is a meaningful quantity. This requires an 'instantaneous equilibrium' of the condensate which then also should be characterized by an instantaneous value of the order parameter (or the energy gap) leading to an instantaneous value of $v_{\underline{K}}^2$, the single particle momentum occupation probability in the superconducting state. A lower limit for a characteristic time constant may be given by the 'pairing time' (that is the time between the emission of the virtual phonon by one electron of the Cooper pair and the reabsorption by the second electron) which may be estimated by $\xi_0/v_F\sim10^{-12}-10^{-13}$ s. Therefore, if the quasiparticles are assumed to come into equilibrium among themselves (for which τ_E is a characteristic time constant), this will easily be able for the condensate. It will be discussed in section 5.10 that the TDGL theory is based on Gorkov's Green's function equations which have to be assumed to hold as long as one works with a TDGL theory. In the other case already an instantaneous value of $\Delta(\underline{r},t)$ or $\psi(\underline{r},t)$ appears to be a doubtful quantity. Naturally, there remains the question whether there can occur processes in a superconductor that are faster than the pairing time at all.

In the TDGL theory the time evolution of the order parameter is obtained by the solution of the TDGL equation. The TDGL theory can give a description of the dynamics of the order parameter throughout the filament containing a phase-slip center, especially also in the core region. Within the TDGL theory one can ask for the phase-slip process itself and (as will be discussed below) it turns out that periodical phase-slip solutions are a consequence of the theory. This is an important difference to phenomenological models. In these models the phase-slip event has to be grafted onto the problem as an additional assumption: In the RSM model free-energy arguments are invoked as a criterion for the occurrence of the phase-slip event. In the SBT model the time behaviour of the phase is a consequence of the assumption about μ_p. In the KSS model the time evolution of the phase depends on the assumption about the current phase relation and the behaviour of charge imbalance waves in the vicinity of the core region of the phase-slip center.

To describe the nonequilibrium properties of a phase-slip center, the TDGL theory has to consider the behaviour of the quasiparticle excitations. This is not contained in the version given above, but has been included more than a decade later.

Before we proceed, we remark that Schmid's result as given in eq. (164) can be written in the following way

$$\tilde{D}^{-1}\left(\frac{\partial}{\partial t} - i\frac{2e}{\hbar}\tilde{\Phi}\right)\hat{\psi} + \xi^{-2}(|\hat{\psi}|^2 - 1)\hat{\psi} + \left(\frac{\nabla}{i} + \frac{2e}{\hbar}\underline{A}\right)^2\hat{\psi} = 0 \qquad (165)$$

where $\tilde{\Phi} = (-1/e)(\mu - \mu_F)$, $\hat{\psi} = \psi/\psi_0$, and $\tilde{D}^{-1} = \pi\hbar/\xi^2 8k(T_{c0} - T)$. In the dirty limit ($\ell \ll \xi_0$) where the GL coherence length ξ is given by $\xi_D = 0.853\,\xi_0^{1/2}\ell^{1/2}[T_{c0}/(T_{c0} - T)]^{1/2}$, it is $\tilde{D} = D$ where $D = \ell v_F/3$ is the diffusion constant. To obtain eq. (165) we introduced the equilibrium value, $\psi_0 = (-\alpha/\beta)^{1/2}$, of the order parameter and $\xi^2 = -\hbar^2/2m'\alpha$ from section 5.1. and used $e' = -2e$.

In a similar way, but using now the definition of the magnetic penetration depth, $\lambda(T)$, and the result for $\underline{j}_s$ from section 5.1, the total current density of Schmid's theory is given by $\underline{j} = \underline{j}_s + \underline{j}_n$, where

$$\underline{j}_s = -\frac{\hbar}{4e\mu_0\lambda^2}\left[\hat{\psi}^*\left(\frac{\nabla}{i} + \frac{2e}{\hbar}\underline{A}\right)\hat{\psi} + \hat{\psi}\left(-\frac{\nabla}{i} + \frac{2e}{\hbar}\underline{A}\right)\hat{\psi}^*\right] \qquad (166)$$

and

$$\underline{j}_n = \sigma(-\underline{\nabla}\tilde{\Phi} - \dot{\underline{A}}) \qquad (167)$$

The first derivation of a TDGL equation from the microscopic theory has been done by Gorkov and Eliashberg [292]. The derivation of a nonlinear TDGL equation of any generality is very difficult, because pure superconductors and also alloys usually have a gap in the excitation spectrum. The resulting singularity in the density of states leads to slowly decaying oscillations in the time domain and the various quantities cannot be expanded in powers of Δ [2, 292]. Therefore, the derivation has been performed for a superconductor with paramagnetic impurities which act as pair breakers and round off the singularity in the density of states. For a sufficient concentration of paramagnetic impurities the excitation spectrum becomes gapless. Gorkov and Eliashberg performed their calculations for the gapless limit.

Already at a first view the terms which additionally enter the equation compared to the time independent case [293] seem to be those of Schmid. The analogy between the Gorkov-Eliashberg result and Schmid's result (for the dirty limit where $\tilde{D} = D$) becomes completely visible if the microscopic equations are also written in the same normalized form chosen for eqs. (165) - (167), adopted by Hu and Thompson [2, 294, 295]. The gapless limit result is given by eqs. (165) - (167) if $\hat{\psi}$ is replaced by the normalized order parameter of the microscopic theory $\hat{\mathbf{\Delta}}_s = \mathbf{\Delta}_s / \mathbf{\Delta}_{0s}$, where $\mathbf{\Delta}_{0s} = \pi k [2(T_{c0}^2 - T^2)]^{1/2}$ is the equilibrium order parameter in the absence of currents and fields but calculated for the critical temperature in the presence of magnetic impurities [2, 293, 295]. Furthermore, $\xi(T)$ and $\lambda(T)$ have to be replaced by their values in the presence of magnetic impurities and depend on the spin flip scattering time. The expressions are given in the cited literature. In addition, Gorkov and Eliashberg give an expression for the charge density (as also Schmid does) and remark that the Maxwell equations coupling Φ and $\underline{A}$ to the charge and current density have to be taken to complete the set of equations.

From these results the conclusion has been drawn that Schmid's TDGL equations are only valid in the gapless limit. This statement is established by the work of Eliashberg [296] who extended the theory to a metal with lower concentration of magnetic impurities as in ref. 292. This is expected to have much more in common with the properties of ordinary superconducting alloys. However, he got very different results: Two new functions of energy had to be introduced to describe the behaviour of the 'normal' electrons [296, 297]. These functions are closely related to the even and odd part of the distribution function describing the occupation probability of the excitation spectrum with nonequilibrium quasiparticles (see section 5.5). Kinetic equations for these distribution functions for a finite gap superconductor have then been seeked. The work of Schmid and Schön [79, 81] and Larkin and Ovchinikov [298 - 300] brought great success in this task.

It is now about ten years ago that Kramer and Watts-Tobin then derived generalized TDGL equations, valid for dirty superconductors in the vicinity of T_{c0} with a finite energy gap, which now consider the properties of the quasiparticle excitations [301]. Their work takes into account inelastic electron-phonon scattering only. Schön and Ambegaokar [302, 303] derived essentially the same set of TDGL equations which, however, include various pair breaking mechanisms characterized by the parameter Γ given in section 5.4. There has been a critical reexamination of both works by Hu [304], who proposed some improvements. Finally, Watts-Tobin, Krähenbühl, and Kramer [305] derived general equations for the dynamic behaviour of dirty superconductors in the Ginsburg-Landau regime, $|T_{c0} - T| \ll T_{c0}$, from the microscopic theory. In the direct vicinity of the critical temperature a local equilibrium approximation then leads to the generalized TDGL equations already given in ref. 301. These TDGL equations are [305]

$$\frac{\pi\hbar}{8k|T_{c0}-T|}\left[1+\left(\frac{2\tau_E}{\hbar}|\mathbf{\Delta}|\right)^2\right]^{-1/2}\left[\frac{\partial}{\partial t}-\frac{2ie}{\hbar}\tilde{\Phi}+\frac{2\tau_E^2}{\hbar^2}\frac{\partial|\mathbf{\Delta}|^2}{\partial t}\right]\mathbf{\Delta} = \xi_D^2\left(\underline{\nabla}+\frac{2ie}{\hbar}\underline{A}\right)^2\mathbf{\Delta}+\left(1-\frac{|\mathbf{\Delta}|^2}{\mathbf{\Delta}_0^2}\right)\mathbf{\Delta} \tag{168}$$

for the order parameter, and

$$\underline{j} = -\frac{\pi\sigma}{4ekT_{c0}}|\Delta|^2\underline{Q} - \sigma\left(\underline{\nabla}\tilde{\Phi} + \frac{\partial\mathbf{A}}{\partial t}\right) \tag{169}$$

for the total current density.

Here*[1], $\underline{Q} = (2e/\hbar)\underline{A} + \underline{\nabla}\varphi$, $\Delta_0 = [(8\pi^2k^2T_{c0}^2/7\zeta(3))(1-T/T_{c0})]^{1/2}$, and $\tilde{\Phi}$, ξ_D as defined above. However, now charge imbalance actually contributes to $\tilde{\Phi}$. The equilibrium order parameter, Δ_0, is obtained from $\psi_0=(-\alpha/\beta)^{1/2}$ by inserting the relation between ψ and Δ and the microscopic expressions for α and β as given in section 5.1.

The result for the total current density given in eq. (169) is equal to that one given above: The second term is the normal part written down in eq. (167). The first term is nothing else but the supercurrent density of the time independent GL theory given in section 5.1, however, inserting $\psi(\underline{r},t)=|\psi(\underline{r},t)|\exp(i\varphi(\underline{r},t))$ and using the relation between ψ and Δ from eq. (4), assuming the dirty limit. Thus, the superconducting contribution to the total current density is equal to eq. (166).

To compare the equation for the order parameter, eq. (168), with eq. (165), one should multiply eq. (168) by $1/\xi_D^2\Delta_0$ and use that $\Delta/\Delta_0=\psi/\psi_0=\hat{\psi}$ due to eq. (4), to see that all terms, except those containing the inelastic collision time, τ_E, are already present in Schmid's theory. The weight of these contributions scales with the parameter γ, given by

$$\gamma = 2\tau_E\Delta_0/\hbar = 6.13(kT_{c0}/\hbar)\tau_E[(T_{c0}-T)/T_{c0}]^{1/2} \tag{170}$$

sometimes called 'pair-breaking parameter of the TDGL theory'. It is $\gamma=\Delta_0/\hbar\Gamma$, if only the inelastic electron phonon contribution of Γ as given in eq. (60) is considered. Actually $1/\gamma$ indicates the pair-breaking strength, because for $\gamma \ll 1$ the singularity of the density of states at the gap edge is totally smeared out [305], leading to strong pairbreaking and the gapless limit [2,305,306]. In the gapless limit eq. (168) yields exactly Schmid's result given in eq. (165).

The TDGL equations of Watts-Tobin, Krähenbühl, and Kramer are valid for dirty superconductors ($\ell \ll \xi_0$) in the Ginsburg-Landau regime ($|T_{c0}-T| \ll T_{c0}$). Moreover, the local equilibrium approximation has to be fulfilled, that means all quantities are assumed to vary in time and space slowly over the inelastic collision time τ_E and the diffusion length $\Lambda_E=(D\tau_E)^{1/2}$, respectively. Then gradients and time derivatives of the odd and even part of the quasiparticle distribution function (see section 5.5) can be neglected in the kinetic equations for the distribution function. In other words, the quasiparticle distribution function adjusts to local instantaneous values of Δ and $\tilde{\Phi}$.

---

*[1] Note that it is $\Delta = |\Delta|\exp(i\varphi)$ and $e>0$ in the present work, while it is $\Delta=|\Delta|\exp(-i\tilde{\Theta})$ and $e<0$ in ref. 305.

The local equilibrium approximation does not lead to the 'local equilibrium' which we have introduced in section 5.3. The approximation leads to deviations from that 'local equilibrium', which are characterized by a distribution function adjusted to an instantaneous value of $\tilde{\Phi}$, that means to a situation in which a chemical potential is a senseful quantity. Nevertheless, we will not introduce a new name, because the nomenclature is used throughout the literature.

In a homogeneous material both conditions for the local equilibrium approximation are fulfilled very close to T_{c0}, where $\Lambda_E \ll \xi_D(T)$. For $\Lambda_E = \xi_D(T)$ it is $\Delta T_E = |T_{c0} - T_E| = \pi\hbar/8k\tau_E$ and the local equilibrium approximation may, thus, be used for $|T_{c0} - T| \ll \Delta T_E$. The break-down of the local equilibrium approximation should at least occur for $|T_{c0} - T| = \Delta T_E$, i.e. at $\gamma^2_{max} = 4\pi^3 k T_{c0} \tau_E / 7\zeta(3)\hbar$. Moreover, it may be useful to introduce the quantity $\Delta T_{GL} = |T_{c0} - T_{GL}|$, which is $\gamma = 1$ for, characterizing the upper limit for the use of the 'simple' (i.e. gapless) version of the TDGL theory. Due to the definition of γ we get $\Delta T_{GL} = 7\zeta(3)\hbar^2/32\pi^2 k^2 T_{c0} \tau_E^2$ so that $\gamma^2_{max} = \Delta T_E/\Delta T_{GL}$.

Usually, the TDGL equations are used in reduced units. For this purpose one has to introduce the quantities

$$t_0 = \frac{\xi_D^2}{Du} = \frac{\pi\hbar}{8k|T_{c0} - T|u} \quad ; \quad u = \frac{\pi^4}{14\zeta(3)} = 5.79$$

$$\tilde{\Phi}_0 = \frac{\hbar}{t_0} \quad ; \quad j_0 = -\frac{\pi\sigma\mathbf{\Delta}_0^2}{4ekT_{c0}\xi_D(T)} = -\frac{\sigma\hbar}{2et_0\xi_D(T)} \tag{171}$$

We remark that $t_0 = \tau_{GL}/u$ and $|j_0| = (1/0.385)\, j_c$, where j_c is the GL result for the critical current of a quasi-one-dimensional superconductor as given in section 5.1, assuming the dirty limit. Furthermore, we note that $t_0 = \tau_{GL}/u = \tau_{OR}$, where τ_{OR} is the dirty limit result for the supercurrent response time as introduced in section 5.7, inserting $\mathbf{\Delta}_0$ for the gap.

Then multiplying eqs. (168) and (169) by $1/\mathbf{\Delta}_0$ and introducing the quantities $\hat{t} = t/t_0$, $\underline{\hat{j}} = \underline{j}/j_0$, $\hat{\psi} = \mathbf{\Delta}/\mathbf{\Delta}_0$, and $\hat{\mu} = -2e\tilde{\Phi}/\tilde{\Phi}_0$, yields

$$\frac{u}{(1+\gamma^2|\hat{\psi}|^2)^{1/2}}\left(\frac{\partial}{\partial\hat{t}} + i\hat{\mu} + \frac{1}{2}\gamma^2\frac{\partial|\hat{\psi}|^2}{\partial\hat{t}}\right)\hat{\psi} = \xi_D^2\left(\underline{\nabla} + \frac{2ie}{\hbar}\underline{A}\right)^2\hat{\psi} + (1-|\hat{\psi}|^2)\hat{\psi} \tag{172}$$

$$\underline{\hat{j}} = \xi_D\left[|\hat{\psi}|^2\underline{Q} - \left(\underline{\nabla}\hat{\mu} - \frac{2e}{\hbar}\frac{\partial\underline{A}}{\partial\hat{t}}\right)\right] \tag{173}$$

This normalized form of the TDGL equations is used in the literature. Usually, the '^' is omitted and sometimes it is set $\hbar = k = 1$ and distances are measured in units of $\xi_D(T)$. This will, however, not be done in the present work.

Although these TDGL equations are usually used to describe the phase-slip state, it is a very important question of current interest in which way properties change if the local equilibrium approximation is not valid. This will be the case for somewhat lower temperatures, especially if the inelastic scattering time, τ_E, of a material is large. Then the quasiparticle distribution cannot adjust to the local instantaneous values of Δ and $\tilde{\Phi}$.

Recently Baratoff [307, 308] performed calculations in the gapless limit but without assuming the local equilibrium approximation. Applying the resulting equations to the problem of a current carrying filament he obtained remarkable changes in the behaviour of a phase-slip center, compared to the local equilibrium case. These results will be discussed later. Baratoff argued that the resulting charge imbalance is insensitive to the gapless approximation because the main contribution of the deviation of the distribution function from its equilibrium value comes from energies $E_K \gg \Delta$.

Moreover, Krähenbühl did some work outside the local equilibrium approximation [309].

More recently, Rangel and Kramer developed a theory of periodically driven current carrying superconducting filaments [127, 310, 311]. Parts of these calculations were carried out exactly, without using the local equilibrium approximation [127, 311]. A comparison with experiment shows that the exact results agree exactly with the measurement while the local equilibrium expressions deviate significantly [128, 311].

We now discuss the work which has been done on the description of the dissipative phase-slip state in a current carrying filament by applying the TDGL theory to the problem:

An early attempt has been done by Rieger, Scalapino, and Mercereau (RSM) [265], using the 'simple' TDGL equations as given in eqs. (164) - (167). In their representation the phase-slip event is grafted onto the problem by a free energy argument. We have discussed the 'RSM model' in much detail in section 5.8.2.

Another early calculation based on the simple TDGL theory has been carried out by Fink and Poulsen [312 - 315]. In that model it is assumed that at the critical current the superconductor enters a new superconducting state. This state is assumed to be spatially periodic with an order parameter vanishing at certain values of the coordinate x. Between the adjacent zeros the superconductor is regarded as singly connected with a constant μ_p. It is assumed that there is a step-like increase of μ_p along the filament. Thus, the phases of the singly connected regions have a different time evolution leading to a continuous phase slip at the zeros of the order parameter. The problem of this model is that the order parameter cannot reestablish after the phase-slip event which will turn out to be characteristic for the behaviour of a phase-slip center. In more recent work on the dissipative phase-slip state [316] the type of solution described has been considered again, but now to give a static approximation for high γ values and large total current densities.

The most systematic and successful analysis of the solutions of the TDGL theory for the case of a current carrying quasi-one-dimensional superconductor has been started a decade ago by Kramer and Baratoff [317, 318]:

Their numerical solutions of the 'simple' TDGL theory show a very small current region below the critical current which 'spontaneous' localized phase-slip occurs in even in a completely homogeneous wire [318]. This result is very important, because it shows that the phase-slip state is a consequence of the theory and has not to be grafted onto the problem by additional assumptions.

Then they extended their investigations by including the coupling to nonequilibrium quasiparticles, but still remaining in the gapless limit [319]. This shifts the value of the current below which phase-slip solutions exist towards larger values closer to the critical current.

The current range which these phase-slip solutions exist in becomes larger if the calculations are not carried out with the simple TDGL theory, but with the generalized TDGL equations given above, considering the properties of the quasiparticle excitations and being valid for a finite-gap superconductor [301, 305, 311, 316, 320]. It is most remarkable that the extension of the current range can lead to the existence of phase-slip solutions above the critical current of the time-independent GL theory.

In the TDGL theory no fluctuations are considered. Therefore, within this theory, also in the phase-slip process there are no fluctuations envolved. It is a completely continuous process: In the core of the phase-slip center the order parameter continuously decreases with time, becomes zero at one point and then increases again. While the order parameter becomes smaller, the gradient of the phase of the order parameter increases in the core region. This must be the case for impressed current, because the supercurrent is proportional to the absolute value of the order parameter and to the gradient of its phase. Therefore, the windings of the complex order parameter become closer and closer. The space for one winding shrinks to zero extent in the limit of vanishing absolute value of the order parameter. In this way, a phase difference of 2π vanishes at the moment where the order parameter becomes zero.

The rate $\nu_{PSC} = 1/\tau_{PSC}$ which the phase-slip event (that means the vanishing of the phase difference 2π) repeats at is related to the time averaged voltage drop across the phase-slip center by the Josephson relation $\langle V \rangle = (h/2e)\,\nu_{PSC}$, where (compare eq.(54)) the voltage has to be measured by superconducting probes [305].

Now we will discuss the phase-slip solutions of the generalized TDGL equations (eqs. (172) and (173)) somewhat more in detail [301, 305, 311]:

The results are obtained by numerical methods using a computer. For the calculations a periodic array of phase-slip centers is assumed with a periodic length, d_p, that means the distance of the phase-slip centers in the array is d_p. It is looked for solutions which are periodic in space with period

d_p. Therefore, the TDGL equations are solved in a range $x=0$ to $x=d_p/2$ with periodic boundary conditions. The conditions required are[311] $\hat{\psi}(x)=\hat{\psi}^*(-x)$, $\hat{\mu}=0$ at $x=0$, $\hat{\psi}(d_p/2+x)\exp(i\tilde{\varphi})=\hat{\psi}^*(d_p/2-x)\exp(-i\tilde{\varphi})$, where $\tilde{\varphi}=-\int_0^{\hat{t}} d\hat{t}'\hat{\mu}(d_p/2,\hat{t}')$ or, with a different gauge of the electrochemical potential, $\hat{\psi}(d_p/2+x)=\hat{\psi}^*(d_p/2-x)$, $\hat{\mu}=0$ at $x=d_p/2$, $\hat{\psi}(x)\exp(-i\tilde{\varphi}')=\hat{\psi}^*(-x)\exp(i\tilde{\varphi}')$, where $\tilde{\varphi}'=-\int_0^{\hat{t}} d\hat{t}'\hat{\mu}(0,\hat{t}')$. In the first case, where $\hat{\mu}(0)=0$, the position of the periodic phase slip (that means of the core of the phase-slip center) is usually chosen to be at $x=d_p/2$, whereas in the second case, where $\hat{\mu}(d_p/2)=0$, it is chosen to be at $x=0$. The voltage $\hat{V}$ generated by a unit of length d_p is then the difference of $\hat{\mu}$ across the periodic length and is, thus, given by $\hat{V}(\hat{t})=\hat{\mu}(-d_p/2)-\hat{\mu}(d_p/2)$, for a phase-slip center at $x=0$, being equal to $\hat{V}(\hat{t})=2\hat{\mu}(0,\hat{t})$.

The question is whether the time average of this voltage (converted to physical units) is measured at the ends of our sample across the superconducting contacts (which measure μ_p). For an isolated phase-slip center, with $d_p=\infty$, this seems to be the case, because for such a large distance from the phase-slip center there are no charge imbalance effects present, implying $\mu=\mu_p$ at $\pm d_p/2$ and the equality of $\hat{\mu}$ with the normalized electrochemical pair potential. For a periodic array of phase-slip centers charge imbalance effects from two phase-slip centers compensate each other at the half distance between two cores, so that again $\mu=\mu_p$ at $\pm d_p/2$. Therefore, also in this case the voltage mentioned is related to the change of the electrochemical pair potential across the periodic length. The gauge $\hat{\mu}(d_p/2)=0$, thus, does not mean anything else but in $\tilde{\Phi}$ to measure μ relative to $\mu(d_p/2)=\mu_p(d_p/2)$ instead of measuring μ relative to μ_F.

For a periodic array of phase-slip centers, the calculated voltage across one center should be measured by superconducting (or normal conducting) voltage probes at $\pm d_p/2$. In a sample with strongly superconducting contacts a periodic array of phase-slip centers cannot establish. Nevertheless, enhancing the number of phase-slip centers in a sample (by increasing the total current) is qualitatively the same as decreasing the periodic length in an array of phase-slip centers. Thus, qualitative comparisons of several phase-slip centers in a whisker and of a phase-slip center in a periodic array should be allowed.

In Fig. 17a the absolute value of the order parameter, $|\hat{\psi}|$, and the potential $\hat{\mu}$ are plotted against the distance from the core at $x=d_p/2$ at two different times. These are the instant of the phase-slip when $\hat{\psi}(d_p/2,\hat{t})=0$ and the instant in between two phase slips when $|\hat{\psi}(d_p/2,\hat{t})|$ reaches its maximum. In Fig. 17b the order parameter $|\hat{\psi}(d_p/2,\hat{t})|$ and the normal part of the total current $\hat{j}_n(d_p/2,\hat{t})$ as occurring at the core situated in the middle of the periodic length are plotted against the time. At the instant of the phase slip it is $|\hat{\psi}(d_p/2,\hat{t})|=0$ and $\hat{j}_n=\hat{j}$, so that $\hat{j}_s=0$. Just after the phase slip it is $\hat{j}_n>\hat{j}$, because a small amount of supercurrent is flowing opposite to the direction of the total current.

In Fig. 18 the time averaged spatial behaviour of both quantities is given (for a different set of parameters and the core of the phase-slip center at

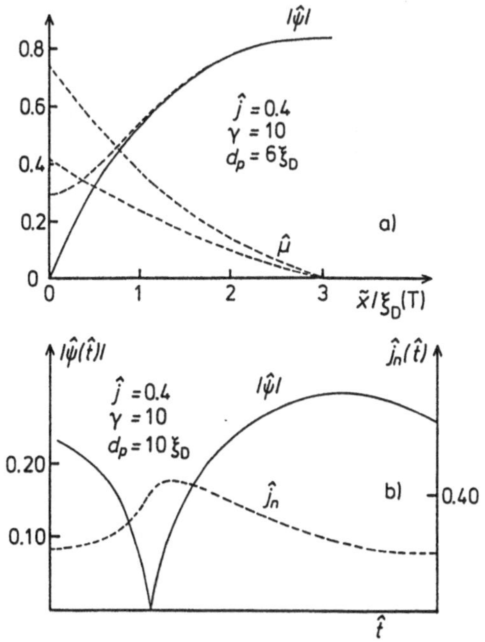

Fig. 17: *Phase-slip behaviour in a quasi-one-dimensional superconductor calculated from the generalized TDGL theory [305]*

a) Spatial dependence of $|\hat{\psi}|$ and $\hat{\mu}$ at the times of their minima and maxima during the phase-slip cycle, respectively. Here, $\tilde{x} = d_p/2 - x$ denotes the distance from the core of the phase-slip center, situated at $x = d_p/2$, i.e. at $\tilde{x} = 0$. The gauge for $\hat{\mu}$ is $\hat{\mu}(x=0) = 0$, so that $\hat{\mu}(\tilde{x} = d_p/2) = 0$. As discussed in the text, this means, that $\mu = \mu_p$ for $x = 0$, i.e. for $\tilde{x} = d_p/2$.

b) Time dependence of $|\hat{\psi}|$ and $\hat{j}_n$ at the core.

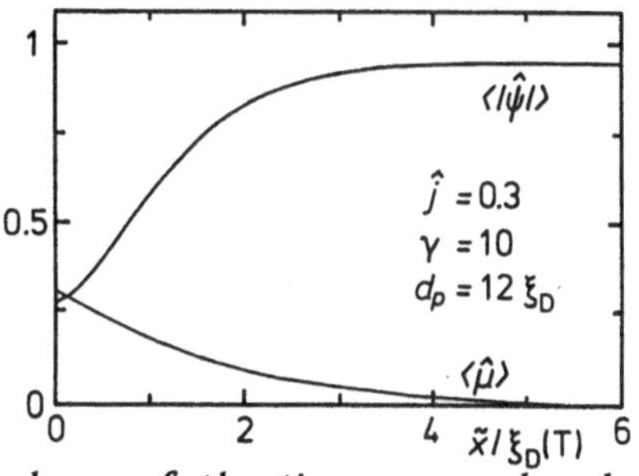

Fig. 18: *Spatial dependence of the time averaged order parameter $\langle|\hat{\psi}|\rangle$ and the potential $\langle\hat{\mu}\rangle$ in the phase-slip state as calculated from the generalized TDGL theory [316].*

Here, $\tilde{x} = x$ denotes the distance from the core of the phase-slip center, situated at $\tilde{x} = x = 0$. The gauge of $\hat{\mu}$ is $\hat{\mu}(\tilde{x} = x = d_p/2) = 0$.

The periodic length d_p and the total current $\hat{j}$ differ from the values chosen in Fig. 17.

x=0). The time averaged absolute value of the order parameter is reduced in the core region and $\langle\hat{\mu}\rangle$ decays continuously into the surrounding superconductor.

The voltage-current characteristics of the filament have been studied in much detail by Kramer and Rangel (KR) for homogeneous filaments and for the case of inhomogeneities [311, 316]. KR perform calculations for arrays of phase-slip centers but give also results for the isolated phase-slip center ($d_p=\infty$).

An example for a homogeneous filament is plotted in Fig. 19 showing the voltage drop per spatial period. The normalized voltage $\hat{V}$ denotes differences of $\hat{\mu}$ and, thus, the voltage V in physical units is related to the normalized quantity by $\hat{V}=-2eV/\tilde{\Phi}_0$. The solution exists between $\hat{j}_{min}$ and $\hat{j}_{max}$. The dissipative state is entered at the critical current of the time independent GL theory. The characteristic shows an intrinsic hysteresis and an extrapolated zero voltage intercept $\hat{j}_0$.

The interval for the existence of the oscillatory phase-slip solution is very small and below $\hat{j}_c$ for $\gamma=0$. For increasing γ the range increases to higher as well as to lower currents. For $\gamma>\gamma_c\approx 5.5$ it is $\hat{j}_{max}=\hat{j}_c$ for the isolated phase-slip center ($d_p=\infty$). The current $\hat{j}_{min}$ is rather insensitive to the distance d_p between the phase slip centers, but $\hat{j}_{min}\rightarrow 0$ for $\gamma\rightarrow\infty$. The maximum current, $\hat{j}_{max}$, depends strongly on d_p. It first increases with decreasing d_p and then decreases again if d_p becomes very small. The increase is much stronger for $\gamma=20$ compared to $\gamma=10$. These results imply

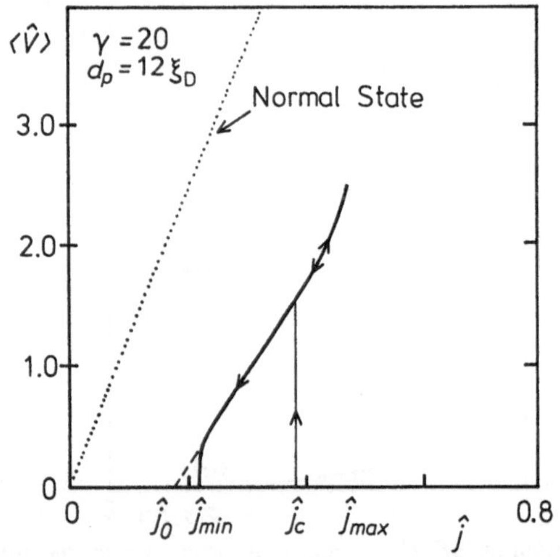

Fig. 19: *$\langle\hat{V}\rangle$-$\hat{j}$ characteristics for a phase-slip solution for a homogeneous filament according to the generalized TDGL theory [316]. Here, $\langle\hat{V}\rangle$ is the time averaged voltage in normalized units across one spatial period. Furthermore, $\hat{j}_c$ is the GL critical current in normalized units, $\hat{j}_{min}$ and $\hat{j}_{max}$ give the current range which the solution exists in, and $\hat{j}_0$ is the extrapolated zero voltage intercept.*

that the solutions for an isolated phase-slip center are restricted to the current range below or equal to j_c and that a periodic array of phase-slip centers is needed to exceed the GL critical current.

It is remarked that for $\gamma_c \approx 5.5$ it follows $\Lambda_{Q^*in} \approx \xi_D(T)$. This can be directly seen by writing Λ_{Q^*in} in the representation $\Lambda_{Q^*in} = (\gamma/u)^{1/2}\xi_D(T)$.

The idea leading to a characteristic with a regular voltage step structure is that for increasing current the filament remains in the state of lowest possible number of phase-slip centers (that means largest possible periodic length) as long as the solution is stable. Then the filament changes to the solution with one additional phase-slip center.

The numerically calculated $\langle\hat{V}\rangle - \hat{j}$ characteristics, as for example shown in Fig. 19, exhibit significant almost linear portions which may be characterized by

$$V(I) = (\rho_n \Lambda_{KR}/A)(I - \beta_{KR} I_c) \tag{174}$$

We have converted the result given in ref. 305 into physical units and dropped the brackets indicating the time average. Moreover, currents have been introduced instead of current densities.

Thus, the voltage developed by a phase-slip center along the linear portion is just that of a normal conducting region of the filament of length Λ_{KR} which a normal current $I - \beta_{KR} I_c$ is flowing through. In this sense Λ_{KR} is an 'equivalent normal length' and $\beta_{KR} I_c = I_0$ is an 'excess current' which does not create any voltage.

The equation is similar to the result of the SBT model, where the equivalent normal length (or briefly 'normal-like length') is $2\Lambda_{Q^*}$ and $I_0 = \langle I_s(X_{PSC})\rangle$ is the time averaged supercurrent in the core of the phase-slip center. As we have already discussed in section 5.8.4, the SBT result holds if the charge imbalance generated by the phase-slip center in time average decays exponentially into the surrounding superconductor with the decay length Λ_{Q^*}, and $\langle I_s(X_{PSC})\rangle$ is voltage independent. It will turn out from the TDGL theory that the interpretation of the normal-like length as two times the quasiparticle diffusion length Λ_{Q^*} is only valid in a certain limit.

In general it is found that

$$\Lambda_{KR} = C(\gamma)\, \xi_D(T) \tag{175}$$

where the function $C(\gamma) = \Lambda_{KR}/\xi_D(T)$ is tabulated in Tab. I of ref. 316 for several values of γ. Moreover a graphical representation is given in that work which is redrawn in Fig. 20 of the present work adding results for $\Lambda_{KR}/\xi_D(T)$ in the range $\gamma < 10$ which were obtained very recently by Rangel [311]. This plot gives a complete overview of the behaviour of $\Lambda_{KR}/\xi_D(T)$ for γ ranging over several orders of magnitude.

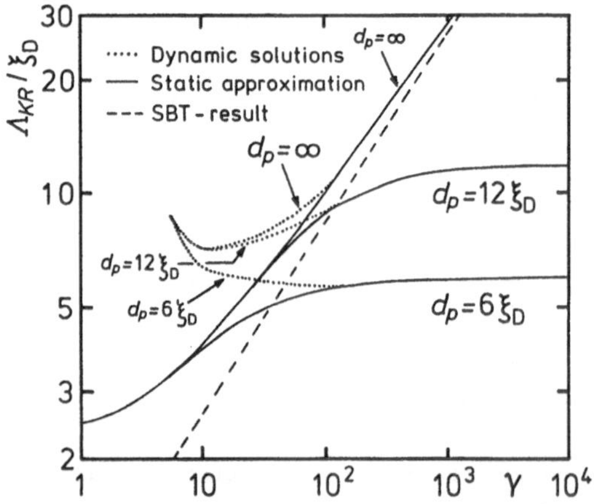

Fig. 20: *Equivalent normal length ('normal-like length') Λ_{KR} normalized by $\xi_D(T)$, as a function of the pair-breaking parameter, γ, according to refs. [311, 316]. Here, $\xi_D(T) = 0.853\,\xi_0^{1/2}\,\ell^{1/2}\,[T_{c0}/(T_{c0}-T)]^{1/2}$ and $\gamma = 6.13\,(k\,T_{c0}/\hbar)\,\tau_E\,[(T_{c0}-T)/T_{c0}]^{1/2}$. The result of the SBT model for quasiparticle relaxation due to inelastic electron-phonon scattering, $\Lambda_{KR}/\xi_D = 2\Lambda_{Q^*in}/\xi_D$, is also plotted.*

Before we explain this figure in detail it is remarked that KR found that for large γ the properties of a phase-slip center can be described by a static approximation where the order parameter remains permanently zero at the core. KR argue [311, 316] that the reason is that the relaxation time of the order parameter increases with increasing γ so that the order parameter cannot relax for large γ during a phase-slip cycle. Therefore the behaviour of the order parameter becomes more and more static and $\hat{\psi}$ remains zero in the core for $\gamma \to \infty$.

In Fig. 20 results for Λ_{KR}/ξ_D are shown for different values of the periodic length d_p, including $d_p=\infty$ (i.e. the isolated phase-slip center), for a homogeneous filament. Results are given for the dynamic solution of the TDGL equations and for the static approximation. For $\gamma \lesssim 100$ the dynamic solutions differ from the static approximation. For $\gamma \gtrsim 100$ the dynamic solutions become equal to the static approximation. To trace the behaviour of $\Lambda_{KR}/\xi_D(T)$ with increasing γ one, therefore, first has to follow the dotted line and then the corresponding full line. Concerning the static approximation it can be seen that all solutions approach the result for $d_p=\infty$ as soon as they become different from the dynamic behaviour.

As can be seen from Fig. 20, Λ_{KR} seems to be nearly independent of d_p for $\gamma \lesssim 30$ as long as Λ_{KR} is clearly smaller than d_p. Nearly no change seems to occur by changing d_p from $12\xi_D$ to ∞ in this γ range. Furthermore, β_{KR} is nearly independent of d_p for not too small periodic lengths [301, 305, 321] so that the values of β_{KR} for $d_p = 12\xi_D$, given in Tab. I of ref. 316 for γ between 10 and 80, may also be applied for the case of an isolated phase-slip center.

For an isolated phase-slip center in a homogeneous filament and γ in the neighbourhood of $\gamma=10$, it is $\Lambda_{KR}/\xi_D \approx 7$, nearly independent of γ. Thus, for $\gamma \approx 10$ it is, for the isolated phase-slip center (and also for a periodic array of phase-slip centers as long as Λ_{KR} is clearly smaller than d_p),

$$\Lambda_{KR} \approx 7\xi_D(T) \sim \Delta T^{-1/2} \tag{176}$$

where $\Delta T = T_{c0} - T$. With increasing γ the ratio Λ_{KR}/ξ_D increases. For the isolated phase-slip center the ratio reaches a value of about 10 for $\gamma=100$. For $\gamma \approx 100$ the results for the dynamic and the static approximation in the case of an isolated phase-slip center are equal to each other. The static results for $d_p=\infty$ are for $\gamma > 30$ well approximated by (see eq. (18) of ref. 316)

$$\Lambda_{KR}/\xi_D = 2(\Lambda_{Q^*in}/\xi_D)[1+0.9803(u/\gamma)^{1/2} - 0.6840(u/\gamma)+...] \tag{177}$$

For $\gamma=100$ this equation yields $\Lambda_{KR}=1.28 \cdot 2\Lambda_{Q^*in}$, so that for $\gamma \approx 100$ it is, for the isolated phase-slip center,

$$\Lambda_{KR} \approx 2.6\Lambda_{Q^*in}(T) \sim \Delta T^{-1/4} \tag{178}$$

Since for a fixed sample (where τ_E and T_{c0} do not change) γ increases with increasing ΔT, that means with decreasing temperature, KR predict a change in the temperature behaviour of Λ_{KR}, for an isolated phase-slip center, from a $\Delta T^{-1/2}$ behaviour close to T_{c0} to a $\Delta T^{-1/4}$ behaviour at lower temperatures.

For $\gamma < 10$ an increase of Λ_{KR} with decreasing γ is predicted. Thus, very close to T_{c0} the temperature dependence of Λ_{KR} is predicted to be stronger than a $\Delta T^{-1/2}$ law. Below $\gamma=6$ the numerical results seem to approach a line where $\Lambda_{KR}/\xi_D(T) \sim \gamma^{-1}$, as well for the isolated phase-slip center as also for a periodic array of phase-slip centers. More explicitly, the line is roughly given by $\Lambda_{KR}/\xi_D(T) \approx 47.6/\gamma$, as may be concluded from Fig. 20. In this case it is $\Lambda_{KR} \sim \xi_D/\gamma \sim \Delta T^{-1}$.

The generalized TDGL theory, thus, predicts, for an isolated phase-slip center in a homogeneous quasi-one-dimensional superconductor, a change of the temperature dependence of the normal-like length from a ΔT^{-1} temperature law very close to T_{c0}, to a $\Delta T^{-1/2}$ law at somewhat lower temperatures and then to a $\Delta T^{-1/4}$ law at lower temperatures. The SBT result is expected to hold for $\gamma \gtrsim 1000$. We add that the ratio $I_0/I_c = \beta_{KR}$, as tabulated in Tab. I of ref. 316, is predicted to decrease with increasing γ, that means with decreasing temperature.

Since $\gamma \sim \tau_E \Delta T^{1/2}$, the properties of a phase-slip center cannot only be changed by changing the temperature but also by making the sample from a different material. This is very important for the performance of experiments: Effects which may never be observed in one material, because the temperature range which they are predicted to occur in is, for instance, so close to T_{c0} that the transition is governed by fluctuations, may be observable in a material with smaller τ_E within a realistic temperature range.

The change of $\Lambda_{KR}/\xi_D(T)$ with increasing γ is a consequence of the two different length scales of the problem:

The relaxation length of the order parameter is $\xi_D(T)$. For the normal current (or for the electric field, or for the charge imbalance) in a superconductor being in a homogeneous superconducting state with $|\hat{\psi}|=1$, an exponential decay with the decay length Λ_{Q^*in} is predicted by the TDGL theory [311, 316].

These different length scales are important for the behaviour of a phase-slip center. For large γ, where Λ_{Q^*in} is several times ξ_D, the dominant portion of the electric field decays outside the core region of the phase-slip center, where the superconducting state is nearly homogeneous, i.e. $|\hat{\psi}|$ is close to unity. This leads to a normal-like length proportional to Λ_{Q^*in}. For small γ it is $\Lambda_{Q^*in} < \xi_D$ and the dominant part of the electric field decays in the core region yielding a normal-like length proportional to $\xi_D(T)$.

KR also investigated the influence of inhomogeneities, considering 'T_{c0}-type' weak regions and reduced mean free path 'MFP-type' weak regions. For T_{c0}-type weak regions KR obtain that the situation is shifted toward the 'static limit'. Thus (see Fig. 20), the cross-over of Λ_{KR}, for an isolated phase-slip center, from a $\Delta T^{-1/2}$ to a $\Delta T^{-1/4}$ law is shifted to lower values of γ, that means to temperatures closer to T_{c0}. As a lower limit the static approximation may be considered. The influence on $\beta_{KR}=I_0/I_c$ is not quite clear. There seems to be evidence that β_{KR} is decreased by T_{c0}-type weak regions. For MFP-type weak regions a reduction of the strength of the temperature dependence of Λ_{KR} is observed. The quantity β_{KR} is predicted to decrease with decreasing electron mean free path of the weak region.

It is important to keep in mind that the results of the generalized TDGL theory are only valid for the local equilibrium case where $\Lambda_E \ll \xi_D(T)$, i.e. $\gamma \ll \gamma_{max}$, and, therefore, the theory may not be applicable for filaments with large τ_E or for temperatures not close enough to T_{c0}. For the case that the local equilibrium approximation does not hold, Baratoff calculated the behaviour of phase-slip centers in a gapless quasi-one-dimensional filament [307, 308, 322]. He found that the time average of $\Phi_B=(\mu-\mu_F)+\hbar\dot{\varphi}/2$ shows an unusual spatial dependence. Since $\Phi_B=-e\,\Phi_{Q^*}$, if we assume that $\mu_{c,q}=\mu_F$, the results just describe the spatial dependence of the time averaged behaviour of charge imbalance decay:

Far away from the core of the phase-slip center the (time averaged) charge imbalance decays exponentially with a decay length Λ_{Q^*in}. Closer to the core there is a nonexponential drop within matching regions that decouple the core region with its rapid variations dynamically from the asymptotic wings of exponential decay. It is concluded that in addition to Λ_{Q^*in} there is a continuous spectrum of decay lengths extending to roughly $\Lambda_E=(D\tau_E)^{1/2}$.

For temperatures not too close to T_{c0} it is expected that most of the quasiparticle charge relaxes in the matching regions. Then the normal-like length corresponding to the phase-slip center is given by $2s_B^2\Lambda_E$. Here s_B^2 is a factor which has to account for the nonexponential decay in the two matching regions around a phase-slip center and for core contributions. There is not any analytic estimation available for this factor. Only for an exponential charge imbalance decay with Λ_E and if core effects can be neglected (see the SBT model) the factor s_B^2 would be unity. Nevertheless, the normal-like length and, thus, the differential resistance of the V-I characteristics is expected to be temperature independent in the sense that it is nondivergent for T approaching T_{c0}. One may evaluate an estimate for τ_E from the measurement by setting $s_B=1$, keeping in mind that the result describes an effective time, corresponding to an equivalent exponential charge imbalance decay that would give the same differential resistance.

Very close to T_{c0} the charge imbalance decay is expected to be dominated by the wings, and the normal-like length is equal to $2\Lambda_{Q^*in}$ which diverges like $\Delta T^{-1/4}$. The temperature range of the divergent behaviour at least extends from T_{c0} to a temperature T given by the condition $\tau_{GL}=\tau_E$ which is nothing else but the condition $\Lambda_E=\xi_D$.

5.10. Final and Summarizing Remarks

The overview of theories given in the present chapter represents the fundamental framework for the description of a nonequilibrium superconducting state.

In the section dealing with the (time independent) GL theory the basic quantities are introduced. These are the order parameter, $\psi(\underline{r})$, and its phase, $\varphi(\underline{r})$, the GL coherence length, $\xi(T)$, the magnetic penetration depth, $\lambda(T)$, the GL parameter, $\varkappa$. Furthermore, the decay length of the probability density of a Cooper pair is given at zero temperature ('BCS coherence length', ξ_0) and at the critical temperature ($\xi_0(T_{c0})$), respectively. The connection of the phenomenological GL equations for the order parameter and the supercurrent density with Gorkov's result derived from the microscopic many-particle theory is discussed. An important result of the GL theory is the critical current density, j_c, of a quasi-one-dimensional superconductor.

Next the basic ideas of the BCS theory are mentioned, introducing the BCS ground state and its quasiparticle excitations. The nature of a quasiparticle and its effective charge, $Q_{\underline{K}}$, is discussed in detail, because quasiparticle excitations are an important ingredient of a nonequilibrium superconducting state. Then there is a section which there are introduced the two chemical potentials in, $\mu_{c,p}$ and $\mu_{c,q}$ (and the corresponding electrochemical potentials μ_p and μ), of Cooper pairs and quasiparticles, respectively. Their values in equilibrium and in the presence of a charge imbalance are calculated as well as the influence of a time-dependent phase of the order parameter.

There is a section dealing with charge imbalance relaxation containing results for charge imbalance relaxation times in several physical situations. In this context a detailed discussion of calculations and experimental values for the inelastic electron-phonon collision time, τ_E, is needed, including the influence of the electron mean free path on τ_E and the role of electron-electron scattering.

An arbitrary quasiparticle disequilibrium envolves the excitation of two modes, the 'energy mode' and the 'charge imbalance mode'. The energy mode changes the magnitude of the energy gap and may in certain cases be described in terms of an effective quasiparticle temperature T^* which is larger than the bath temperature. Moreover, collective excitations of the whole Cooper pair and quasiparticle system may be excited. The only collective excitation which is well established is the Carlsson-Goldman mode the properties of which are given somewhat in detail.

Then the dynamics of charge imbalance is examined envolving charge imbalance relaxation and collective excitations in a situation where the chemical potential of the Cooper pairs is not necessarily independent of time. The result is a charge imbalance wave equation. In the general case this equation describes propagating damped dispersive waves of charge imbalance. One limiting case is the Carlsson-Goldman collective mode and another limit is a diffusive motion of quasiparticles with an exponential spatial decay governed by the 'quasiparticle diffusion length' (or 'charge imbalance decay length') called Λ_{Q^*}. In this context we also discuss the interaction of electrons with collective excitations ('electron-bogolon' scattering).

Then it is explained why phase-slip processes are needed in a superconductor with a potential difference between its ends. Subsequently the three main models for a phase-slip center in a current carrying quasi-one-dimensional superconductor are described. The RSM model deals with the dynamics of the order parameter applying a 'simple' TDGL equation, but does not take into account the role of nonequilibrium quasiparticles. The SBT model considers the relaxation of nonequilibrium quasiparticles, but assumes an exponential spatial charge imbalance decay. The KSS model considers the dynamics of charge imbalance and contains the SBT result as a limit.

The only description of the dissipative phase-slip state which is directly based on the microscopic many-particle theory is the TDGL theory. After reviewing the development of this approach, the generalized TDGL equations describing a finite gap superconductor are given and the results of their application to the phase-slip problem.

It is clear that the theoretical overview cannot be complete and some references may be given leading to additional work done in this field [6 - 9, 323 - 327].

Finally, all theoretical results concerning the superconducting state which are given in the 'Overview of Theories' (chap. 5) of the present work are (with some exceptions) throughout only valid for weak-coupling [2] superconductors.

This is clear for the BCS theory and all descriptions developed in this framework widely used in chap. 5. The phenomenological version of the GL equations may be valid more generally. However, the microscopic derivation of these equations is based on the Gorkov Green's function equations [46, 47, 285, 328, 329] which are only valid for the weak-coupling case. The Gorkov Green's function equations are also the basis of the theory of Schmid and Schön concerning the relaxation of charge imbalance and of the energy mode [79]. Also the generalized TDGL equations are derived starting from Gorkov's equations [305].

In the work of Kaplan et al. [103] most results for the quasiparticle lifetimes are valid for weak-coupling superconductors, but there are also results given for the strong-coupling materials Hg and Pb.

A special case is the Josephson relation which reflects a general property of a quantum mechanical state characterized by a macroscopic wave function. This relation seems to be exact although concluded from a theory containing approximations and it may be applied for all kinds of (equilibrium) superconductors.

Most experiments dealing with nonequilibrium superconductivity have been carried out with weak-coupling superconductors such as aluminium and tin. However, also superconductors with a strong electron-phonon coupling were studied [14, 25, 40, 93, 138]. An example is the strong-coupling material lead , which experimental results will also be given for in the present work. As usually done, we consider the weak-coupling limit results for an explanation but introduce material parameters for the strong-coupling material such as the energy gap $\Delta(0)$ which differs from the weak-coupling BCS result [14, 40, 138, 330].

Indeed the slope of the reduced energy gap, $\Delta(T)/\Delta(0)$, as a function of the reduced temperature, T/T_{c0}, for most strong-coupling superconductors is in a good approximation given by the BCS weak-coupling limit result [2, 331]. Moreover, the single particle momentum occupation probability in the superconducting ground state (called $v_{\underline{k}}^2$ in the BCS theory) is similar to the BCS result. Also for strong coupling superconductors the occupation probability shows a rapid but continuous change at the Fermi energy. The only difference is that this change is less steep in the strong-coupling case and the occupation probability is somewhat smaller than for a weak-coupling superconductor below the Fermi energy and somewhat larger above [331].

Another basic feature is the probability of transferring energy and momentum to the groundstate of the system by injecting or extracting an electron. For a BCS superconductor this probability is very large at the gap edge due to the singularity in the quasiparticle density of states (see Fig. 9 for a sketch of the BCS quasiparticle excitation spectrum). Then the probability decreases with increasing quasiparticle energy, until the quasiparticle density of states has become equal to the normal state density of states, which is nearly constant close to the Fermi border in a free electron model. Also in a strong-coupling superconductor the transfer probability is large at the gap edge where the excitation spectrum is close to the BCS result. For increasing energy the situation becomes more involved (see Fig. 50 of ref. 331 and consider the particle-hole symmetry discussed in that work to get the complete information about the transfer probability). The transfer probability also decreases, but the excitation spectrum deviates more and more from the BCS spectrum. For even higher energies the spectrum of a strong-coupling superconductor approaches again the BCS result while the transfer probability increases again. The transfer probability is small in regions where the excitation spectrum of the strong-coupling superconductor deviates substantially from the BCS spectrum.

These facts in mind, the weak-coupling limit theories may be applied also for the interpretation of results obtained for strong-coupling superconductors in the sense of a (more or less) rough approximation.

In the framework given in the overview of theories we will try to understand our experimental results discussed in the following chapters. This will be possible in many cases, but there are situations which additional phenomenological models have to be developed in for the description of the phenomena observed, and there are experimental results which are not understood at present.

6. Equilibrium Properties

Before one can start with the experimental study of nonequilibrium effects in a quasi-one-dimensional superconductor, it is necessary to investigate the equilibrium properties of its superconducting state. Information is needed about the critical current, I_c, and the critical temperature, T_{c0}, in order to decide whether the voltage step structure in the transition characteristics appears below I_c or at overcritical currents.

We identify the current related to the first onset of voltage at the sample with the critical current and, thus, assume the voltage step structure to be an overcritical phenomenon. In the following sections it is shown that this choice leads to reasonable results for the critical current and the extrapolated critical temperature.

6.1. Critical Current

The dependence of the critical current, I_c, on the transition temperature at this current, T_c, of a pure tin whisker with a diameter of about 1.3 μm has been plotted in Fig. 21 together with the prediction of theoretical calculations. We added measurements of the 'jump-back current', I_R, at which the whisker recovers the superconducting state for decreasing current [332]. The vertical and horizontal distance between both experimental curves gives the width of the hysteresis in current and temperature, respectively, which will be discussed in detail in chap. 11.

Very close to the critical temperature, $I_c^{2/3}(T_c)$ follows the straight line predicted by the GL theory for a quasi-one-dimensional superconductor (see section 5.1.). It is remarkable that the measurements follow this straight line also in a temperature region where strictly the quasi-one-dimensionality condition is violated because the diameter of the sample is still smaller than $\xi(T)$ but already several times larger than $\lambda(T)$.

For lower temperatures a deviation from the straight line occurs. In this temperature range the critical current follows Silsbee's rule. This rule is valid for a thick wire of a type I superconductor with a diameter much larger than $\lambda(T)$ and $\xi(T)$, where the supercurrent only flows in a thin skin at the surface of the wire. If the magnetic field generated by the supercurrent at

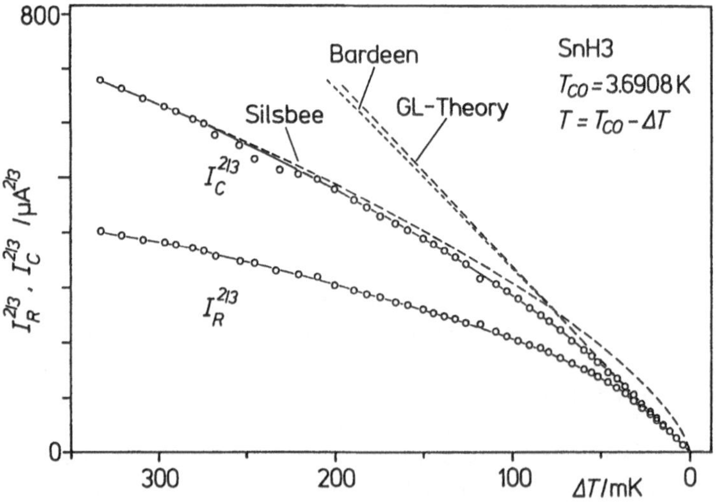

Fig. 21: *Critical current, $I_c^{2/3}$, and jump-back current, $I_R^{2/3}$, for a tin whisker (sample Sn H 3 of ref. 332) as a function of the bath temperature, T, which is equal to the transition temperature, T_c. Here, T_{c0} is the critical temperature. The full line is the estimated slope through the measured values. The dashed lines are the results of theoretical calculations for the critical current due to the GL theory for a quasi-one-dimensional superconductor, Bardeen's formula, and Silsbee's rule, respectively.*

the surface of the sample becomes equal to the critical magnetic field, the superconductor changes into the dissipative intermediate state. Thus, the critical current is determined by the condition [1]

$$B_{cth}(T_c) = \mu_0 I_c / 2\pi r_w \tag{179}$$

The thermodynamical critical magnetic field is given by [333]

$$B_{cth}(T) = B_{cth}(0)[1-(T/T_{c0})^2 + D(T)] \tag{180}$$

where D(T) denotes a slight deviation of the critical field from a parabolic law. This deviation has been predicted by theory [48] and there are detailed experimental investigations of this quantity for weak-coupling and strong-coupling superconductors [331, 333 - 339]. Then it follows

$$I_c(T_c) = I_c(0)[1-(T_c/T_{c0})^2 + D(T_c)] \tag{181}$$

with $I_c(0)=2\pi r_w B_{cth}(0)/\mu_0$, where r_w is the radius of the whisker assuming a circular cross sectional area. We determined $I_c(0)$ at the lowest measuring temperature where the approximation of a thick wire should be best satisfied. Using $D(T_c)$ as measured for bulk material [333], we calculated $B_{cth}(0)$ for tin and found satisfactory agreement with the bulk material value [332]. Moreover, $I_c(T_c)$ as given in eq. (181) leads to a good description of the critical current over a wide temperature range. For temperatures close to T_{c0}, where $I_c^{2/3}(T_c)$ is a straight line the critical current is overestimated. It is remarked that Silsbee's result as drawn in Fig. 21 has been calculated without considering the correction $D(T_c)$.

In Fig. 21 we have also drawn the result of Bardeen's phenomenological expression for a filament which is so thin that it remains quasi-one-dimensional at lower temperatures [106, 332]

$$j_c(T_c) = j_c(0)\,[1-(T_c/T_{c0})^2]^{3/2} \tag{182}$$

where $j_c(0)=K_{GL}/2^{3/2}$, and the constant K_{GL} has been defined in section 5.1. in connection with the GL critical current. Bardeen's formula is not flat enough to describe our experimental results. The reason is that our sample at lower temperatures seriously violates the quasi-one-dimensional condition due to the decrease of $\xi(T)$ and $\lambda(T)$ with decreasing temperature and, therefore, cannot be described by this expression.

Romijn et al. [340, 341] investigated narrow, dirty ($\ell \ll \xi_0$) evaporated thin aluminium strips which the current density is uniform for over the cross section even far below T_{c0}, because the widths of their samples are smaller than the magnetic penetration depth $\lambda(T)$ (which increases with decreasing ℓ) for all measuring temperatures. The GL coherence length (which decreases with decreasing ℓ, but ξ_0 of Al is rather large) is still large enough compared to the width, so that no vortices can exist in the strips and, therefore, no vortex flow can occur. Qualitatively the shape of the results of Romijn et al. is similar to that of our measurements, but their critical currents, measured down to $T_c/T_{c0} \approx 0.2$, are reasonably well described by Bardeen's formula and well described by the theory of Kupriyanov and Lukichev [342]. Moreover, Neumann and Kao [343] found a $[1-(T_c/T_{c0})^2]^{3/2}$ temperature law for very narrow indium strips down to $T_c/T_{c0} \approx 0.4$.

The conclusion is that the tin whisker of Fig. 21 behaves like a quasi-one-dimensional superconductor very close to T_{c0} and like a three-dimensional wire far away from the critical temperature. There is an intermediate region where $I_c^{2/3}(T_c)$ deviates from a straight line but does not follow Silsbee's rule.

Our measurements are usually carried out close to T_{c0} where we found the $I_c^{2/3}(T_c)$ straight line predicted by the GL theory for whiskers made of various materials and alloys, such as Sn [19, 22, 23, 37, 332], <u>Sn</u>-In alloys [15, 28], In [16, 17], <u>In</u>-Sn alloys [17], the In-Pb alloy system [25], Zn [20], and <u>Zn</u>-Ag alloys [20].

An example for a measurement of an indium whisker is shown in Fig. 22. The straight line is the estimated slope through the results of the measurements, with a value of $dj_c^{2/3}/dT_c$ which differs only by 4% from the

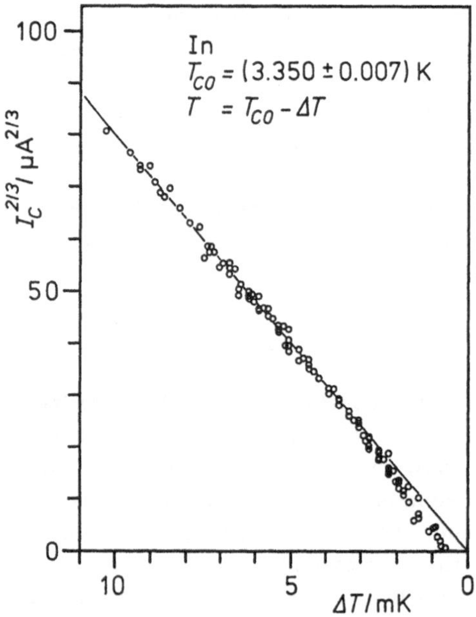

Fig. 22: *Critical current, $I_c^{2/3}$, as a function of the bath temperature, T, which is equal to the transition temperature, T_c, for an indium whisker (sample In of ref. 16) at temperatures close to the critical temperature, T_{c0}. The full line is the estimated slope through the measured values.*

prediction of the GL theory [16]. The deviation of the experimental results from the straight line for temperatures very close to T_{c0} is probably caused by fluctuation effects.

We compared the measured values for $d j_c^{2/3}/d T_c$ with the prediction of the GL theory (see eq. (17) of section 5.1) and always found agreement between experiment and theory over a wide range of electron mean free paths. As an example the results for Sn-In alloys [15] are given in Fig. 23. The asymmetric error bars at the experimental values are caused by the inaccuracy of the determination of the length of the whisker, L, with the light microscope. The squeezed Wood metal of the contacts hinders the view on the actual locus of contact. Therefore, there is a much smaller tolerance of L toward a shorter length than toward a longer length. Similar plots are obtained for whiskers made of In and In-Sn alloys [17], the whole In-Pb alloy system [25], and of Zn and Zn-Ag alloys [20]. There is no fitting parameter in the theory. The critical temperature, T_{c0}, which is extrapolated from the straight line, $I_c^{2/3} \sim (T_{c0} - T_c)$, for vanishing current actually does not act as a fitting parameter because it is not chosen to give agreement with the theoretical value of the slope of the $I_c^{2/3}(T_c)$ straight line.

The agreement with the GL theory implies that the material parameters used and the determination procedure of the electron mean free path should be reasonable for our samples. Furthermore, it justifies the identification of

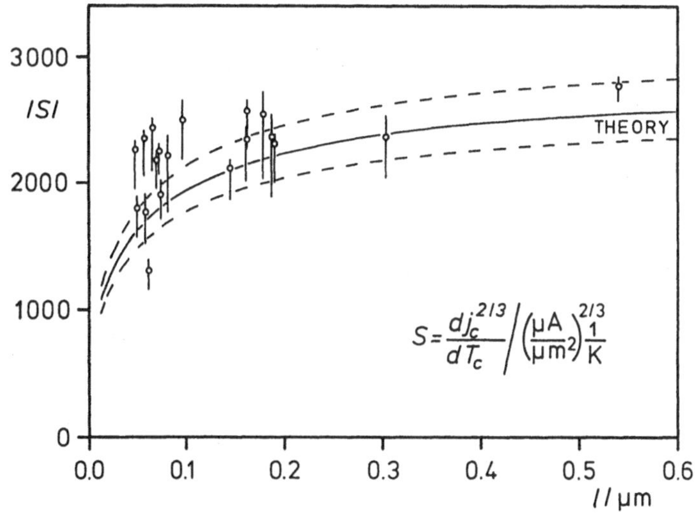

$$S = \frac{dj_c^{2/3}}{dT_c} \Big/ \left(\frac{\mu A}{\mu m^2}\right)^{2/3} \frac{1}{K}$$

Fig. 23: *Slope of the $j_c^{2/3}(T_c)$ straight line for the Sn-In alloy whiskers of ref. [15] versus the electron mean free path ℓ. Here, j_c is the critical current density. The full curve has been calculated from the GL theory (eq. (17) of section 5.1.), considering the change of the critical temperature as a function of ℓ in an anisotropic superconductor according to the theory of Markowitz and Kadanoff [89]. For details of this calculation see appendix 6 of ref. [15]. The dashed lines give the inaccuracy of the theoretical curve due to the inaccuracy of the material parameters entering the calculation.*

the critical current with the current which the first onset of voltage occurs at. Such an agreement would never be obtained for the current which the fully normal conducting state is reached at. That current shows a very different behaviour, as, for instance, plotted in refs. 19, 28, and 37.

The $(T_{c0} - T_c)$ temperature dependence of $I_c^{2/3}$ close to T_{c0} has also been reported for long, narrow evaporated thin films (microbridges) made of Sn [32, 344], In [343], and Al [35, 73, 322, 340, 341].

6.2. Critical Temperature

The critical temperature, T_{c0}, is the transition temperature T_c in the absence of currents and fields. The values given in the present work are obtained by extrapolating the $I_c^{2/3}(T_c)$ straight line to vanishing critical current. This procedure neglects the influence of fluctuations on the superconducting transition [38, 42 - 44] which suppress the critical current below the $I_c^{2/3}(T_c)$ straight line for temperatures extremely close to T_{c0} [38].

In superconductors with anisotropic properties of the crystal lattice the strength of the effective electron-electron interaction via virtual phonons ('electron-phonon interaction') and, thus, the energy gap is different in different crystallographic directions. For pure bulk material the critical temperature is governed by the direction with the largest energy gap. In the presence of (nonmagnetic) impurities the electrons are scattered at these irregularities, thus, traveling through regions of different strength of electron-phonon interaction. As we have already discussed in section 5.4 in connection with the charge imbalance relaxation due to gap anisotropy, the pairing is now between reversed exact eigenstates of the doped normal metal which are less able to take advantage of the crystal anisotropy because they are a superposition of plane waves with different wave number vectors [89, 90, 91]. The consequence is a reduction of the critical temperature of an anisotropic superconductor containing a small amount of impurities (called 'mean-free-path effect' or 'anisotropy effect' [1, 89]). For higher concentrations the superconductor enters a region where the changes in T_{c0} are more and more determined by properties of the specific impurity (called 'valence-effect' [1, 89]).

Tin is an anisotropic material [15, 345] and, therefore, samples of Sn-In alloys show the effects described above [15] as demonstrated in Fig. 24.

A theoretical expression for the reduction, $\delta T_{c0} = T_{c0} - T_{c0r}$, of the critical temperature, T_{c0}, below the critical temperature T_{c0r} of the clean matrix, has been calculated by Markowitz and Kadanoff [89], yielding

$$\delta T_{c0} = K^i \tilde{\chi}_i + \langle a^2 \rangle T_{c0r} \tilde{I}_1(\tilde{\chi}_i) \tag{183}$$

Here, $\langle a^2 \rangle$ characterizes the anisotropy of the gap, $\tilde{\chi}_i$ is proportional to the density of impurities, and K^i is a parameter characterizing all effects besides the anisotropy effect. The function $\tilde{I}_1(\tilde{\chi}_i)$ is given by eq. (47) of ref. 89 and is plotted in Fig. 4 of that work. In the range $1 < \tilde{\chi}_i < 100$ there is an analytical approximation for this function, so that

$$\delta T_{c0} = (K^i - 0.36 \langle a^2 \rangle T_{c0r}) \tilde{\chi}_i + 0.078 \langle a^2 \rangle T_{c0r} \tilde{\chi}_i \ln \tilde{\chi}_i \tag{184}$$

Setting for tin with indium impurities $\tilde{\chi}_i = C \rho_{nSnIn\perp}$, where C is constant, and using results for C, $K^i C$, and $\langle a^2 \rangle$ as determined experimentally for bulk tin single crystals with indium impurities [346] leads to the full curve in Fig. 24. Details are given in appendix 5 of ref. 15. Here, $\rho_{nSnIn\perp}$ is the residual resistivity of tin with indium impurities, perpendicular to the tetrad axis of the tetragonal elementary cell of 'white tin'.

For whiskers we obtain the maximum lowering of the critical temperature at the same resistivity as in the case of bulk material. For small resistivities there is a stronger reduction of T_{c0} in the case of whiskers. This effect is probably caused by specular scattering at the surface of the whisker which does not contribute to the residual resistance but may cause a reduction of the gap anisotropy. This conclusion would explain the reduction of the critical temperature below the bulk material for pure tin whiskers.

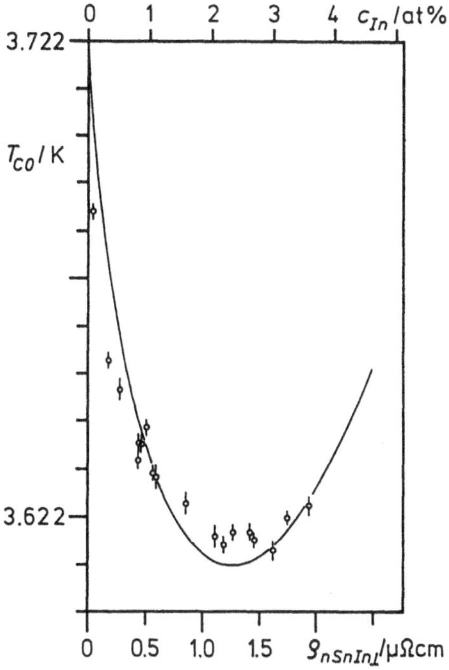

Fig. 24: *Critical temperatures, T_{c0}, of whiskers from Sn-In alloys as a function of the resistivity $\rho_{nSnIn\perp}$, respectively as a function of the indium concentration [15]. The full curve is valid for bulk single crystals.*

The measuring point at the smallest residual resistivity in Fig. 24 is obtained from a pure Sn whisker showing the typical reduction of the critical temperatures for these specimens.

Due to their hexagonal crystal structure[18] also Zn and Zn-Ag alloys show an anisotropic behaviour of their physical properties[347 - 349]. Also the energy gap is anisotropic[351 - 353]. A critical discussion of energy gap measurements is given by Auluk[354]. The critical temperature of whiskers made of Zn and Zn-Ag alloys is reduced below the bulk material value of pure zinc[20]*¹. The results for the alloys are well described by the curve of Markowitz and Kadanoff as given in eq. 184 setting $\tilde{\chi}_i$ proportional to the residual resistance ratio, $\rho^* = R_n/(R_{298K} - R_n)$, and inserting parameters valid for bulk material alloys [20, 355]. The experimental values obtained for the critical temperature of pure Zn whiskers are situated alongside the curve, because for a given ρ^* their T_{c0} is much more depressed than predicted by the Zn-Ag alloy curve. For alloy whiskers a scattering of conduction electrons at impurities dominates. For pure Zn whiskers the dominating contribution to the residual resistance is diffuse surface scattering. However,

*¹ See Fig. 5 of ref. 20. The numbers indicating the division of the temperature scale should have negative signs.

specular scattering at the surface is also present [356], and as in the case of tin whiskers we conclude that this effect although not visible in ρ^* leads to a gap anisotropy reduction.

Indium is a slightly anisotropic material [14, 347], also showing anisotropy effect if being alloyed with a small amount of impurity atoms [89, 357, 358]. Also our pure In whiskers show a reduction of their critical temperature below the bulk material value of pure indium [16, 17]. For increasing residual resistance ratio (that means decreasing ℓ) the critical temperature first decreases and then increases again (see Tab. 2 of ref. 17). Since we did not dope the material by a certain impurity, the change in ρ^* and T_{c0} is probably caused by surface scattering effects or fortuitous impurities. In principle a decrease of T_{c0} and a subsequent increase can also occur if ρ^* is changed by increasing the number of lattice defects by plastic deformation [1]. This effect is not very probable in the case of whiskers. Even submitted to a strong bending a straight whisker becomes straight again after removing the bending force, indicating that the filament was deformed elastically.

Our In-Sn alloy whiskers have critical temperatures above the value of pure bulk indium [17] and systematically fit the results of Seraphim et al. [358] performed for In-Sn alloy bulk material. Also indium doped with lead impurities should lead to an anisotropy effect [89, 357, 358]. We have, however, not investigated In-Pb whiskers in the low concentration region so that our samples have already critical temperatures above the pure indium bulk material [14, 40].

Also the critical temperature of pure Pb whiskers is reduced below that of pure bulk material [14, 40, 359]. Adding small concentrations of In leads to critical temperatures below the critical temperature T_{c0r} of pure bulk Pb, while small amounts of Bi may depress T_{c0} below T_{c0r} or (for somewhat larger concentrations) enhance T_{c0} to values above T_{c0r} [359]. Thus, there is evidence for an anisotropy effect in lead. However, the data are not sufficient to make a fit using eq. (184) and to get K^i and $\langle a^2 \rangle$ from this fit.

Also measurements of other specimens seem to indicate an anisotropy of the energy gap in lead. Doping bulk Pb with impurity concentrations between 1 and 10 at% in several cases leads to a linear decrease of T_{c0} [360, 361]. This linear decrease does, however, not start from zero impurity concentration but extrapolates to a temperature below the critical temperature of pure bulk lead. Thus, for concentrations below about 1 at% there must be a much steeper decrease of T_{c0}. This decrease has indeed been measured by Gamari-Seale and Coles [362] and has been interpreted as anisotropy effect. This interpretation has been adopted by the other authors cited [360, 361]. There is evidence for the anisotropy of the energy gap in lead also from other investigations, although a reliable value of the anisotropy parameter $\langle a^2 \rangle$ is not available from the literature [363 - 366].

7. Fundamental Properties of Phase-Slip Centers

7.1. Introductory Remarks

After the experimental discovery of a voltage step structure in the characteristics of tin whiskers and its first systematic study (see chap. 4), Skocpol, Beasley, and Tinkham (SBT) interpreted the phenomenon as a sequence of localized phase-slip centers activated one after the other when the measuring current is increased (see section 5.8.3). We then started a series of experiments performed with Sn and In whiskers (including Sn-In and In-Sn alloys) to prove the fundamental assumptions and predictions of this model.

The first test concerns the statement that the voltage developed at a phase-slip center is governed by the diffusion and relaxation of nonequilibrium quasiparticles. For this purpose we calculated the quasiparticle diffusion length from the differential resistance of a single voltage step. Usually we used the first step which represents an isolated phase-slip center.

Next, the properties arising from the dynamics of the order parameter in the core of the phase-slip center were investigated. For this problem we determined the time averaged supercurrent given by the extrapolated zero voltage intercept of the V - I characteristics after the first voltage step. Furthermore, we subjected the whisker to high-frequency radiation to prove whether a cyclical high-frequency process at Josephson frequency is associated with the appearance of a phase-slip center.

To answer the question of a localization of the phase-slip centers in the sample, measurements with several potential probes have been performed where the V - I characteristics of several sections of the sample could be measured at the same run. Multi-potential-probe samples are also used to explore the reason for the distance in current between the voltage steps which turns out to be the consequence of a stabilization phenomenon caused by an active phase-slip center.

Finally, some measurements of the electrochemical potentials of Cooper pairs and quasiparticle excitations are reported in the present chapter. The results indicate a modification of the slope proposed by SBT. For an interpretation of these measurements a model calculation is presented which is based on the assumption of an excitation of collective excitations by the phase-slip center.

In the following sections the experiments will be discussed somewhat in detail. It turns out that the SBT model seems to contain most of the essential physics needed for the description of the voltage-step phenomenon.

7.2. Quasiparticle Diffusion Length and Time Averaged Supercurrent

The V - T transition curves and the V - I characteristics of Sn, Sn-In, and In whiskers look very similar for the different materials. For Sn we showed some curves in chap. 4 and we refer to the cited literature, where both kinds of curves are plotted for the other metals [15, 16].

The first voltage step of a measured V - I characteristic of these samples is sketched in Fig. 25. It is characterized by a straight line with differential resistance $(dV/dI)_1$ following a voltage jump of height V_1 at the critical current I_c and by an extrapolated zero voltage intercept I_0. Furthermore, the jump-back current I_R and voltage V_R characterize the hysteretic behaviour. The first voltage step is, thus, empirically described by

$$V(I) = (dV/dI)_1(I - I_0) \tag{185}$$

Now we assign a 'normal-like' length, L_{An1}, to $(dV/dI)_1$, as the whisker length, L, is assigned to the residual resistance, R_n, by equating $L_{An1}/(dV/dI)_1 = L/R_n$, or

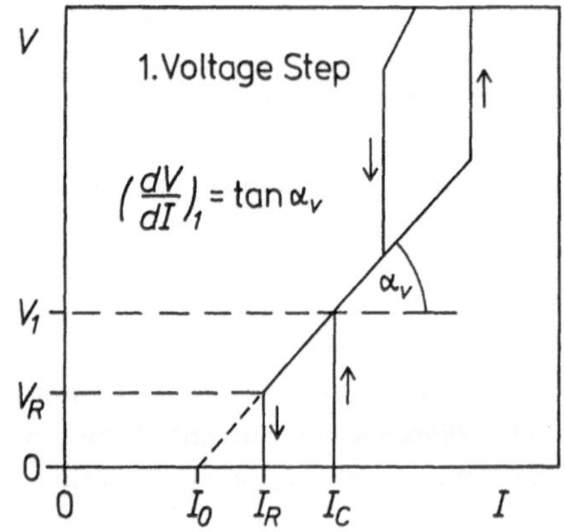

Fig. 25: *Sketch of the first voltage step (and the lower part of the second voltage step) in the V-I characteristics of a quasi-one-dimensional superconductor. The first voltage step is related to the appearance of an isolated phase-slip center in the filament. The quantities are explained in the text.*

$$L_{An1} = (L/R_n)(dV/dI)_1 \tag{186}$$

Introducing this length into the expression for the voltage and using $R_n = \rho_n L/A$, where ρ_n is the residual resistivity and A the cross-sectional area of the sample, yields

$$V(I) = (\rho_n L_{An1}/A)(I - I_0) \tag{187}$$

This expression shows that the voltage can be described as if generated by a normal conducting region of length L_{An1} which a normal current $I - I_0$ is flowing in.

For the materials mentioned the V-I characteristics of the first voltage step is a straight line. In other words, the differential resistance and the quantity I_0 do not depend on the total current, I. This is also the case in the SBT model and we, therefore, can directly compare eq. (187) with eq. (138) getting

$$L_{An1} = 2\Lambda \tag{188}$$

and

$$I_0 = \langle I_s(x_{PSC}) \rangle \tag{189}$$

where the quasiparticle diffusion length, $\Lambda = ((1/3)\ell v_F \tau_2)^{1/2}$, has to be identified with the charge imbalance relaxation length, $\Lambda_{Q*} = ((1/3)\ell v_F \tau_{Q*})^{1/2}$, and $\langle I_s(x_{PSC}) \rangle$ is the time (or phase angle) averaged supercurrent flowing in the core of the phase-slip center located at x_{PSC}.

The KSS model yields that the SBT result only holds in the high voltage dc limit where $I \gg I_c$. In general, the time average of the supercurrent in the core is not equal to its measuring current independent phase angle average, but changes along the V-I characteristics.

The TDGL theory leads to $L_{An1} = \Lambda_{KR}$ and $I_0 = \beta_{KR} I_c$, where Λ_{KR} becomes equal to Λ_{Q*in} only for large values of the pair-breaking parameter, γ.

These arguments in mind, a comparison of experimental results with the SBT model should be regarded as a first step to clarify the nature of the phenomena observed and to come to a quantitative analysis of the measurements. The basic problems to deal with are the dependence of L_{An1} on ℓ, the interpretation of the evaluated scattering time τ_2 as a charge imbalance relaxation time, and the order of magnitude of the time averaged supercurrent.

For whiskers of Sn and Sn-In alloys the differential resistance $(dV/dI)_1$ and, thus L_{An1}, and the ratio I_0/I_c are usually found to be temperature independent throughout the investigated temperature range [15, 16, 19, 23, 367]*1.

---

*1 For tempertures very close to T_{c0} there seems to be a slight temperature dependence in the data for the pure Sn whisker of Fig. 6 (see ref. 16 for an evaluation of $(dV/dI)_1$). We shall come back to this problem in chap. 12.

In this case one may average results obtained for different measuring temperatures.

The length L_{An1} is calculated for each sample from its averaged differential resistance [15] and then plotted in Fig. 26 as a function of the electron mean free path ℓ of the sample, using a double logarithmic scale. The experimental values follow a straight line with slope 1/2 which means that $L_{An1} \sim \ell^{1/2}$ as predicted by the SBT model.

The measuring points for $\ell < 1\,\mu m$ are obtained from <u>Sn</u>-In alloy whiskers while those for $\ell > 1\,\mu m$ are from pure Sn whiskers. For the point most to the right, the electron mean free path ℓ is larger than the diameter of the sample, d_w, as calculated assuming a circular cross sectional area. This indicates that specular scattering of electrons at the surface of the whisker occurs which does not contribute to the residual resistance which the electron mean free path is calculated from. In section 6.2 we argued that specular surface scattering may cause a reduction of gap anisotropy. If we would assume now that a specular scattering event also influences the relaxation behaviour of nonequilibrium quasiparticles, so that ℓ is limited by d_w, this would lead to a position of the measuring point which is closer to the solid line in Fig. 26. Also results from other pure tin whiskers [22, 367] systematically fit into this figure and for samples with $\ell > d_w$ the measuring points are closer to the solid line if ℓ is replaced by d_w.

We also evaluated the differential resistance of the second voltage step for several Sn and <u>Sn</u>-In alloy whiskers. In analogy to eq. (186) we define a length L_{An2} by $L_{An2} = (L/R_n)\Delta(dV/dI)_2$, where

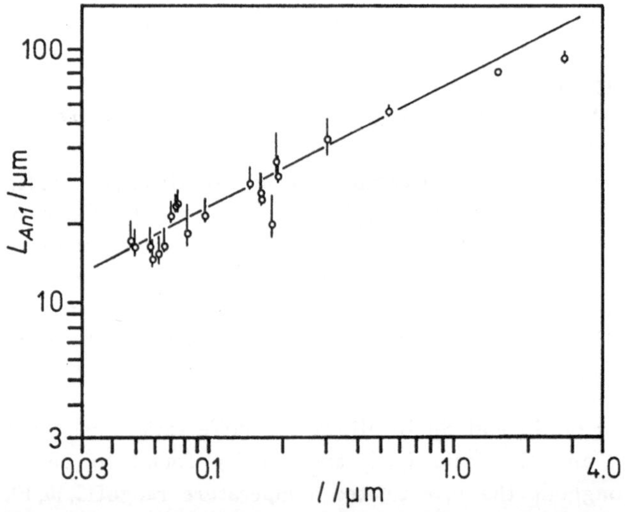

Fig. 26: *Normal-like lengths, L_{An1}, of Sn and <u>Sn</u>-In alloy whiskers as a function of the electron mean free path ℓ of each sample, using log-log coordinates [15]. The solid line has the slope 1/2 and, thus, represents the dependence $L_{An1} \sim \ell^{1/2}$.*

$\Delta(dV/dI)_2 = (dV/dI)_2 - (dV/dI)_1$ is the increase of the differential resistance due to the appearance of the second voltage step. The results for L_{An2} follow the same straight line as those for L_{An1}. This indicates that the second voltage step is generated by a repetition of the phenomenon leading to the first step.

From the measurements of L_{An1} the quasiparticle relaxation time τ_2 can be calculated by $\tau_2 = 3L^2_{An1}/(4\ell v_F)$. The result depends on the value inserted for the Fermi velocity [15, 17 remark 6]. According to ref. 89 it is $v_F = 6.84 \cdot 10^5$ m/s for tin. Assuming that this value also holds for our Sn-In alloy whiskers with their small In impurity concentration it follows $\tau_2 = 6.14 \cdot 10^{-9}$ s for each point of the solid line in Fig. 26. This result (called τ_{OSnW} in the following) represents a graphically averaged value for all Sn and Sn-In samples measured. Since the time is independent of temperature it cannot be related to one of the steady-state charge imbalance relaxation times given in section 5.4.

We also averaged the ratio I_0/I_c for each sample [15] and plotted the results as a function of the residual resistance ratio ρ^* of each sample in Fig. 27. Most results are obtained from Sn-In alloy whiskers. There are three measuring points, for $\rho^* < 0.01$, which belong to pure Sn whiskers and represent the typical range of I_0/I_c for these specimens. In most cases pure tin whiskers show values for I_0/I_c in the upper half of this range and their I_0/I_c may even somewhat exceed the value 0.8 [22, 23, 332, 367]. For increasing alloy content (that means increasing ρ^* or decreasing ℓ) the ratio I_0/I_c firstly decreases to values below 0.5 and then slightly increases again. Although the special slope of I_0/I_c with changing alloy content cannot be understood within the SBT model, which predicts $I_0/I_c = 0.65$ for all

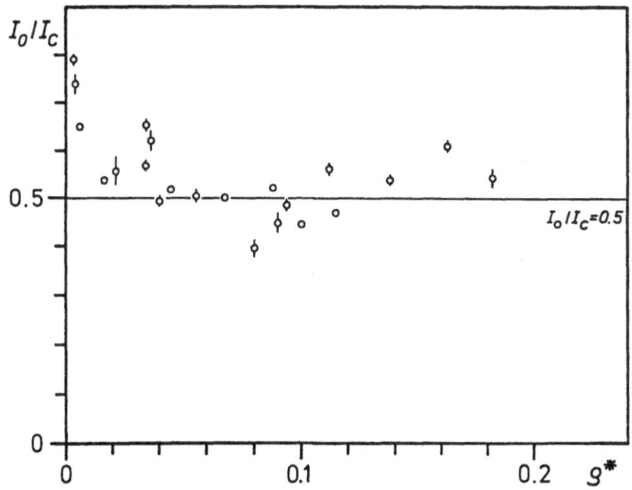

Fig. 27: *Ratio I_0/I_c versus the residual resistance ratio, ρ^*, for Sn and Sn-In alloy whiskers [15].*

specimens, the experimental values are not far away from its prediction. There is also a region where I_0/I_c is obtained close to 0.5 as the RSM model yields.

For In whiskers $(dV/dI)_1$ and I_0/I_c are only independent of temperature for temperatures not too close to T_{c0}. Very close to the critical temperature, however, In whiskers show a differential resistance and a ratio I_0/I_c which is strongly temperature dependent[16, 17]. Experimental results of $(dV/dI)_1$ and I_0/I_c as a function of the temperature for an In whisker are shown in Fig. 28. For $\Delta T > 7$ mK both quantities do not depend on temperature, but there is a rapid increase if the temperature approaches the critical temperature. It is remarkable that, nevertheless, the height of the first voltage jump, $V_1 = V(I_c)$, follows a straight line[16] as in the case of Sn and Sn-In whiskers. Since $V_1 = (dV/dI)_1 (1 - I_0/I_c) I_c$, this observation indicates that

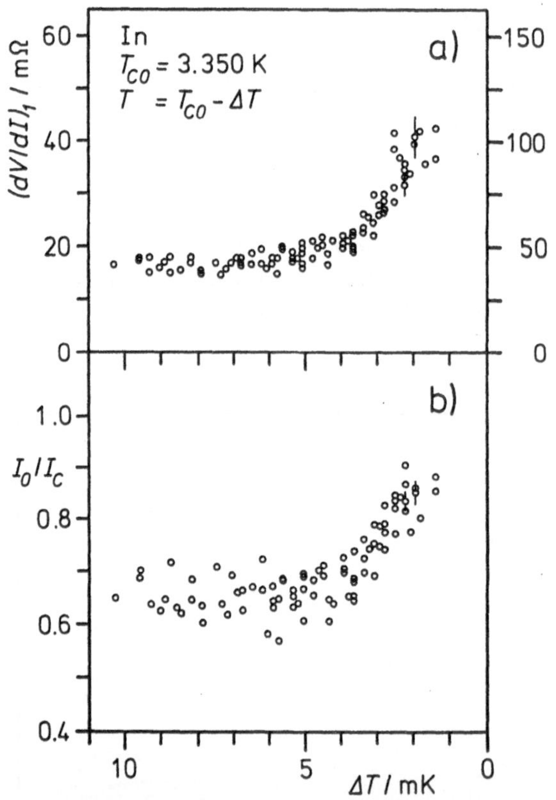

Fig. 28: *Differential resistance $(dV/dI)_1$ and ratio I_0/I_c of the first voltage step in the V-I characteristics of an indium whisker (sample In of ref. 16) as a function of the temperature, $T = T_{c0} - \Delta T$. The size of the normal-like length, $L_{An1} \sim (dV/dI)_1$, is indicated by the right hand scale in part a) of the figure.*

for In whiskers the temperature dependences of $(dV/dI)_1$ and $(1-I_0/I_c)$ compensate each other so that their product is independent of the temperature and, therefore, is constant for different critical currents.

For In whiskers (and In-Sn whiskers) we averaged $(dV/dI)_1$, $\Delta(dV/dI)_2$, and I_0/I_c only in the region where the quantities do not depend on temperature. The resulting 'low temperature values', $(L_{An1})_{0InW}$, $(L_{An2})_{0InW}$, and $(I_0/I_c)_{0InW}$, behave very similar to the corresponding quantities of Sn and Sn-In whiskers[17]. Both normal-like lengths are proportional to $\ell^{1/2}$, following the same straight line of slope 1/2 in a double logarithmic plot which a temperature independent quasiparticle relaxation time $\tau_2=\tau_{0InW}=2.23\cdot10^{-9}$ s is obtained from (using $v_F=7.20\cdot10^5$ m/s valid for In[89]). The results for $(I_0/I_c)_{0InW}$ scatter around a value of about 0.68 which is close to the SBT prediction[17]. There seems to be a very slight increase in this data for increasing ℓ (that means decreasing ρ^*), but the results remain below those obtained for Sn whiskers. It is, however, remarked that even the purest In whiskers have residual resistance ratios which are much larger than for a tin whisker[15, 17, 22, 367].

Next we discuss the evaluation and interpretation in the temperature range where the characteristic properties, $(dV/dI)_1$ - and thus L_{An1} - and I_0/I_c, of a voltage step increase for T approaching T_{c0}. While the behaviour of I_0/I_c cannot be understood within the SBT model there is a way to interpret the divergent behaviour of L_{An1}. The idea is to examine the experimental data for $(dV/dI)_1$ or L_{An1} concerning their temperature dependence and then to look for a charge imbalance relaxation process with a temperature dependent steady state charge imbalance relaxation time, τ_{Q^*}, which yields the same temperature law for Λ_{Q^*}. A comfortable way to do this is to plot $(dV/dI)_1$ or L_{An1} as a function of $\Delta T=T_{c0}-T$ using log-log coordinates.

For the In whisker of ref. 16 the temperature dependence of L_{An1} seems to change from a $\Delta T^{-1/4}$ to a $\Delta T^{-1/2}$ and a ΔT^{-1} law if the temperature approaches T_{c0} (see Fig. 29). For this sample the change between the first two temperature laws is not distinct. This effect is, however, clearly observable for sample In 17 in Fig. 30 which does only show these two temperature dependences. In the same figure we plotted the differential resistance of sample In 20 showing a $\Delta T^{-1/4}$ law over the whole temperature range where its differential resistance changes with temperature.

The only temperature dependence of L_{An1} which can be understood within the framework of the SBT model in the described manner is the $\Delta T^{-1/4}$ law which would be obtained by assuming charge imbalance relaxation due to inelastic electron-phonon collisions. Then $\tau_{Q^*in}=\tau_{Q^*in}(0)(T_{c0}/\Delta T)^{1/2}$ as given in section 5.4 has to be inserted into the charge imbalance relaxation length leading to $L_{An1}=2\Lambda_{Q^*in}\sim\Delta T^{-1/4}$. This identification leads to an experimental value for $\tau_{Q^*in}(0)\sim\tau_E$ which can be calculated from the measured L_{An1} at a given ΔT value. The experimental values for $\tau_{Q^*in}(0)$ are in reasonable agreement with those predicted by the theory invoking Tinkham's estimate of

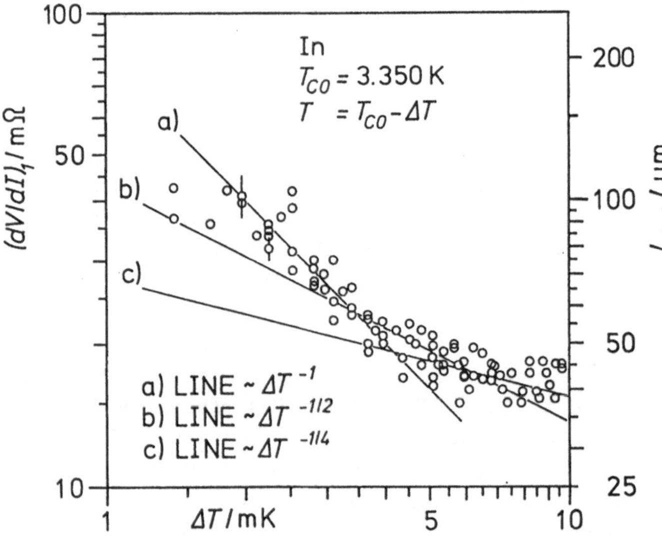

Fig. 29: *Differential resistance $(dV/dI)_1$ of the first voltage step and normal-like length L_{An1}, respectively, as a function of the temperature using log-log coordinates for sample In of ref. 16. The straight lines, a) - c), have a slope (-1), (-1/2), and (-1/4), thus representing a temperature dependence of $(dV/dI)_1$ - or L_{An1} - proportional to ΔT^{-1}, $\Delta T^{-1/2}$, and $\Delta T^{-1/4}$, respectively.*

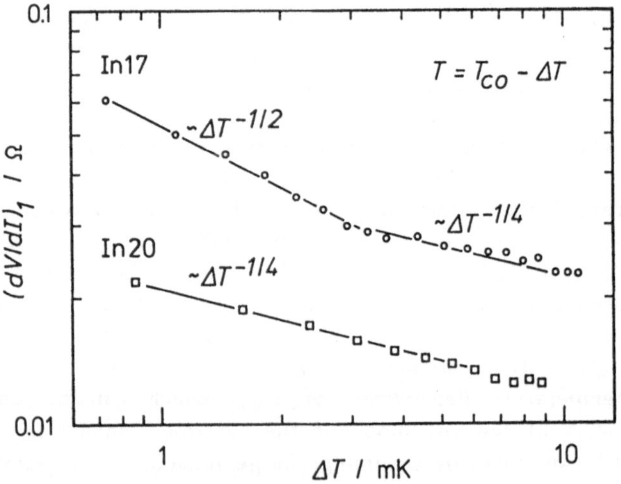

Fig. 30: *Differential resistance $(dV/dI)_1$ of the first voltage step as a function of the temperature for two In whiskers of ref. 17, sample 17 with T_{c0} = 3.3557 K and sample 20 with T_{c0} = 3.3673 K.*

eq. (72) for a prediction of τ_E (see Tab. I of ref. 40 *[1]). Moreover, experimental results for τ_E can be derived from $\tau_{Q^*in}(0)$ as obtained from the measurements which systematically fit into Fig. A2, given in the appendix, which concerns the dependence of τ_E on the electron mean free path.

In principle also the stronger temperature laws $L_{An1} \sim \Delta T^{-1/2}, \Delta T^{-1}$ can be obtained by inserting the steady-state charge imbalance relaxation time related to the relaxation of charge imbalance by elastic scattering in the presence of an anisotropic gap, τ_{Q^*el}, into the expression for Λ_{Q^*}. For quasiparticle energies which are large compared to the averaged value of the gap it is $\tau_{Q^*el} \sim \langle \mathbf{\Delta} \rangle^{-2} \sim \Delta T^{-1}$ somewhat further away from T_{c0}, changing to $\tau_{Q^*el} \sim \langle \mathbf{\Delta} \rangle^{-4} \sim \Delta T^{-2}$ close to T_{c0}. Explicit expressions for τ_{Q^*el} are given in section 5.4.

The problem is, however, that the stronger temperature laws are measured closer to T_{c0} than the $\Delta T^{-1/4}$ law related to inelastic electron-phonon processes: Already in section 5.4 we pointed out that charge imbalance relaxation due to elastic processes becomes more and more negligible compared to the relaxation by inelastic electron-phonon scattering for T approaching T_{c0}. The reason is that τ_{Q^*el} shows a stronger increase with increasing temperature than τ_{Q^*in}. Thus, τ_{Q^*in} is smaller than τ_{Q^*el} for all temperatures above the temperature where both times are equal and, thus, elastic processes cannot dominate at these temperatures.

If we, nevertheless, want to interpret the measured normal-like length in terms of τ_{Q^*el}, we have to postulate an (unknown) mechanism which 'switches off' the inelastic electron-phonon processes if the temperature is very close to T_{c0}. Then $L_{An1} = 2((1/3)\,\ell\, v_F\, \tau_{Q^*el})^{1/2}$, where, assuming $\underline{E_K} \gg \langle \mathbf{\Delta} \rangle$, it is $\tau_{Q^*el} \approx \hbar^2 \underline{E}_K^2 / 4\tau \langle a^2 \rangle_0 \langle \mathbf{\Delta} \rangle^4$ due to the fact that $\hbar / 2\tau \langle \mathbf{\Delta} \rangle \gg 1$ for our In whiskers at temperatures $\Delta T \lesssim 5\,\mathrm{mK}$. Thus, $L_{An1} \sim \Delta T^{-1}$ is predicted as observed for sample In very close to T_{c0} (see Fig. 29). Here we estimate $\langle \mathbf{\Delta} \rangle$ by $\mathbf{\Delta}$ as given in eq. (69).

In more detail, it is $\tau_{Q^*el} = \tau_{Q^*el}(0)\,(T_{c0}/\Delta T)^2$ with the prediction $\tau_{Q^*el}(0) = \hbar^2 \underline{E}_K^2 v_F / 351.8 \langle a^2 \rangle_0\, \ell\, (k T_{c0})^4$. Inserting the material parameters for sample In ($\ell = 0.738\,\mu m$, $\langle a^2 \rangle_0 = 0.021$ [16]) the theoretical prediction for $\tau_{Q^*el}(0)$ can be evaluated. On the other hand, an experimental result for $\tau_{Q^*el}(0)$ is obtained by applying the expression for L_{An1} mentioned above at $\Delta T = 2\,\mathrm{mK}$, where $L_{An1} = 100\,\mu m$ as can be seen from Fig. 29. By comparing both results we get $\underline{E}_K$ which has to be regarded as an averaged energy of the relaxing quasiparticles, because the experimental $\tau_{Q^*el}(0)$ represents an average over these quasiparticle energies. The problem is that we find $\underline{E}_K \approx \mathbf{\Delta}$, contradictory to $\underline{E}_K \gg \langle \mathbf{\Delta} \rangle$ as assumed for the expression of τ_{Q^*el} given above.

This discussion shows that the interpretation of L_{An1} in terms of τ_{Q^*el} is no appropriate method. In this context it is remarked that we also in refs. 16 and 17 made an attempt to understand the stronger temperature laws

---

*[1] In this context we refer to ref. 40 because the evaluations in refs. 15 and 17 were carried out considering Tinkham's early calculation for a branch imbalance relaxation time $\tau_{Q\,in}$, discussed in section 5.4.

of L_{An1} in indium whiskers by elastic relaxation processes. For this purpose we applied Tinkham's result (discussed in section 5.4) for a branch imbalance relaxation time τ_{Qel}. We assumed that the energies of the nonequilibrium quasiparticles are large compared to the averaged gap and, furthermore, neglected averaging effects which change $\langle a^2 \rangle$ from its clean bulk material value $\langle a^2 \rangle_0$. The resulting expression for τ_{Qel} is equal to that one obtained for τ_{Q^*el} in section 5.4 in the same limit ($\hbar / 2\tau \langle \Delta \rangle \ll 1$ and $E_{\underline{K}} \gg \langle \Delta \rangle$) and leads to $L_{An1} \sim \Delta T^{-1/2}$. In the cited work, moreover, an averaged quasiparticle energy has been introduced by characterizing the energy of the quasiparticles creating the branch imbalance by an effective temperature so that τ_{Qel} had already the meaning of an averaged relaxation time. Assuming this temperature to be about T_{c0} (implying a quasiparticle energy of kT_{c0}) we got a nice agreement between experiment and theory. Nevertheless, this agreement remains a puzzle, because it should not be allowed to ignore the influence of averaging effects on the mean square gap anisotropy, and the mean quasiparticle energy assumed should be too high. To comment the latter, the time averaged voltage generated by the phase-slip center is of order $1\,\mu V$ (see Fig. 3 of ref. 16) and so an order of magnitude estimate for the quasiparticle energy may be $1\,\mu eV$ which is of order $10^{-3}\,kT_{c0}$ for sample In.

Now we return to the problem of the temperature independent quasiparticle relaxation times τ_{OSnW} and τ_{OInW}. As mentioned above, no steady-state charge imbalance relaxation time is known which is independent of temperature.

There are several arguments to identify these times with the inelastic electron-phonon collision time τ_E:

At the end of section 5.4 and in discussing eq. (107) we pointed out that in non-steady-state experiments (with a time dependent chemical potential of the pairs) charge imbalance decay should be governed by τ_E rather than by τ_{Q^*}.

Moreover, there are Baratoff's calculations performed outside the local equilibrium approximation (which would lead to the generalized TDGL equations), predicting matching regions which decouple the rapid variations in the core region of a phase-slip center from the asymptotic wings of exponential charge imbalance decay. We have discussed his theory at the end of section 5.9. For temperatures not too close to T_{c0}, where $\tau_{GL} \le \tau_E$, most of the quasiparticle charge is expected to relax in these matching regions leading to a temperature independent normal-like length $L_{An1} = 2 s_B^2 ((1/3) \ell v_F \tau_E)^{1/2}$. The factor s_B^2 accounts for core effects and nonexponential charge imbalance decay in the matching regions. An estimate for τ_E may be evaluated from the measured L_{An1} by setting $s_B = 1$. This estimate characterizes an equivalent exponential charge imbalance decay. The condition $\tau_{GL} < \tau_E$ leads to the prediction that a temperature independent normal-like length should be observed if ΔT becomes larger than 8.4 mK for tin and larger than 11.9 mK for indium. To obtain these values we used $\tau_E = 3.56 \cdot 10^{-10}$ s for Sn and $\tau_E = 2.52 \cdot 10^{-10}$ s for In, as given in Table A1 for polycrystalline pure bulk materials after Tinkham's estimate. Using τ_E as evaluated from the $\Delta T^{-1/4}$ dependence of L_{An1} for the In whiskers mentioned in the present work (see the second to fourth line concerning the material In in Tab. A1 and identify the different samples by their T_{c0} values) results in temperature borders of about 8 - 13 mK below T_{c0} for a change to a temperature independent normal-like length. While for Sn a temperature independent behaviour is usually found up to temperatures much closer to T_{c0}, these temperature borders sound quite reasonable for In whiskers.

Nevertheless, it is not possible to identify the times τ_{OSnW} and τ_{OInW} with the inelastic electron-phonon collision time τ_E of the material of the whisker, because τ_E is an order of magnitude smaller than these times (compare the values given above and see also Tab. A1). Thus, there also is not any interpretation of the temperature independent normal-like lengths of Sn and In whiskers within the SBT model. However, there is a remarkable empirical rule which connects the times τ_{OSnW} and τ_{OInW} with the scaling time τ_0 introduced by Kaplan et al. [103] and discussed in section 5.4: It is $\tau_0(\mathrm{Sn}) = 2.30 \cdot 10^{-9}$ s and $\tau_0(\mathrm{In}) = 0.799 \cdot 10^{-9}$ s, so that we find $\tau_{OSnW}/\tau_0(\mathrm{Sn}) = \tau_{OInW}/\tau_0(\mathrm{In})$ within less than 5%. Thus, the scaling time for Sn and In has to be multiplied with the same factor to give the measured temperature independent quasiparticle relaxation times for these materials. Since $\tau_0 = 8.4\,\tau_E$ the same kind of rule is valid for the inelastic electron-phonon scattering times derived from this scaling time.

In section 5.7 we discussed the work of Kulik, who introduced a temperature independent characteristic length $\Lambda_{Ku}(0)$ for the calculation of the penetration of the electromagnetic field into the superconductor in the presence of electron-bogolon scattering. In the following we will show, how this length may be connected with the normal-like length of a phase-slip center:

We start from eq. (119), the equation for the scalar potential $\Phi(\underline{r}, t)$ of the electromagnetic field. Inserting the expression for $\Lambda^2_{Ku}(0)/\Lambda^2_{Ku}(\omega_{Ku})$ given in eq. (120) yields an equation which is nothing else but the following differential equation

$$\Lambda^2_{Ku}(0)\,\underline{\nabla}^2\Phi = \tau_{0R}\tau_{eb}\ddot{\Phi} + (\tau_{0R} + \tau_{eb})\dot{\Phi} + \Phi \qquad (190)$$

This result is in formal analogy to the charge imbalance wave equation (eq. (113)), compared to which it is replaced Q^* by Φ, $D\tau_{Q^*} = \Lambda^2_{Q^*}$ by $\Lambda^2_{Ku}(0)$, and τ_α by τ_{eb}.

As done by KSS (see section 5.8.4) also this equation can be compared with the telegraph equation leading to the series resistance $\tilde{R}_{Ku} = 1/\sigma A$, the leakage conductance $\tilde{G}_{Ku} = \sigma A/\Lambda^2_{Ku}(0)$, the series inductance $\tilde{L}_{Ku} = \tau_{0R}/\sigma A$, and the leakage capacitance $\tilde{C}_{Ku} = (\sigma A/\Lambda^2_{Ku}(0))\tau_{eb}$.

The telegraph equation for Φ can be obtained from the following system of equations

$$-\partial I_n/\partial x = \tilde{G}_{Ku}\,\Phi + \tilde{C}_{Ku}\,\partial\Phi/\partial t \tag{191}$$

$$-\partial\Phi/\partial x = \tilde{R}_{Ku}\,I_n - \tilde{L}_{Ku}\,\dot{I}_s \tag{192}$$

To see this, one has to calculate $\partial\dot{I}_n/\partial x$ from the time derivative of eq. (191) and $\partial\dot{I}_s/\partial x$ from the space derivative of eq. (192) and then consider that for the total current, $I = I_s + I_n$, it is $\partial I/\partial x = 0$, so that $\partial\dot{I}_n/\partial x + \partial\dot{I}_s/\partial x = 0$. Into the resulting equation one inserts $\partial I_n/\partial x$ from eq. (191).

The impedance $Z_{Ku} = \Phi/I_n$ can be calculated from these two equations [368] by assuming that the space and time dependence of Φ and I_n are equal (the expression for $\Phi(\underline{r},t)$ is given in eq. (122)), yielding

$$Z_{Ku}(\omega_{Ku}) = \left(\frac{\tilde{R}_{Ku} - i\,\omega_{Ku}\,\tilde{L}_{Ku}}{\tilde{G}_{Ku} - i\,\omega_{Ku}\,\tilde{C}_{Ku}}\right)^{1/2} = \frac{\Lambda_{Ku}(0)}{\sigma A}\left(\frac{1 - i\,\omega_{Ku}\,\tau_{QR}}{1 - i\,\omega_{Ku}\,\tau_{eb}}\right)^{1/2} \tag{193}$$

If we try now to apply these results to a phase-slip center in analogy to KSS (see especially chap. 4 of ref. 158) there appear some problems: One has to know the voltage at the core region (shrinked to one point in the KSS model) which may be expressed in terms of the values of μ_p or Φ_{Q*} on both sides of the core region yielding $V_{-+}(t) = \mu_p(x=0^-,t)/(-e) - \mu_p(x=0^+,t)/(-e)$ or $V_{-+}(t) = \Phi_{Q*}(x=0^+,t) - \Phi_{Q*}(x=0^-,t) = 2\,\Phi_{Q*}(x=0^+,t)$. Here we applied eq. (45) together with $\Phi_{Q*} = Q^*/2e^2N_0$ as introduced in section 5.3 and considered that μ changes continuously in the origin and, therefore, does not differ at $x=0^+$ and $x=0^-$. On the other hand, the potential Φ as introduced in section 5.3 is given by $\Phi = \Phi_{Q*} + \Phi_{\dot{\varphi}}$, where $\Phi_{\dot{\varphi}} = (1/2e)\hbar\dot{\varphi}$. Since $\mu_p = \mu_F - e\,\Phi_{\dot{\varphi}}$, it follows $\Phi = Q^*/2e^2N_0 - (1/e)(\mu_p - \mu_F)$ or, using $\mu - \mu_p = Q^*/(-2eN_0)$ from eq. (45), we get $\Phi = \mu - \mu_F$. Thus, Φ does not only contain contributions from the charge imbalance but also from nonequilibrium effects in the condensate. Only in the limit where no charge imbalance waves are excited ($\omega_{Ku} = 0$), but where there is an exponential spatial charge imbalance decay, $\dot{\varphi}$ can be assumed to be constant outside the core region and may be set to zero at one side of the phase-slip center so that $V_{-+} = 2\,\Phi\,(x=0^+,t)$. This situation is present in the high voltage dc limit which reproduces the SBT result. In analogy to eq. (144) it is the voltage developed by the phase-slip center

$$V = 2Z_{Ku}(\omega_{Ku}=0)\,[I - \langle I_s(0,t)\rangle] \tag{194}$$

where the time average of the supercurrent is equal to its phase-angle average leading to the normal-like length $L_{An1} = 2\Lambda_{Ku}(0)$. In section 5.7 we showed that $\Lambda_{Ku}(0)$ may be expressed as $\Lambda_{Ku}(0) = (3^{1/2}Dm/\hbar)\,\xi_0(T_{c0})$, where $D = \ell v_F/3$ and $\xi_0(T_{c0}) = \hbar v_F/2\pi k T_{c0}$.

The length $\Lambda_{Ku}(0)$ does not depend on temperature for a given sample which T_{c0} has a fixed value for. However, $\Lambda_{Ku}(0) \sim \ell$ is in disagreement with the observed $\ell^{1/2}$ dependence of L_{An1} and the absolute value of $\Lambda_{Ku}(0)$ is more than an order of magnitude too large to explain the measured values of L_{An1} in the case of Sn and In: With material parameters for pure bulk material and $\ell \approx 1\,\mu m$ as typical for a pure whisker, it is $\Lambda_{Ku}(0)$ equal to 761 μm and 922 μm for Sn and In, respectively.

This discussion shows that the electron-bogolon mechanism is too weak to account for the phenomena observed and, so, is not the key for an explanation of a non-diverging normal-like length.

The features reported for Sn, Sn-In, and In whiskers are representative for the phenomena which may be also observed for whiskers made of other materials: While $(dV/dI)_1$ - or L_{An1} - and I_0/I_c do not depend on temperature for whiskers of Pb, Zn, and Zn-Ag alloys, the whole variety of temperature laws has been observed for In-Pb whiskers. The results for these materials (including Pb-In and Pb-Bi alloys) will be given in chapters 8 and 9.

Next, we discuss the experimental work done with long thin film microbridges evaporated on substrates:

For Sn microbridges SBT measured a temperature independent normal-like length which changes proportional to $\ell^{1/2}$ [32]. Assuming $v_F = 1 \cdot 10^6\,m/s$, they got from their experiments $\tau_2 = 8 \cdot 10^{-10}\,s$ which transforms to $\tau_2 = 1.16 \cdot 10^{-9}\,s$ with our value for v_F of tin. SBT proposed to identify their value for τ_2 with the inelastic electron-phonon scattering time, τ_E, or with the characteristic time for quasiparticle recombination [2, 32].

Subsequently Kadin, Skocpol, and Tinkham [84] observed that the differential resistance of a voltage step generated by a phase-slip center in an Sn microbridge changes proportional to $\Delta T^{-1/4}$ over a range of about 50 mK. For this microbridge the conduction of heat out of the film is expected to be optimized and the authors argued that deviations from this behaviour may be heating induced effects. This microbridge does not contain a notch or something else frequently used in that work to fix the locus of the phase-slip center and to reduce its critical current.

Direct determinations of the quasiparticle diffusion length (or charge imbalance relaxation length) are pioneered by Dolan and Jackel who performed spatially resolved measurements of the pair and quasiparticle electrochemical potentials near a phase-slip center in a tin microbridge [70]. For this purpose they applied two sets of voltage probes to the sample, one set being superconducting and the other normal conducting. The probe/sample interfaces are oxide barrier tunnel junctions. As we have discussed in section 5.3, it is thus possible to measure the (time averaged) values of μ and μ_p. The spatial dependence obtained for $\langle\mu\rangle$ and $\langle\mu_p\rangle$ is just that one proposed by the SBT model (see Fig. 13). The behaviour of $\langle\mu\rangle - \langle\mu_p\rangle$ can be fitted by an exponential decay on either side of the phase-slip center with a decay length which is proportional to $\Delta T^{-1/4}$ and which can be identified with Λ_{Q^*in}. The τ_E value arising from this identification is given in Tab. A1. Moreover, Dolan and Jackel calculated $L_{An1}/2$ from the differential resistance (according to eq. (186)). They found that there is only a restricted temperature range which this quantity changes proportional to $\Delta T^{-1/4}$ in, its absolute value still being larger than the measured Λ_{Q^*in} by about 20%. Very close to T_{c0} and further away from T_{c0} the deviation becomes larger. The temperature dependence of $L_{An1}/2$ is stronger than a $\Delta T^{-1/4}$ dependence in the higher temperature region and weaker in the lower temperature range. Dolan and Jackel argue that in both regions the evaluation of the differential resistance does not lead to reliable results, because their V-I characteristics are not really straight. They propose intrinsic effects of the phase-slip center to be responsable for the curvature of the characteristics at high temperatures while heating effects are invoked in the low temperature range.

It is interesting to note that Dolan and Jackel also generated a charge imbalance by injecting quasiparticles into the strip through one of the tunnel junctions between a voltage probe and the sample, leading to the same decay length as in the case of an active phase-slip center. This indicates that the same quantity is envolved in both kinds of experiments.

Aponte and Tinkham [72] also performed spatially resolved measurements of $\langle\mu\rangle - \langle\mu_p\rangle$ near a phase-slip center in long tin microbridges and they also found an exponential charge imbalance decay with the characteristic length $\Lambda_{Q^*in} \sim \Delta T^{-1/4}$. The corresponding τ_E values are given in Tab. A1.

In the case of In, Jillie [33] obtained a temperature dependent normal-like length $L_{An1}/2 \sim \Delta T^{-1/4}$ from measurements of the differential resistance generated by a phase-slip center in a long microbridge. The identification $L_{An1}/2 = \Lambda_{Q^*in}$ leads to a reasonable value for τ_E (see Tab. A1).

Recently Weissbrod et al. [369] investigated phase-slip centers in In microbridges, observing that $L_{An1}/\xi(T) \sim \gamma^{1/2}$ which means $L_{An1} \sim \Delta T^{-1/4}$. Here, γ is the pair-breaking parameter of the TDGL theory.

The history of experimental investigations of phase-slip centers in long Al microbridges sounds somewhat similar to the tin case:

The early V-I measurements of Klapwijk et al. [34, 35] yield a temperature independent normal-like length changing proportional to $\ell^{1/2}$. Also in this case the application of the SBT model leads to an inelastic relaxation time (τ_2 = $1.9 \cdot 10^{-7}$ s) and the inelastic electron-phonon scattering time τ_E was invoked for an interpretation of this time.

Then Stuivinga et al. [73] performed spatially resolved measurements of the quasiparticle electrochemical potential near a phase-slip center, yielding that also in the case of aluminium $\Lambda_{Q^*in} \sim \Delta T^{-1/4}$ governs the spatial decay of the charge imbalance.

This is, however, not the end of the story, because recently Liengme et al. [136] again measured V-I characteristics of Al microbridges and derived a value for L_{An1} from their differential resistance. They found that $L_{An1} \sim \Delta T^{-1/4}$ very close to T_{c0} and then changes to a temperature independent behaviour as predicted by Baratoff's theory (see section 5.9. and this section). The temperature range of the $\Delta T^{-1/4}$ law depends on the homogeneity of the sample. For very homogeneous samples the range is very small and an early crossover to a temperature independent behaviour is observed. For inhomogeneous samples the divergent behaviour extends to lower temperatures until the change to the temperature independence occurs. In the temperature independent regime it is $L_{An1} \sim \ell^{1/2}$ and according to the SBT model an inelastic relaxation time, $\tau_2 = 43.5 \cdot 10^{-9}$ s, is evaluated which is very close to calculated τ_E values for this material.

It is usually believed that in those microbridges which do not show a temperature dependent normal-like length, heating effects mask the weak $\Delta T^{-1/4}$ temperature dependence of the differential resistance [8, 73, 84, 94, 260], although there is no completely convincing explanation why they should lead to an apparently temperature independent slope [8].

For whiskers, heating effects are negligibly small in the region of phase-slip behaviour and cannot be considered to mask a $\Delta T^{-1/4}$ temperature law [40, 370].

The situation seems to be rather envolved: Normal-like lengths, L_{An1}, calculated from the differential resistance of measured V-I characteristics show a variety of temperature dependences. For whiskers and microbridges L_{An1} may be independent of temperature. Moreover, in both kinds of samples normal-like lengths are observed which diverge proportional to $\Delta T^{-1/4}$. In addition, whiskers may show a $\Delta T^{-1/2}$ or even a ΔT^{-1} dependence for L_{An1}. In whiskers L_{An1} may be temperature independent over the temperature range investigated or start with a temperature independent behaviour at lower temperatures and then pass one or more temperature laws for T approaching T_{c0}. The behaviour of microbridges is usually somewhat different. They show either a normal-like length which is only independent of temperature or which follows a $\Delta T^{-1/4}$ law over the whole temperature range. Only the Al microbridges of Liengme et al. change from a temperature independent L_{An1} to $L_{An1} \sim \Delta T^{-1/4}$ if T becomes very close to T_{c0}. On the other hand, all spatially resolved measurements of the charge imbalance decay close to a phase-slip center deliver a decay length which changes proportional to $\Delta T^{-1/4}$ and which can be identified with Λ_{Q^*in}.

In our opinion an understanding of the measurements cannot be achieved by stating that $L_{An1} = 2\Lambda_{Q^*in}$ (leading to $L_{An1} \sim \Delta T^{-1/4}$) and then looking for a disturbance (like heating effects) if this relation is not fulfilled. Rather one should apply this relation only if a $\Delta T^{-1/4}$ law is observed for L_{An1} and then prove whether the resulting value for τ_E has a reasonable magnitude and, thus, charge imbalance relaxation by inelastic electron-phonon processes dominates. In temperature ranges where L_{An1} follows a different temperature law one can formally apply the relation $L_{An1} = 2\Lambda$ and then try to identify the resulting relaxation time τ_2 with a charge imbalance relaxation mechanism. If the latter does not appear to be successful, there are three possibilities. The first one is that the charge imbalance relaxes by an unknown mechanism, the second one is that a known mechanism is masked by a disturbance , but the third reason can also be that the SBT result $L_{Anl} = 2\Lambda$ does not hold in the experimental situation considered.

The third possibility seems to be most remarkable. As mentioned in the beginning of this section, as well the KSS model as also the TDGL theory predict that the SBT result does not hold in general. The TDGL theory finds that $L_{Anl} = 2\Lambda_{Q^*in} \sim \Delta T^{-1/4}$ only for large γ. For decreasing γ this theory expects a change to $L_{Anl} \sim \xi_D \sim \Delta T^{-1/2}$ and then to $L_{Anl} \sim \xi_D/\gamma \sim \Delta T^{-1}$. For a fixed sample γ decreases with increasing temperature and, therefore, L_{Anl} should run through the different temperature laws if T approaches T_{c0}. A detailed comparison of experimental results with the prediction of the TDGL theory will be done in chap. 10. The TDGL theory does not predict a temperature independent normal-like length. A temperature-independent L_{Anl} results from Baratoff's calculations performed outside the local equilibrium approximation of the TDGL theory, leading to reasonable agreement not only in the case of the Al microbridges of Liengme et al. [136] but also for Zn and Zn-Ag whiskers (see chap. 8).

An important feature is the influence of inhomogeneities. The TDGL theory predicts that they influence the properties of a phase-slip center (see section 5.9). Liengme et al. [136] found that inhomogeneities extend the temperature range which $L_{Anl} \sim \Delta T^{-1/4}$ in. Compared to whiskers, microbridges may be rather inhomogeneous samples with variations of 20 - 30 mK of the critical temperature along the bridge [32, 70, 84]. Moreover, a widely used technique is to fix the locus of the phase-slip center in a microbridge by a pinning center such as, for instance, a notch [84, 369]. This method is especially comfortable for spatially resolved measurements of the charge imbalance decay [70, 72, 73]. Chapter 12 will deal with phase-slip centers appearing at a weakly superconducting point with continuously adjustable strength ('tunable weak link'). Among other effects evidence is found for an influence of the strength of superconductivity in the tunable weak link on the temperature dependence of the normal-like length, L_{Anl}, of the phase-slip center appearing at the tunable weak link.

We add some remarks on the predictions of the KSS model concerning $(dV/dI)_1$ and I_0/I_c outside the SBT limit. The approximate solutions of this model are plotted in Fig. 16 with τ_E/τ_{QR} as a parameter which decreases from the left to the right, indicating that for a fixed sample T becomes closer to T_{c0}. The curves are not straight. In the non-hysteretic case they are still bent at the critical current. For a qualitative discussion we may fit a straight line to the straight part of each curve. The slope of this line as well as its extrapolated zero voltage intercept increase along the set of curves for decreasing τ_E/τ_{QR}, that means for increasing temperature. Thus, the KSS model predicts a differential resistance which increases somewhat stronger than proportional to $\Delta T^{-1/4}$ (which is the temperature dependence of R_{eff}). Moreover, the ratio I_0/I_c is predicted to increase with increasing temperature.

This is in qualitative agreement with the experiments. A quantitative straight-foreward comparison of the approximate KSS solutions with measured V-I characteristics is not successful. A trial can be made for materials with temperature dependent differential resistance such as indium. The problem is that the change into the hysteretic regime is observed closer to T_{c0} in the experiment than predicted by KSS. Thus, there is a sharp voltage jump followed by a straight line in the experimental curve, but a continuous transition with slowly changing slope in the theoretical curve. The reason for the different hysteretic behaviour are contributions to the hysteresis in addition to the charge imbalance wave mechanism (see chap. 11). Up to now we did not work through the problem of calculating V-I characteristics according to KSS which are modified by these contributions. It would be most desirable to do this not only for approximate but also for exact solutions of the model. In the present article we will restrict our discussions to qualitative comparisons with the approximate KSS solutions. An exception is the discussion in chap. 11, where we quantitatively calculate the contribution of charge imbalance waves to the hysteresis width from these curves.

7.3. The Influence of High-Frequency Radiation

The aim of investigations in an electromagnetic high-frequency radiation (HF) field is to prove the assumption of a cyclical high-frequency process at Josephson frequency to be associated with the voltage steps in the transition curves of a quasi-one-dimensional superconductor.

The HF radiation from a signal source with variable attenuator was applied to the whisker with a coaxial cable. An inductive loop was used to couple the radiation to the sample. In most cases this antenna was placed in a small distance above the two superconducting blocks of the squeeze contact holding the whisker [16, 24, 40, 371]. Very recently we performed some measurements with the loop situated in the space between the contact blocks [372]. In this case the loop had only a diameter of about 1 mm and was made of a thin superconducting wire turned around the whisker. The second method is of advantage, because it does not need the careful shielding to minimize the influence of the HF radiation on the carbon resistor thermometer as was necessary in the first case. The reason is that the coupling of the radiation is more efficient, so that smaller HF-power levels could be used. Moreover, the radiation is expected to be captured between the superconducting contacts.

The first measurements in a HF field were performed with a tin whisker [24, 371]. With applied radiation, steps of zero slope ('current steps') appear in the V-I characteristics in addition to the voltage step structure (Fig. 31).

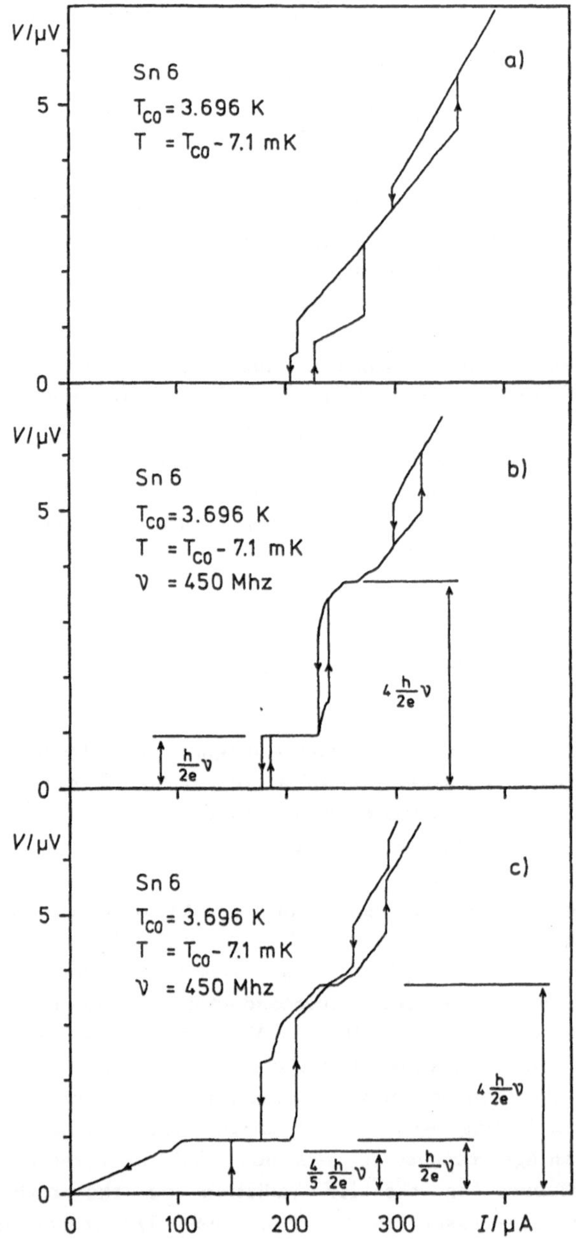

Fig. 31: *V-I characteristics of an Sn whisker for increasing and decreasing current [24, 371].*
(a) Without HF radiation
(b),(c) Applied radiation $\nu = 450\,MHz$*, where the HF power input level is larger in c) than in b).*
Current steps are observed at $V_{\tilde{n}} = \tilde{n}\,(h/2e)\,\nu$ *with* $\tilde{n} = 1,\ 4,$ *and* $4/5$*, respectively. (The sample is called 'Sn 6' in ref. 24 but 'Sn' in ref. 371.)*

These current steps are observed for increasing and decreasing current. The voltage V_{J1} of the main step is related to the frequency ν of the applied HF field by the Josephson relation, $V_{J1} = (h/2e)\nu$. Moreover, harmonic and subharmonic steps are observed at $V_{J\tilde{n}} = \tilde{n}\, V_{J1}$, with $\tilde{n}$ a positive integer or fractional number, respectively.

We performed measurements at different fixed temperatures and with different frequencies of the HF field. Plotting the measured V_{J1} as a function of ν results in a straight line with slope $h/2e$, indicating that the voltages which the main current step appears at are given by the Josephson relation in the whole frequency range investigated (240 MHz - 900 MHz).

The main current step may extend over the whole current range of the first voltage step or occur within the first step (similar to the current step at V_{J4} in Fig. 31). The critical current is reduced by the HF radiation. Usually, the hysteresis width becomes smaller, except at higher input power levels, where the voltage for decreasing current does not disappear until the current has become zero (Fig. 31 c).

We also performed measurements in a HF field using whiskers of In [16, 24, 372], In-Pb alloys [40], and Pb [372] and obtained similar results. Measuring curves for Pb whiskers (which do not show subharmonic current steps) are given in chap. 9.

Current steps similar to those observed for our whiskers were predicted by Josephson [68, 263] for the characteristics of tunnel junctions subjected to a microwave field. The phenomenon is usually called 'inverse ac Josephson effect' and first was observed by Shapiro [373]. Meanwhile the appearance of current steps is a familiar effect occurring in various weakly coupled superconducting systems (that means systems with a weakly superconducting region) which are placed into a HF-radiation field [68, 279, 374]. Typical specimens are tunnel junctions [373, 375], point contacts [376, 377], thin film bridges [161, 378], short microbridges [33, 35, 379 - 382], and proximity effect bridges [383]. Also wide microbridges show inverse ac Josephson-like effects [384, 385]. Moreover, these effects are observed in the long tin microbridges of SBT in the phase-slip regime [32]. Series arrays of hundreds of proximity effect bridges respond cooperatively to the applied radiation to produce 'current steps' at $n_{ar} \cdot V_{J1}$, where n_{ar} is the number of bridges in the array [386]. The same is the case for large arrays of tunnel junctions which are used to get a high precision voltage standard [387 - 389].

In many cases the current step phenomenon can be described as inverse ac Josephson effect or by the analysis of the influence of HF radiation on a system with a current phase relation which is similar to that one derived by Josephson [68, 279, 374, 390, 391]. For proximity effect bridges a phase-slip model with included HF radiation gives the best agreement with the measurements [266, 383, 392].

We made attempts to understand the 'current step' phenomenon observed in the V-I characteristics of our whiskers within a phase-slip model with included HF radiation [24, 371, 372, 393]. The idea of these attempts is as follows: The first voltage step can be described by the empirical formula, $V(I) = (dV/dI)_1 (I - I_0)$, of section 7.2. It is assumed that the differential resistance, $(dV/dI)_1$, does not change if a HF field is applied. Therefore, the quantity I_0 is assumed to be responsible for the effect. The expectation is that in a HF field I_0 may vary to compensate changes of the total current, I, so that the voltage V(I) remains constant over a certain current range. In all models for a phase-slip center discussed in section 5.8, the quantity I_0 is interpreted as a time averaged supercurrent $\langle I_s(t) \rangle$, so that the problem reduces to the calculation of $I_s(t)$ in a HF field.

Our first calculation [24, 371] was performed within the RSM model which has been developed to describe the phase-slip behaviour in a weak link. The RSM model results in an expression for the time averaged voltage $\langle V(t) \rangle (I)$ across the weak link, eq. (130), which is similar to our empirical formula for the first voltage step. In the RSM expression the residual resistance of the weak region appears instead of $(dV/dI)_1$, because the model does not consider the relaxation of nonequilibrium quasiparticles. Furthermore, $\langle \overline{I}_s(t) \rangle$ appears instead of I_0, where $\overline{I}_s(t)$ is the time dependent supercurrent, spatially averaged over the entire weak superconductor. For sufficiently large total currents, I, the $\langle V(t) \rangle (I)$ characteristic becomes straight and shows an extrapolated zero voltage intercept $\langle \overline{I}_s(t) \rangle = (1/2) I_c$.

Although this model cannot quantitatively describe the V-I characteristics of our samples (because it does not take into account the quasiparticles), it was hoped that it nevertheless leads to the observed behaviour of the time averaged supercurrent in a HF field. For our calculations we used the analytical approximation given by RSM for the current-phase relation of the supercurrent, eq. (131), and included the influence of the HF radiation by adding a radiation induced voltage $v_a \cos(\omega_a t + \varphi_a)$ in the equation which describes the time development of the phase difference $\varphi_{12}(t)$ across the weak link (eq. (127)). To model a straight line behaviour of $\langle V(t) \rangle (I)$ in the absence of the HF field, we assumed that the voltage generated by the phase-slip center is nearly constant and, thus, the actual time dependent voltage in the equation for $\varphi_{12}(t)$ was replaced by its average over one phase-slip period. (This leads to $\varphi_{12}(t) = \varphi_{12}(0) + \omega_{PSC} t$ in the absence of the radiation, yielding $\langle \overline{I}_s(t) \rangle = (1/2) I_c$.) Then we calculated the time averaged supercurrent for an amplitude v_a of the radiation-induced voltage which is small compared to the time averaged voltage generated by the phase-slip center. If the angular frequency ω_a of the external field is equal to ω_{PSC}, the resulting expression for $\langle \overline{I}_s(t) \rangle$ in addition to the term $(1/2)I_c$ has a nonvanishing term depending on the difference of $\varphi_{12}(0)$ and the phase angle of the radiation, φ_a. Thus, the superconductor has the possibility to adjust $\langle \overline{I}_s(t) \rangle$, that means I_0, over a certain current range, as required for a 'current step'.

Our first approach only delivered the main current step. Recently we generalized the calculation so that it contains predictions for harmonic and subharmonic steps. Moreover, a critical discussion of our procedure is given including arguments from the SBT and the KSS model [372, 393]. Argumentation and calculation are discussed below:

Again we start from the empirical formula for the first voltage step and assume that only I_0 is affected by the HF field. In the SBT model and in the KSS model I_0 is interpreted as time averaged supercurrent $\langle I_s(X_{PSC}, t)\rangle$ in the core of the phase-slip center located at X_{PSC}. In the SBT model $\langle I_s(X_{PSC}, t)\rangle$ is constant throughout the V-I characteristic. KSS show that this is only valid in the high voltage dc limit (at sufficiently large measuring currents), where the voltage across the core region is nearly time independent. The voltage V(I) of our empirical formula arises from a time-averaged difference $\langle \Delta\mu_p\rangle$ of the electrochemical pair potentials far away from the core of the phase-slip center (measured with superconducting contacts between the ends of the sample). In the SBT model (and in the dc limit of the KSS model) the potential μ_p is actually time independent further away from the core of the phase-slip center. In the general case KSS show that $\Delta\mu_p$ contains an ac component (eq. (158)).

The modeling of the core region and, thus, the calculation of $I_s(X_{PSC}, t)$ is different in both models. SBT assume that the core region with strong variations of the order parameter extends over some $\xi(T)$. They give an expression for the space dependence of μ_p in this region which they calculate the supercurrent in the core from, using the Josephson relation and the expression for the supercurrent density of the GL theory. In the KSS model the core region is represented by a Josephson oscillator of zero extent. The supercurrent is calculated from the phase difference of the order parameter on both sides of the Josephson element via a current-phase relation which is a free parameter of the model.

In the SBT model, μ_p becomes space independent outside the core region and equal to its value measured at the contacts far away from the phase-slip center. KSS showed that the time average of the difference of the electrochemical potentials across their core region (that means the Josephson oscillator) is equal to the time-averaged difference of μ_p as measured across the contacts. In a time resolved picture, however, the difference of μ_p across the core region may be different from the one measured at the contacts.

These problems in mind, we performed the following calculation for the time averaged supercurrent $\langle I_s(X_{PSC}, t)\rangle$ in the core of a phase-slip center submitted to HF radiation: We model the core region by a Josephson oscillator characterized by a suitable current-phase relation which relates the phase difference across the core region with the supercurrent in the core. The core region with the borders $\underline{r}_1$ and $\underline{r}_2$ may have zero extent as in the KSS model, but this is not essential for our calculations. The time development of the phase difference $\varphi_{12}(t) = \varphi(\underline{r}_1) - \varphi(\underline{r}_2)$ across the core region is calculated from the Josephson relation. For this purpose we assume that the difference $\mu_{p,12} = \mu_p(\underline{r}_1) - \mu_p(\underline{r}_2)$ of the electrochemical pair potential across the core region is not much different from the value detected across the contacts and does only sligtly vary in time. So we replace the actual time dependent $\mu_{p,12}$ by its time averaged value $\langle\mu_{p,12}\rangle$ which is related with the time averaged voltage across the contacts by $V=\langle\mu_{p,12}\rangle/(-e)$. The influence of the HF radiation is included by adding a radiation induced voltage.

Due to the Josephson relation the time evolution of $\varphi_{12}(t)$ in the absence of the radiation is given by

$$\dot{\varphi}_{12}(t) = (2e/\hbar)V \tag{195}$$

Now the HF radiation is assumed to modulate the voltage leading to

$$\dot{\varphi}_{12}(t) = (2e/\hbar)(V + v_a \cos(\omega_a t + \varphi_a)) \tag{196}$$

with the frequency of the HF field ω_a and φ_a a phase angle.

Integrating over the time yields

$$\varphi_{12}(t) = \tilde{\varphi}_{12}(0) + \omega_{PSC}\, t + (\omega_{PSC}/\omega_a)(v_a/V)\sin(\omega_a t + \varphi_a) \tag{197}$$

where $\omega_{PSC} = (2e/\hbar)V$ and $\tilde{\varphi}_{12}(0) = \varphi_{12}(t=0) - (\omega_{PSC}/\omega_a)(v_a/V)\sin\varphi_a$.

For the description of the supercurrent in the core we propose the following current-phase relation which is periodic in $\varphi_{12}(t)$:

$$I_s(X_{PSC}, t) = a I_c (1 + \sum_m b_m \cos(m\varphi_{12}(t))) \tag{198}$$

The coefficients b_m determine the amplitudes of the time dependent terms, where 'm' is an index throughout this section (and not the electon mass, as usually) which may run from 1 to ∞. If there is no HF radiation ($v_a = 0$), the time average over one phase-slip period yields $\langle I_s(X_{PSC}, t)\rangle = a I_c$ so that the factor 'a' is nothing else but I_0/I_c.

In the special case where $a = 1/2$ and the sum only contains the term $\cos\varphi_{12}(t)$, that means only b_1 is non-zero and equal to unity, our current-phase relation reduces to that proposed by RSM.

Now we insert $\varphi_{12}(t)$ into the current-phase relation and expand $I_s(X_{PSC}, t)$ into a Fourier Bessel series. The procedure is similar to the case of tunnel junctions [279] and in detail described in the cited literature [372, 393], yielding

$$I_s(X_{PSC}, t) = a I_c \{1 + \sum_m b_m \sum_{n=-\infty}^{+\infty} J_n\left(m \frac{\omega_{PSC}}{\omega_a}\frac{v_a}{V}\right) \cos[m\tilde{\varphi}_{12}(0) + n\varphi_a + (m\omega_{PSC} + n\omega_a)t]\} \tag{198a}$$

Here $J_n(\ldots)$ are Bessel functions of the first kind of integer order [394].

The time average $\langle I_s(X_{PSC}, t)\rangle$ is only different from $a I_c$, if the sum contains time independent contributions and, thus, does not vanish in time average. In the experiment ω_a is fixed and ω_{PSC} is varied by changing the total transport current, I (and thus V). If a situation is reached where there is (at least) one term (m, n) with $m\omega_{PSC} + n\omega_a = 0$, that means $\omega_{PSC}/\omega_a = -n/m$, this term is a time independent contribution and $\langle I_s(X_{PSC}, t)\rangle$ differs from

$a I_c$. The magnitude of this contribution depends on $m\tilde{\varphi}_{12}(0)+n\varphi_a$. If now I is varied further, the superconductor has the possibility to hold its voltage, and thus ω_{PSC}, constant by changing $m\tilde{\varphi}_{12}(0)+n\varphi_a$, which can be done by readjusting $\varphi_{12}(t=0)$. This leads to the occurrance of a current step in the V - I characteristic.

The main current step is predicted to be observed for $\omega_{PSC}=\omega_a$. In this case the sum contains m time independent contributions. If there is only one $b_m \neq 0$, only harmonic current steps are obtained. These steps appear if ω_{PSC} becomes an integer multiple of ω_a. In the other case, where there are several $b_m \neq 0$, also subharmonic current steps should occur for rational values of ω_{PSC}/ω_a as soon as $m\,\omega_{PSC}+n\,\omega_a=0$. Thus, m accounts for the substeps. These predictions for $\langle I_s(X_{PSC},t)\rangle$ are similar to those of the RSJ (resistivity shunted junction) model of Gundlach and Kadlec [390].

For systems which besides the main step show harmonic and subharmonic steps, there are several $b_m \neq 0$, so that the current-phase relation is not simply sinusoidal. If only the main step and its harmonics are observed, one expects only $b_1 \neq 0$ and, thus, a pure sinusoidal current-phase relation.

For a pure sinusoidal current-phase relation with $b_1 \neq 0$ only, it is $m=1$ and $n=-1$ for the main current step occurring at $\omega_{PSC}=\omega_a$. Throughout the step the voltage V(I) is equal to V_{J1} and the time averaged supercurrent is given by

$$\langle I_s(X_{PSC},t)\rangle = a I_c\{1-b_1 J_1(v_a/V)\cos[\tilde{\varphi}_{12}(0)-\varphi_a]\} \tag{199}$$

Here we used $J_{-n}(\ldots)=(-1)^n J_n(\ldots)$ for the Bessel functions [394]. The maximum width ΔI of this current step is obtained, if $\tilde{\varphi}_{12}(0)-\varphi_a$ varies between 0 and π, leading to $\Delta I = 2aI_c b_1 J_1(v_a/V)$. For small arguments, v_a/V, it is [394], $J_n(v_a/V)\approx(v_a/2V)^n/\Gamma(n+1)$, where $\Gamma(n+1)=n!$ with Γ the Gamma function (and not a pair-breaking parameter, as usually) and thus $J_1(v_a/V)\approx v_a/2V$, leading to

$$\langle I_s(X_{PSC},t)\rangle \approx a I_c\{1-b_1(v_a/2V)\cos[\tilde{\varphi}_{12}(0)-\varphi_a]\} \tag{200}$$

If we assume that $a=1/2$, $b_1=1$, and $\tilde{\varphi}_{12}(0)\approx\varphi_{12}(t=0)$, as done in refs. [24, 371], this expression agrees with our early calculations performed within the RSM model [24, 371].

For whiskers of Sn, In, and In-Pb, we observed the main current step, harmonic, and subharmonic steps. In the case of Pb whiskers we never observed subharmonic steps, but only the main step at $\omega_{PSC}=\omega_a$ and the first harmonic step at $\omega_{PSC}=2\omega_a$. This gives experimental evidence for a non-sinusoidal current-phase relation in Sn, In, and In-Pb whiskers and for a pure sinusoidal current-phase relation in Pb whiskers.

It is remarked that in Pb whiskers the main and the first harmonic current step both occur while there is only one phase-slip center present in the sample. For the other materials, harmonic steps are observed if more than one phase-slip center is activated. See for instance Fig. 31, where V_{J4} appears in the presence of two phase-slip centers. The observation of current steps in the presence of more than one active phase-slip center may have to do something with an interaction between the phase-slip centers and will be discussed at the end of this section.

In the radiation field the critical current for a given bath temperature is depressed below its zero-field value for our samples. Mechanisms which may reduce the critical dc current are Joule heating due to radiation induced quasiparticle currents, amplitude modulation of the supercurrent, pair breaking by the absorption of photons from the electromagnetic field and by pair-breaking effects of the magnetic field of the HF radiation. For Pb whiskers we performed detailed investigations of the reduction of the critical current with increasing HF input power and found that the measurements are well described by the pair-breaking effect of the magnetic field of the HF radiation [372], while already rough estimates of the other effects show that they should be of minor importance [393].

In principle electromagnetic high-frequency radiation can lead to a stimulation of superconductivity, as discussed in section 5.5. In this case the critical current would be enhanced in a HF field. In our experiments we did not observe HF stimulated superconductivity. The reason is that the frequency of the applied radiation field was probably too low.

There is a low frequency border, ν_c, of microwave radiation stimulated superconductivity [144 - 146, 191]. Following Eliashberg [191] we calculated this border for the Sn whisker of ref. 371 and the In whisker of ref. 16, inserting Tinkham's estimate of τ_E as given in eq. (72) for the energy relaxation time of the excitations [24]. The frequency border depends on the material and increases with decreasing temperature for a given sample. For the Sn whisker it is $\nu_c(\Delta T = 1\,\text{mK}) \approx 2\,\text{GHz}$ and $\nu_c(\Delta T = 10\,\text{mK}) \approx 3.1$ GHz, while for the In whisker ν_c changes from about 2.4 GHz to 3.7 GHz in the same temperature range. These frequencies are much larger than our maximum measuring frequency of 900 MHz.

An expression for the lowest frequency that can lead to an enhancement of superconductivity is given by Mooij [144] to be $\omega_{c,min} = 1.73\,\tau_E^{-1}$, where $\omega_{c,min} = 2\pi\nu_{c,min}$. Using τ_E values from Tab. A 1 it follows that $\nu_{c,min}$ is of order 1 GHz for Sn and In but of order 10 GHz for Pb.

Now we come back to the phenomenon that harmonic current steps are observed after a second phase-slip center has become active in the whisker. This requires a cooperative behaviour of both phase-slip centers. One possibility is that the individual V-I characteristic of each phase-slip center shows a current step and that there is a certain current range which both current steps are present in. The current step at V_{J4} in Fig. 31 may be the consequence of a current step at V_{J2} in the individual characteristics of both phase-slip centers, but also other combinations (for instance V_{J1} and V_{J3}) are possible where also subharmonic steps may be involved.

The presence of current steps in the individual characteristics with overlapping current ranges may be due to a fortuitous interaction with the external radiation. However, it is also possible that this is the consequence of an interaction between the phase-slip centers.

Naturally, a current step in the V-I characteristic of two phase-slip centers can also be explained without assuming a current step in each individual characteristic. In this case it must be stated that both phase-slip centers cooperatively change their voltages as a function of the measuring current so that the resulting voltage developed by both phase-slip centers remains constant. This may for instance be achieved by a rise of the voltage developed by one phase-slip center and a corresponding reduction of the voltage developed by the other phase-slip center.

Information what really happens can only be obtained from a measurement where the total voltage across the whisker and the individual characteristics of the phase-slip centers are measured at the same run in a microwave field. We have not done this experiment up to now.

There are, however, similar effects reported in the literature. As mentioned above series arrays of proximity effect bridges in a HF field show a synchronization of their voltages so that a current step appears at $n_{ar}V_{J1}$ [386]. The same effect was observed in series arrays of In microbridge Josephson junctions (each junction is a short thin film microbridge which connects two wider film areas) if being submitted to HF radiation [33, 395]. In small arrays of these microbridges with $n_{ar} \leq 7$ also harmonic and subharmonic current steps at $n_{ar}m_iV_{J1}$ and $n_{ar}V_{J1}/m_i$ were observed, where m_i is an integer number [33]. Also pairs of variable-thickness microbridges (each junction being a short narrow bridge connecting two wider and thicker banks) show a current step at $2V_{J1}$ [396].

Dai, Yeh, and Kao [397] investigated a special arrangement of two closely spaced indium thin film microbridges. Their sample consists of a short thin film bridge (typically only 1.4 µm long, and 0.6 µm wide, and 0.12 µm thick) between two wide electrodes. This weak-link region contains two microbridges, in the sense that one half of the narrow film between the electrodes has a smaller width than the other part. An additional voltage probe (about 0.5 µm wide, probing μ_p) is positioned in the middle of the weak-link region and attached to the microbridge with the larger width. Thus, the V-I characteristics of the whole weak-link region and of the individual microbridges could be measured at the same run. The microbridges have a different critical current and the shape of the characteristics indicate an interaction between the voltage carrying states of the microbridges. If this system of microbridges is submitted to HF radiation, current steps appear in the V-I characteristics of the whole weak-link region. The remarkable phenomenon is that the individual characteristics of the microbridges do not show a current step although there is a current step observed in the total weak-link voltage. The voltage throughout this current step is held constant due to a V-I curve distortion of the individual microbridges: The voltage across one microbridge continuously increases while the voltage across the second microbridge decreases and vice versa.

Similar V-I curve distortion was observed by Yu and Mercereau for a proximity effect bridge with an additional potential probe in the weak-link region [398] which in this case measured the quasiparticle electrochemical potential [69, 398]. With applied HF field the V-I characteristics across the weak link show a current step, while the individual potential differences between the probe attached to the weak-link region and superconducting probes on both sides of this region show a compensatory distortion.

The experiments of Dai et al. show that a pair of microbridges in a HF field can generate a current step in the V-I characteristics measured across both bridges while there are no current steps in the individual V-I characteristics of both bridges. However, in their experiment both bridges belong to a common weak-link region. It is not clear, whether this result also holds for microbridge Josephson junctions separated by a wide film area or in our whiskers.

There is another cooperative effect observed in coupled Josephson oscillators without a HF field, called 'voltage locking' or 'frequency locking' or 'phase locking'. The experimental observation is that over a certain current range the individual V-I characteristics of two microbridge junctions are pulled together so that their voltages synchronize to the same absolute value, implying that the phase-slips in the two bridges are locked to the same repetition frequency. Voltage locking is also observed in the experiments of Dai et al. [397]. The phenomenon has been widely studied with closely spaced microbridge Josephson junctions [33, 399 - 404] but is also occurring in multiple terminal weak links consisting of several superconductors in connection with a common region via separate weak links [405, 406].

The most direct method to investigate the properties of a Josephson junction is to observe the electromagnetic radiation emitted from this junction [279, 407, 408]. The detected radiation is usually not monochromatic with the Josephson frequency $\nu_J = 2eV/h$, where V is the time averaged voltage across the junction, but has a content of higher harmonics. If two (or several) Josephson junctions lock to synchronize their Josephson oscillations (that means the time evolution of the phase difference and, thus, the voltage across the junction and the time dependent supercurrents in the element) to a 'coherent', that means 'in-phase' behaviour, the power of the radiation is much larger than in the non-coherent case due to the constructive interference of the radiation. For two junctions radiating with the power P_1 and P_2, the total power of the coherent array would be $P = (P_1^{1/2} + P_2^{1/2})^2$. Coherent radiation has been detected from low number arrays of Josephson tunnel junctions [409], large number arrays of proximity effect bridges [410], two closely spaced microbridge Josephson junctions [400, 411], and two microbridge Josephson junctions coupled by a shunt resistor [412].

There are several possibilities for the coupling mechanism leading to a locked state of Josephson junctions [403, 404, 413]: The order parameter in the middle of a microbridge Josephson junction goes to zero periodically with the Josephson frequency, related with a periodic variation of the order parameter within some $\xi(T)$ around this point (see section 5.9). Therefore, the order parameter oscillation in one microbridge directly modulates the order parameter in the other bridge, if the distance between the bridges is less than some $\xi(T)$. This mechanism may be called 'ac-order parameter coupling'. Moreover, such a Josephson oscillator can generate a charge imbalance wave (see section 5.8.4) which may travel to another microbridge and modulate its superconducting properties. The decay length (denoted with $\lambda_{d,KSS}$ in the KSS model) is usually much larger than $\xi(T)$ and depends on Λ_{Q^*in}, τ_{QR}, and τ_E. This mechanism is an 'ac-quasiparticle coupling'. There can also be a 'high-frequency electromagnetic coupling' due to the emission of radiation from one junction which then induces oscillations in another junction. This coupling can act over larger distances than the first two mechanisms. We refer to the reviews cited above for a detailed discussion of calculations envolving the different coupling mechanisms.

The ac-quasiparticle coupling by a charge imbalance wave was proposed by Lindeloff and Bindslev-Hansen [400], and by KSS [158]. Within the KSS transmission line representation, Blackburn [414] found that a locking of the phase-slip voltages occurs if two phase-slip centers in a distance Λ_{Q^*in} in a quasi-one-dimensional superconductor are coupled by a charge imbalance wave. The phase-slip centers are assumed to develop at two points of weakened superconductivity with different weakness.

The recent calculations of Frank et al. [415] demonstrate that a pair of coupled noncapacitive Josephson weak links can mutually lock subharmonically. Subharmonic locking means that it is $n_1V_1 = n_2V_2$, where V_1 and V_2 are the time averaged voltages across the individual junctions and n_1 and n_2 denote two positive prime integers.

Voltage locking could explain a non-fortuitous generation of current steps with overlapping width in the individual V - I characteristics of two phase-slip centers, leading to a current step in the common characteristics of both centers. The idea is that only one phase-slip center synchronizes with the external HF field to give a current step in its V - I characteristics. This phase-slip center then interacts with the other center so that the voltages of both centers are locked to the same value and also the second center generates a current step (at the same voltage) in its characteristics. (Probably also subharmonic locking can be included in the discussion.) The phenomenon would be somewhat similar to an experiment of Lindelof and Bindslev-Hansen [400, 403] performed in the absence of HF radiation with two closely spaced microbridge Josephson junctions. In their experiment the currents through both bridges could be separately adjusted. So they adjusted the current I_1 through the first microbridge to a constant value V_1 and then measured the $V_2 - I_2$ characteristics of the second bridge, observing current steps in the $V_2 - I_2$ characteristics at $V_2 = V_1$ and $V_2 = 2V_1$.

Another effect which may generate a current step in the V - I characteristics of a Josephson element is predicted by Seifert [416] to be a periodic modulation of the maximum Josephson current. This modulation may be achieved by a periodic irradiation of the superconductor, a periodic quasiparticle injection or an irradiation of the superconductor with a HF radiation of periodically varying amplitude. These disturbances excite the energy mode and, thus, lead to a periodic weakening or strengthening of the superconducting state (see section 5.5). A further discussion of this mechanism is given in refs. 417 and 418.

Vanneste et al. [419] observed inverse ac Josephson-like effects in a weak link generated by laser irradiation of a superconducting film and submitted to a microwave field. However, they did not find current steps due to a modulation of the laser irradiation. Recently Logvenov et al. [420] measured current steps in the V - I characteristics of SC/NC/SC junctions when these specimens were periodically irradiated by a laser. However, the appearance of these steps seems to be associated with the phase modulation of the Josephson oscillation due to alternating thermoelectric supercurrents which appear in the junction.

Nevertheless, a periodic stimulation or weakening of the superconducting properties remains an interesting alternative explanation for current steps in the presence of two active phase-slip centers in our whiskers. As we will discuss in section 5.7, an active phase-slip center can stabilize the superconductivity in the sample. If the stabilizing mechanism contains an ac component, there could be a periodic strengthening of superconductivity.

Now, we return to the discussion of a single phase-slip center in a HF field. The analytical description given in this section already explains the significant properties observed in the experiment. However, the calculations do not deal with the envolved space and time dependent processes in the phase-slip center, but model the actual behaviour in the core by describing the core as a kind of 'black box' with a phase difference at its ends and being characterized by a current-phase relation.

For a more detailed understanding the TDGL theory has to be considered: Using Schmid's version of the TDGL equations (see section 5.9), Fjordbøge and Lindeloff [421] performed analog computer simulations to describe the dynamic properties of (short) superconducting microbridges with a combined dc and ac current flowing through the bridge. The calculated V - I characteristics show inverse ac Josephson effect like behaviour with harmonic and subharmonic current steps.

Very recently Kramer and Rangel (KR) developed a TDGL theory of periodically driven current carrying superconducting filaments [127, 310, 311]. While we assumed a modulation of the voltage across the phase-slip center, KR consider the case of a modulation of the current. They predict that for a pure ac current drive the filament can remain in a homogeneous superconducting state with low dissipation or (depending on the ac amplitude) may switch into a dissipative phase-slip state. The nature of this ac-driven phase-slip state is involved. There are periodic and chaotic solutions. In most cases the time-averaged voltage is zero, but there are also solutions where the time-averaged voltage is given by $\langle V \rangle = \hbar\omega_a/2e$.

The existence of an ac phase-slip state is confirmed by Rachford et al. [422], who studied the microwave response of superconducting microbridges. Measuring the reflected power from a cavity containing the microbridge, they found step structures in the reflected-incident power plots. These experiments have already been carried out in the year 1975.

For a superposition of a dc current and an ac current, KR predict inverse ac Josephson like effects with harmonic and subharmonic current steps. These steps belong to a periodic solution of the order parameter, while the V-I characteristic between the current steps is generated by quasi-periodic solutions. On a current step at $\langle V \rangle = (n_1/m_1)(\hbar\omega_a/2e)$ the phase of the phase-slip center and the ac current are locked, in the sense that one has n_1 phase slips every m_1 periods of the ac current. Here, n_1 and m_1 are integer numbers. Moreover, chaotic behaviour [423, 424] is predicted by KR [310, 311] with pairs of forward and backward phase slips occurring in an irregular manner in addition to the n_1 phase slips per m_1 periods of the ac current. If KR for instance decrease the dc current on a wide current step, corresponding to the basic frequency (that means $\langle V \rangle = \hbar\omega_a/2e$) or to a harmonic, they observe a period doubling cascade for the phase-slip period and then a change to a chaotic behaviour still occurring on the step. Enhancing the dc current toward the other end of the current step leads to intermittent type chaos.

The transition from the homogeneous superconducting state with low dissipation into the dissipative phase-slip state is hysteretic [311]. In the presence of an dc and ac current there is a critical value of the dc current up to which the homogeneous state exists. This critical value is smaller than the GL critical current and depends on the angular frequency of the ac current, its amplitude, and the pair-breaking parameter of the material. The main current step exists above and below this critical value of the dc current. For increasing dc current one thus expects that the filament enters the phase-slip state at the critical dc current by a voltage jump to the main current step, and that for decreasing current the homogeneous superconducting state is recovered for a smaller dc current. This behaviour is indeed observed in our experiments (see Fig. 31 b).

Also the case of two phase-slip centers submitted to a combined dc and ac current is briefly discussed within the TDGL theory [311]. On a current step the voltage is then given by $\langle V \rangle = (n_1/m_1 + n_1'/m_1')(\hbar\omega_a/2e)$, where also n_1' and m_1' denote positive integer numbers. Also in this case period-doubling cascades are observed, leading to chaos for each phase-slip center. The special case that n_1/m_1 and n_1'/m_1' are equal is only found at the border, where the filament switches into a state with two phase-slip centers. In the example shown by Rangel [311] this transition occurs on a current step with $n_1 = 2$ and $m_1 = 1$. Also the transition between the state with one and two phase-slip centers is hysteretic, because the solutions have an overlapping dc-current range.

At present it is not possible to observe the predicted time evolution of the phase-slip center in our samples: KR give time resolved results for the normalized (see section 5.9) electrochemical quasiparticle potential $\hat{\mu}(x=0, t)$ and the normalized order parameter $\hat{\psi}(x=0,t)$, where the core of the phase-slip center is assumed to be localized at $x = 0$. As in the dc case (see section 5.9), KR perform their calculations for a periodic array of phase-slip centers with a periodic length d_p and use the gauge $\hat{\mu}(x=d_p/2)=0$. As pointed out in section 5.9, it is $\hat{\mu}(x=0, t) = \hat{V}(\hat{t})/2$, where $\hat{V}(\hat{t})$ is the normalized voltage per spatial period. Converted to physical units the voltage has the meaning of the change of μ_p across each phase-slip center in the array. There are no predictions made for an isolated phase-slip center ($d_p=\infty$) as done in the dc case (see section 5.9). Thus, what is called above, 'a phase-slip center' or 'two phase-slip centers' should be denoted more carefully by 'a phase-slip center in a periodic array with spatial period d_p' and 'two phase-slip centers within d_p'. However, even if we assume that the calculated time dependent behaviour of one or two phase-slip centers in an array is similar to that of one or two phase-slip centers in a whisker (as may be expected from the qualitative agreement of the time averaged results), it is not possible to measure the time evolution of the voltage signal. This voltage signal is small and changes very fast. Its amplitude is only of order $1\,\mu V$ and its frequency of order 1 GHz. Thus, a time resolved measurement would require a resolution of much better than $1\,\mu V$ on a time scale much smaller than 10^{-9} s.

At the moment one must therefore try to get information about chaotic states from the time averaged voltage-current characteristics as measured with a dc current in a HF field. KR predict chaotic behaviour at the borders of a current step, but still within the step, and quasi-periodic solutions between the steps. The calculated characteristic as given in Rangel's thesis [311] in the environment of the main current step looks quite similar to the measured curve, but it is not possible to distinguish between an ac Josephson like or chaotic behaviour within the current step.

In the case of Josephson junctions breaks and discontinuities in the V - I characteristics (similar to those observed in the characteristics of In whiskers in a HF field [16]) are taken as indication for chaos.

Theoretically, chaotic behaviour in Josephson junctions is usually studied by computer calculations or equivalent circuit electronic analog simulations within a periodically driven capacitive resistively shunted junction model with a combined dc and ac current drive. In some cases additional noise currents are introduced and there are some articles which only consider an ac current drive [425 - 446]. Moreover, there are also efforts to study the problem analytically [434, 447 - 449].

Experimental work (which is more scarce in this field) has been performed with Josephson tunnel junctions and with (short) microbridge Josephson junctions by applying a dc current and HF radiation to the sample [426, 438 - 441, 444, 447, 448]. Recently it has been experimentally observed that also picosecond current pulses may induce chaotic states in dc current carrying Josephson tunnel junctions [451].

While the single capacitive Josephson junction (described by a capacitive resistively shunted junction model) displays chaotic behaviour if an ac or combined dc and ac current drive is applied, chaotic states are already observed for a pure dc current drive if the system also has a self-inductance. This has been studied theoretically and experimentally for a Josephson tunnel junction with a self-capacitance, shunted by an external resistor which has a self-inductance [452, 453].

There has also been carried out some theoretical work on chaos in linear arrays of Josephson junctions. Analog simulations are made for a pair of coupled Josephson junctions, described by a normal conducting shunt coupled pair of noncapacitive resistively shunted equivalent circuits. A combined dc and ac current drive could be applied to the system and chaotic states were observed, although a single noncapacitive resistively shunted junction cannot exhibit chaos [454, 455]. For an array of three resistively coupled noncapacitive Josephson junctions chaos is predicted to occur already with a dc current drive only [456]. Moreover, chaotic behaviour is predicted if a pure dc current is applied to a system of three noncapacitive Josephson junctions coupled by passive but reactive circuit elements. Pure inductive or capacitive loads have been studied and also a series inductive-capacitive circuit shunt [457].

It is remarked that theoretical work on long, large-area Josephson tunnel junctions with applied HF radiation or a combined dc and ac current drive also predicts chaotic behaviour [458, 459]. These systems are described by a 'sine-Gordon' equation (see chap. 10 of ref. 279) and may exhibit chaotic fluxon (soliton) motion [459].

This brief discussion of chaotic phenomena already shows, that chaos is an important effect in Josephson junctions. These phenomena are of great basical interest, but they also lead to limitations in the coherent behaviour of coupled junction arrays, considered as technological devices for microwave generation.

Finally we give a summary of this section: The V - I characteristics of our irradiated whiskers exhibit steps of zero slope ('current steps') in addition to the voltage step structure. These current steps can be understood as inverse ac Josephson like effect described by a phase-slip model with included voltage modulation due to the HF-radiation field. This result establishes the assumption that a cyclical high-frequency process at Josephson frequency is associated with the voltage-step phenomenon.

The pair-breaking effect of the magnetic field of the applied radiation depresses the critical current below its zero field value. HF stimulated superconductivity is not observed because the frequency of the radiation is still too low.

The observation of current steps in the presence of two phase-slip centers requires a cooperative behaviour of both centers. Several possible reasons for such a behaviour are discussed, such as overlapping current steps in the individual characteristics of each center or a compensatory change of the individual voltages. Overlapping current steps may be generated by an interaction of both phase-slip centers with the radiation field. Another possibility is that only one center interacts with the radiation field and then synchronizes the phase-slip oscillations by a voltage locking effect or by periodically weakening or strengthening the superconducting state at the locus of the second center.

Next we discussed the predictions of the TDGL theory for the case of a periodically driven current carrying filament. This theory predicts inverse ac Josephson effect like behaviour (current steps) but also an ac phase-slip state and chaotic behaviour. The calculated time averaged voltage-current characteristic in the environment of the main current step looks similar to the measured curve. Also hysteretic behaviour is predicted.

The characteristics of irradiated In whiskers show breaks and discontinuities. In Josephson junctions distortions in the characteristics are indications for chaos and they have been investigated by theory and observed experimentally.

7.4. Localization of Phase-Slip Centers

The investigations discussed in the preceding two sections establish the interpretation that phase-slip centers generate the voltage step structure in the V-I characteristics of a quasi-one-dimensional superconductor. To prove whether these phase-slip centers are localized within a certain region of the sample (and do not move through the specimen) we measured V-I characteristics and V-T transition curves of different parts of a whisker at the same run [21, 24].

For these experiments we used multi-potential-probe contacts as scetched in Fig. 1c. With these contacts it is not only possible to measure the voltage across the Wood metal contacts holding the sample, but there are additional potential probes to measure the voltage across different parts of the whisker. For the present work we adopt the notation introduced in ref. 36] and, thus, denote (from the left to the right) the parts of the sample by the numbers 1 - 4, the potential probes by P1 - P3, and the Wood metal blocks by the capital letters A - E. For samples with two potential probes P1 and P2, and, thus, only three whisker parts, part 4, probe P3, and contact D do not exist. The full variety of the notation will only be needed in section 7.6 to give detailed information about current directions and voltage polarities by capital letter subscripts. In the present and in the following section we do not introduce capital letter subscripts.

It is remarked that in former publications the numbering of whisker parts and voltage probes for samples with two potential probes is the same as in the present work, but that it is different for specimens with three potential probes [21, 22, 24, 460]. In the latter case, the whisker parts in former work were denoted, from left to right, by 3 - 1 - 2 - 4 (instead of 1 - 2 - 3 - 4 in the present work) and the voltage probes by S1 - S3 - S2 or P1 - P3 - P2 (instead of P1 - P2 - P3 in the present work).

Furthermore, in refs. 24 and 460 current directions were characterized by arrows indicating the flow direction of the electrons. In the present work we deal with conventional currents only, flowing from the positive to the negative pole of the battery.

The V - I characteristics of the two central parts of an Sn whisker with three potential probes are shown in Fig. 32. The results of this kind of measurement are [21, 24]:

The onset of voltage in different parts of the whisker occurs at different values of the measuring current. While one part of the sample is still fully superconducting, voltage steps may be present in a different part.

There are voltage steps in the various parts occurring 'synchronously' at the same current (arrows in Fig. 32). The number of these steps differs. It can be altered with a small change of the bath temperature. However, also measuring the characteristics several times at the same fixed temperature leads to different numbers of synchronous voltage steps (compare Figs. 32 a and b). Synchronous voltage steps were observed for increasing and decreasing current. In the total characteristic of both whisker parts a synchronization of voltage steps leads to one large voltage step, occurring at the 'synchronization current'.

We also measured V-T characteristics, yielding that different parts of a whisker become completely normal conducting at the same temperature, although their onset of voltage occurs at different temperatures.

Different samples qualitatively yield the same results, even if the investigated whisker parts have extremely different lengths. In section 7.6 results of sample Sn 2 are shown where one part of the whisker is about a factor of 10 smaller than all the other parts.

These results show that the origin of a voltage step is a localized phenomenon. Together with the results of the preceding two sections, we, thus, conclude that the voltage step structure in the V-I characteristics of our samples is generated by localized phase-slip centers.

Moreover, the experiments discussed in the present section indicate a mutual influence of phase-slip centers localized at different places in the whisker:

The synchronous voltage steps cannot be explained by a dissipative region extending from one part of the sample into the other part. In this case one would always expect the same number of synchronous steps, at least as long as the temperature is not changed (see section 7.6 in this context). We rather assume that there is a mutual activation of phase-slip centers due to an interaction of the centers.

This mutual influence is also apparent in Fig. 32. The characteristics of Fig. 32 a and b for high currents show the same slope. For smaller currents

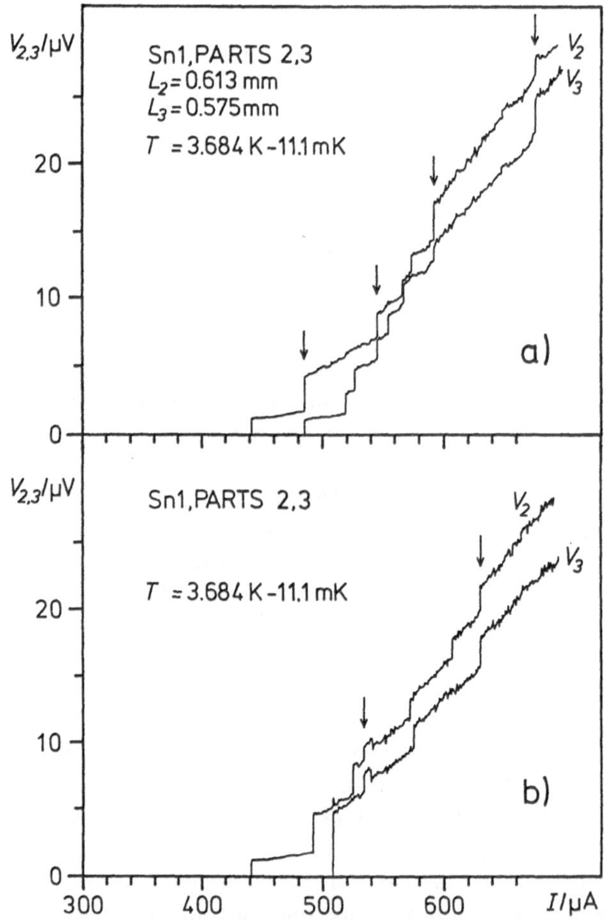

Fig. 32: *a) V-I characteristics of parts 2 and 3 of a tin whisker ('Sn1') with three tin whiskers as additional potential probes, measured at the same run for increasing current I which flows through the whole sample [21, 24]. Arrows indicate voltage steps occurring in different parts of the sample at the same current.*

b) A repetition of the same measurement.

Assuming the onset of voltage in part 2 to be also the onset of voltage across the whole whisker (so that the related current is the critical current of the sample) yields a critical temperature of 3.684 K. The critical temperatures of the potential probes are not known.

The distances of the probes P1 and P3 from the Wood's metal contacts are L_1 = 0.463 mm and L_4 = 0.338 mm, respectively. Further characteristic properties of Sn 1 (called 'Sn' in ref. 21) are given in refs. 22 or 24.

they differ although they were measured at the same fixed temperature. While the first voltage step in part 2 appears at the same current in both figures the further development of $V_2(I)$, $V_3(I)$ in Fig. 32a is different from that one in Fig. 32b. A probable explanation is that in Fig. 32a the first voltage step in part 3 occurs at a smaller current (synchronously with the second voltage step in part 2) than in Fig. 32b and, thus, influences the further development of the characteristics in a way different from that one shown in Fig. 32b.

Finally, also the behaviour of the V-T characteristics of different parts (discussed above) indicate a mutual influence of the phase-slip centers in our whiskers.

These results require a quantitative investigation dealing with the question in how far the presence of a phase-slip center influences the superconducting properties of a filament. This is the subject of the following section, where we also will give a discussion of the interaction effects reported here considering experiments on microbridges and theoretical work.

Now we come back to the localization of the phase-slip centers. In the literature experiments are scarce which demonstrate the localization of phase-slip centers in a quasi-one-dimensional filament not containing any notch or anything else to fix the locus of phase slip. There are the experiments of SBT [32] and Aponte and Tinkham [72] performed with (long) tin microbridges and those of Yeh et al. [461, 462] carried out with (long) indium microbridges.

In these experiments samples with several additional potential probes (made of the same material as the microbridge) were used. However, in the region where a voltage probe is attached to the microbridge, the width of the microbridge is enlarged and in most cases [72, 461, 462] the space between the probes, where the microbridge has a uniform width, is only about 2 µm. In the current carrying state one, therefore, expects a local enhancement of the order parameter in the regions where a superconducting potential probe is connected with the microbridges. This local enhancement of the order parameter should influence the dissipative state developing in the microbridge so that these long microbridges with its potential probes may be regarded as a series connection of several short bridges. This interpretation is established by Yeh et al. [462] who observed that in their indium microbridges with a total length of about 10 µm there is only one phase-slip center in the filament at temperatures close to T_{c0}, if there are no additional potential probes. On the other hand, if superconducting leads are attached to the filament to probe the potential, several phase-slip centers develop in the specimen, one in each region between the leads. In the work of SBT [32] the space between the potential probes is much larger and the influence of the potential probes should be less serious.

Finally, some remarks on the mechanism leading to a localization of phase-slip centers:

In long microbridges commonly variations of the critical temperature of about 20 - 30 mK are observed [32, 70, 84]. Thus, it is assumed that the phase-slip centers in microbridges develop at weakly superconducting regions with a reduced critical current.

Whiskers are much more homogeneous with a very small transition width (in temperature) for small fixed currents. Thus, variations in the critical temperature along the whisker should be very small. The mechanism which leads to a localization in these specimens is finally not known. There may be small variations of the mean free path ℓ of the conduction electrons, yielding local variations of the critical current which depends on ℓ (see section 5.1 and 6.1). In anisotropic materials also T_{c0} depends on ℓ due to the anisotropy effect (see section 6.2) and naturally also T_{c0} enters the expression for I_c. Furthermore, mechanical stress or strain in a whisker fixed between contact blocks cannot be excluded and may lead to local changes of the critical temperature [463 - 467]. Finally, elastically bending a whisker leads to an increase of the resistance at low temperatures. Thus, also bending effects can lead to a local reduction of ℓ and via an isotropy effect also of the critical temperature [468].

7.5. Stabilization of Superconductivity by a Phase-Slip Center

In the preceding section we reported measurements, dealing with multi-potential-probe contacts which do not only indicate a localization of phase-slip centers, but also give evidence for an influence of an active phase-slip center on the superconducting properties of a whisker. We observed voltage steps appearing synchronously at the same current in the V-I characteristics of different whisker parts, V-T characteristics of different sections of the whisker entering the normal state at the same temperature although changing from the superconducting state to the dissipative phase-slip state at different temperatures, and a development of the V-I characteristic of a certain part of the whisker which depends on the appearance of phase-slip centers in an other section of the sample.

These observations are the key for an explanation of the distance in current (or temperature) between the voltage steps in the V-I (or V-T) characteristics of our samples and, thus, for an understanding of a large transition width of nearly homogeneous filaments. The idea is that only the appearance of the first phase-slip center is directly related with the material properties of the whisker, while the appearance of the following phase-slip centers is influenced by those being already present which may stabilize the superconductivity in the whisker so that the next phase-slip center is switched on at a somewhat larger measuring current.

This stabilizing effect has indeed been observed experimentally [22, 24, 460]. For these measurements we had to control the appearance of phase-slip centers in different parts of the sample. This can be done by using multi-potential-probe contacts to impress different currents into different parts of the sample. The measuring technique is described in detail in ref. 24 or 460. In principle a 'measuring current', I_m, flows through the whole sample, while a 'control current', I_k, only flows through one part of the whisker *1. Since both currents are separately adjustable, this enables us to make the effective transport current through this part higher or lower than I_m by impressing a suitable fixed current parallel to I_m (notation briefly '$I_k > 0$' or '+') or in the opposite direction (briefly '$I_k < 0$' or '-'). The exciting result is that phase-slip centers in one part of the sample can enhance the critical current of phase-slip centers in other parts of the whisker to higher values.

The first approach to this problem was made with sample 'Sn 2' with three additional potential probes (also Sn whiskers) [24, 460]. For a sketch of this arrangement see Fig. 1c and consider the remarks made in the beginning of the preceding section. Increasing the measuring current without any control current, usually the first two phase-slip centers in this whisker develop in part 3 and then a phase-slip center appears in part 1.

Now a suitable control current $I_k > 0$ was applied to part 1. The control current being switched on, there are phase-slip centers present in part 1 before the two phase-slip centers in part 3 appear. As a consequence the critical currents of both phase-slip centers in part 3 are shifted to higher values. The magnitude of the critical current enhancement is different for both phase-slip centers, depends on the number of influencing phase-slip centers, and becomes larger for lower measuring temperature, but does not depend on the voltage developed by the influencing phase-slip centers and is within the limits of accuracy the same if the polarities of I_m and I_k are both changed to their opposite.

Investigations of the influence of phase-slip centers in part 3 of Sn 2 on phase-slip centers in part 1 yield a depression of the critical current of the centers in part 1. The depression depends on the polarity of the currents and on the number of influencing phase-slip centers. When their number is increased, the depression becomes smaller and changes into an enhancement.

The results obtained for Sn 2 are given in ref. 460 and in more detail in ref. 24. However, for this sample the control current leaves (or flows into) the sample through a potential probe which is normal conducting during the measurement. Therefore, we cannot exclude an influence of quasiparticle injection or extraction effects on the results. Moreover, the distance between influencing and influenced phase-slip centers may be not very large. The distance is not exactly known, because we do not know the exact position of the phase-slip centers. They must, however, be located in parts 1 and 3,

*1 In former publications [22, 24, 460] the measuring current was called I_1 and the control current was denoted by I_2.

respectively, so that their distance is at least the length $L_2 = 38\,\mu m$ of the separating part 2.

Therefore, sample Sn 3 was prepared with two additional tin whiskers which are superconducting during the measurements and with a separating section of $300\,\mu m$ between the parts where influencing and influenced phase-slip centers are situated in [22, 24]. A sketch of the sample geometry is given in the inset of Fig. 33. The influencing phase-slip centers are situated in part 3 and the influenced phase-slip centers in part 1 at a distance that is at least the length L_2 of part 2.

Under the influence of phase-slip centers in part 3, the critical current of the first phase-slip center in part 1 is shifted to a higher value. Also the critical current of the second phase-slip center is enhanced. Since both critical currents are nearly equally influenced, almost no change of the width between the first and the second voltage step in the $V_1 - I_m$ characteristics is observed. Also in sample Sn 3 the magnitude of the critical current enhancement depends on the number of influencing phase-slip centers. For the first phase-slip center in part 1 it has the largest value if there are four influencing phase-slip centers in part 3. A quantitative evaluation has been carried out for different measuring temperatures. The results are given in detail in ref. 22 or 24.

These measurements show that there is a mutual influence of phase-slip centers even if they are situated far away from each other. Therefore, it appears that a phase-slip center occurring in one part of the whisker can stabilize the superconducting properties of the whole sample, leading to the large transition width observed although the sample is nearly homogeneous.

We can also make some quantitative statements about the homogeneity of our samples by asking, which part of the current difference between two voltage steps is caused by the stabilizing effect discussed above. For this purpose we need information about the critical currents of both voltage steps in the uninfluenced case. If both voltage steps in this case have the same critical current at every time, we can conclude that the whole distance between the steps comes from the stabilizing of superconductivity caused by one or more other phase-slip centers in the sample.

To obtain the desired information in the case of sample Sn 3, we at first only applied the measuring current I_m and determined the $I_c^{2/3}(T_c)$ straight line for the first voltage step of the sample which appears in part 3 of the whisker without being influenced by other phase-slip centers. The first voltage step in part 1 sometimes occurs synchronously at the same current with the first step in part 3. These measurements have not been evaluated. If there is no synchronization between the steps, the first step in part 1 appears at a larger current than the first step of the sample in part 3. Then we applied a control current $I_k < 0$ to part 3 so that the total transport current through this section is reduced and now the first step in part 1 becomes the first voltage step of the sample when the measuring current I_m is increased

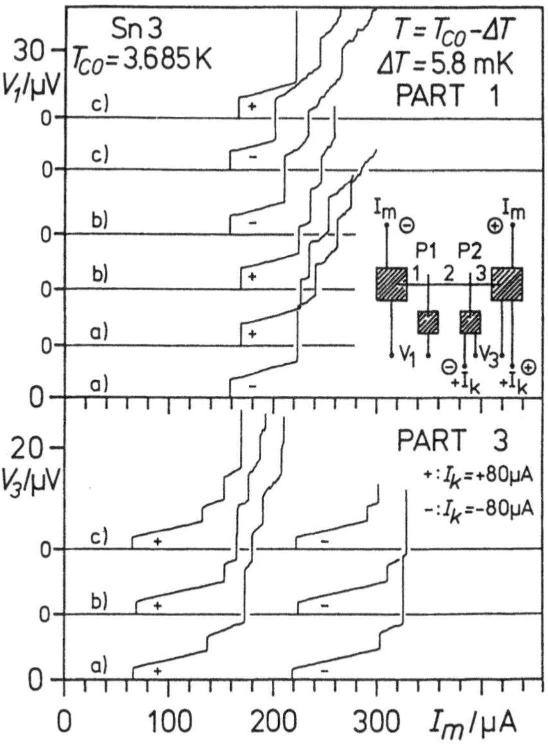

Fig. 33: *The V_1-I_m and V_3-I_m characteristics of parts 1 and 3 of sample Sn 3, a tin whisker with two additional superconducting tin whiskers P1 and P2 as potential probes [22, 24]. The inset shows the sample geometry. The circle minus and plus signs indicate the polarity of the currents. The lengths of the sample parts are L_1 = 0.275 mm, L_2 = 0.300 mm, and L_3 = 0.563 mm.*

For '-' a fixed control current, I_k, is flowing through part 3 of the sample in opposite direction to the measuring current, I_m, while for '+' both currents have the same direction. For '-' the first voltage step (that means the first phase-slip center) in part 1 appears before the occurrance of the first voltage step in part 3, while for '+' there are already voltage steps present in part 3 before the onset of voltage in part 1. Influenced by these voltage steps the occurrance of the first voltage step in part 1 is shifted to higher current values.

In this figure, (a) - (c) are three repetitions of a measurement, measured in the sequence (a) - +, (b) + -, (c) - +. Characteristics marked with the same symbol [for instance, '(a) +'] were measured in the same run.

and, thus, now this step appears without being influenced by other phase-slip centers. In this case its critical current is smaller and follows the same $I_c^{2/3}(T_c)$ straight line as the critical current of the first voltage step in part 3 in the uninfluenced case where only I_m is applied to the whisker. Thus, the whole current difference between these two voltage steps is caused by the stabilizing effect discussed above. For quantitative results of our homogeneity investigations we refer to ref. 22 or 24.

In ref. 367 we stated that influencing phase-slip centers although shifting the critical current of an influenced phase-slip center do not seem to change the dynamics of the phase-slip process and the relaxation behaviour of the nonequilibrium quasiparticles very much. Within our measuring accuracy it has not been observed any change of the differential resistance and the zero voltage intercept I_0. This statement is concluded from our stabilizing experiments with only a few influencing phase-slip centers which cannot be closer than a certain minimum distance to the influenced centers. The statement does not seem to be true in general. See for instance Fig. 32 b, where the voltage jump before the last voltage step in V_2 leads to a small synchronous change of the differential resistance of the $V_3(I)$ characteristic.

We now turn to a brief survey of dc interactions observed between closely spaced microbridges. Subsequently, the problem of the mechanism behind the synchronization, stabilization, and depression effects observed in our whisker experiments will be discussed.

A system that has been widely investigated is a series array of two microbridge Josephson junctions (each junction is a short thin film microbridge which connects two wider film areas) with a very small distance between the bridges. The current through each bridge is separately adjustable and the voltage across each bridge can be measured. As we have discussed in section 7.3, such a system of closely spaced short microbridges shows voltage locking effects. Moreover, this arrangement exhibits dc interactions, that means shifts in the critical currents or changes in the time averaged voltage (which have nothing to do with voltage locking) of one bridge due to an influence of the other bridge.

The first experiments with two closely spaced thin film microbridges have been reported by Jillie et al. [33, 399, 402, 403, 413, 469], using indium microbridges with a length which is smaller than 1 μm and a distance between the bridges which is usually only 2 μm. Several interaction effects are observed in these bridges with a distance which is comparable with the GL coherence length $\xi(T)$ and smaller than the charge imbalance decay length Λ_{Q^*in}:

If a current I_2 is flowing through the second bridge while this bridge remains in the superconducting state ($V_2 = 0$), the critical current I_{c1} of the first bridge is nevertheless reduced below its intrinsic value I^0_{c1} which is measured for a current I_1 flowing through the first bridge only. This depression of the critical current I_{c1} becomes larger for increasing magnitude of I_2 and is generally present, independent of the direction of the currents through the bridges which may be the same ('series case') or opposed ('opposed case').

For a current I_2 which is so large that a phase-slip center develops in the second bridge and, thus, it is $V_2 \neq 0$, the interaction depends on the relative direction of the currents. In the opposed case a further depression of I_{c1} is observed if I_2 is increased. In the series case I_{c1} does not drop further, independent of the magnitude of I_2.

From these experiments it is concluded that the total change, ΔI_{c1}, of the critical current is generated by an interaction which contains a symmetric part, ΔI^s_{c1}, being independent of the relative direction of the currents through the bridges and an asymmetric part, ΔI^a_{c1}, which depends on the relative current direction. The symmetric part decreases the critical current I_{c1} while the asymmetric part only leads to a further decrease of I_{c1} in the opposed case of current flow but acts to increase I_{c1} in the series case. With the assumption that ΔI_{c1} is simply the sum of both contributions, they can be easily separated from a measurement of the critical current I_{c1} as a function of I_2 in the series and opposed case [469].

The mechanisms proposed to explain the experimental results are different in the zero voltage state and in the voltage carrying state of bridge 2, respectively. As long as $V_2 = 0$, there is only a symmetric interaction due to a current induced order parameter depression. In more detail, the current I_2 reduces the order parameter in bridge 2 below its temperature dependent zero current value. The order parameter cannot abruptly increase to its zero current value at the edges of the bridge, but is assumed to recover the zero current value exponentially with a spatial decay length $\xi(T)$. The consequence is a reduction of the order parameter in bridge 1, leading to a depression of its critical current I_{c1}.

In the voltage carrying state with $V_2 \neq 0$, processes are more involved. There remains a symmetric interaction present, but now more factors can reduce I_{c1}. The order parameter in bridge 2 is periodically driven to zero and so its time average is smaller than the value in the zero voltage state and a stronger depression of the order parameter in bridge 1 is expected. Since energy is dissipated in the phase-slip center in bridge 2, Joule heating should occur, enhancing the lattice temperature also in bridge 1. Also a quasiparticle overpopulation may be induced into bridge 1 enhancing the effective quasiparticle temperature in the sense of the T^* effect described in section 5.5.

The asymmetric effect observed for $V_2 \neq 0$ is assumed to be the consequence of a quasiparticle diffusion current generated in the phase-slip center in bridge 2 and flowing through bridge 1. The idea has already been mentioned in section 5.8.3 when discussing the SBT model. Around a phase-slip center the supercurrent is smaller than the total current, because a part of the current is carried as a quasiparticle current. As the distance between the bridges is smaller than Λ_{Q^*}, a quasiparticle current is also flowing through bridge 1. In the series case, where I_1 and I_2 have the same direction, the supercurrent through bridge 1 is smaller than I_1, because a part of the total current is already carried as a quasiparticle current. In the opposed case, where the direction of I_1 is opposed to that one of I_2, the supercurrent through bridge 1 is larger than I_1, because a quasiparticle current which has to be compensated by a supercurrent is flowing through bridge 1 opposite to I_1. Since the change into a dissipative state occurs if the supercurrent becomes too large, the critical current I_{c1} measured will be larger in the series case and lower in the opposed case compared to the value determined by the symmetric effect.

There is another experiment performed with these closely spaced indium microbridges which also demonstrates these interaction effects [33, 469]. In this experiment the current I_2 is fixed to a value where $V_2 \neq 0$ and then I_1 is varied and V_1 and V_2 are measured as a function of I_1. The interesting result is that V_2 changes with I_1 although I_2 is constant. In the opposed case V_2 increases as a function of I_1. The increase is already present while bridge 1 is still in the superconducting state, but it becomes much stronger as soon as a phase-slip center develops in bridge 1. In the series case V_2 behaves very different. Starting from the value for $I_1 = 0$, V_2 slightly drops and then rises again. Close to I_{c1} it is somewhat larger than the initial value. For $V_1 \neq 0$ the voltage V_2 first increases and then decreases again and finally falls below its value at $I_1 = 0$.

Another effect observed in the closely spaced two microbridge junction system is that the critical currents of both bridges clump together to a common value if a current is flowing through both bridges [33, 399, 403, 413]. This common value is depressed below the critical currents of both bridges as measured for a current flowing through the individual bridge only. This 'critical current locking' or 'critical current pulling' effect is observed if the individual critical currents of the bridges are not much different. It is related with a sharpening of the resistive transition. The pulling effect is a consequence of the current induced order parameter depression discussed above and, thus, demonstrates the symmetric interaction.

Attempts were made to describe the interaction effects theoretically. There are estimates for the contributions of the symmetric interaction such as the current induced order parameter depression for $V_2 = 0$ [33, 413, 470] and for $V_2 \neq 0$, heating, and quasiparticle overpopulation effects [33, 469]. There is no quantitative agreement with the measured depression of I_{c1}. Nevertheless, the shift ΔI_{c1}^s as obtained from the current induced order parameter depression effect is of reasonable size [33, 413]. For $V_2 \neq 0$, ΔI_{c1}^s is within a factor of 2 that one calculated due to measured lattice heating [33, 469].

To describe the asymmetric interaction, a resistively shunted junction (RSJ) model is used, into which terms have been included which consider quasiparticle currents injected from one microbridge into the other one [33, 401 - 403, 413, 469]. Comparing the results of this model with measured V-I characteristics of interacting microbridges yields a reasonable qualitative agreement of the general forms [402].

Now we briefly discuss other experimental and theoretical work done with microbridges:

Also Lindelof and Bindslev-Hansen [400] report dc interaction effects between two closely spaced indium (and tin) microbridge Josephson junctions. A heating induced symmetric depression of the critical current is observed and an asymmetric behaviour which increases the critical current in the series case and decreases the critical current in the opposed case. However, only in some cases the asymmetric interaction is induced by nonequilibrium quasiparticles. Usually it is caused by the difference in the current distribution in the region between the bridges which is different in the opposed and series case.

Neuman et al. [471] also report measurements with the closely spaced two indium microbridges system described above. They found a regenerative feedback effect arising from the exchange of quasiparticles between the phase-slip centers in the bridges in the opposed case. The idea is that a phase-slip center which, at a depressed critical current, appears in a microbridge carrying quasiparticle currents generated by a phase-slip center in the other bridge, naturally itself also starts to emit quasiparticles. These quasiparticles migrate to the other center and change its phase-slip behaviour, because they act to enhance the supercurrent through that bridge. The consequence is a different quasiparticle emission of the other phase-slip center toward the center under consideration which reacts with a readjustment of its properties and, thus, with a change of its quasiparticle emission behaviour and so on. This effect results in a step-like increase of the voltage in a bridge carrying a constant current, if a phase-slip center develops in the neighbouring bridge.

Neuman et al. [471] analyzed the interaction effects between the two bridges within a RSJ model with quasiparticle coupling currents. They assumed identical junctions so that the fraction of the quasiparticles generated in one bridge which passes through the other bridge is given by the same coupling constant. In order to include the effect of an ac quasiparticle current generated by feedback effects, the authors do not only perform their calculations within the first order of the coupling constant (as usually done [403]), but also consider second order terms. For the case that there only is a phase-slip center present in one bridge, they calculate the current through the neighbouring bridge. There are three terms in the resulting expression. The first term is the supercurrent through the bridge. The second term is linear in the coupling constant and describes the asymmetric quasiparticle diffusion current effect. The third term is of second order in the coupling constant. It describes a symmetric ac quasiparticle current effect which acts to decrease the critical current of the bridge independent of the relative current direction in the bridges.

Thus, there is also a feedback term in the presence of only one phase-slip center. In principle, such an effect may be expected, because charge imbalance waves should be excited by the active phase-slip center which may act back to the center by the induction of quasiparticle currents. However, we did not succeed in finding a simple qualitative nonartificial argument in terms of a charge imbalance wave picture leading to the symmetric critical current reducing ac quasiparticle current effect.

In a subsequent paper Neuman et al. reported further experiments performed with the two indium microbridges system [472]. In the series case, critical current pulling with a sharpening of the resistive transition is observed for bridges with similar intrinsic critical currents. If the intrinsic critical currents are rather different, the first phase-slip center enhances the critical current of the neighbouring bridge above its intrinsic value. The pulling and sharpening effects are stated to be a consequence of the current induced order parameter depression and the ac quasiparticle feedback interaction while the critical current enhancement is proposed to be caused by the quasiparticle diffusion current effect.

Also the special arrangement of two very closely spaced indium microbridges of Dai, Yeh, and Kao, described in section 7.3, shows quasiparticle induced interaction effects. In their experiments a current I passes through both bridges. The quasiparticle current generated by the phase-slip center developing in the more narrow bridge enhances the critical current of the wider bridge above its intrinsic value. There is a step-like change in the V-I characteristics of the first phase-slip center at the moment when a phase-slip center appears in the second bridge caused by the ac quasiparticle feedback effect. Then the voltage of the first phase-slip center increases with increasing I more slowly due to the quasiparticle diffusion current generated by the center in the second bridge. Moreover, Dai et al. report the observation of critical current pulling in an arrangement of two very closely spaced microbridges which have the same width and, thus, similar intrinsic critical currents.

Also three terminal and four terminal weak links show interaction effects caused by dc and ac quasiparticle currents [405, 406]. The same is the case for the (long) microbridges with several potential probes (described in detail in the preceding section) investigated by Yeh et al. [461, 462] and Aponte and Tinkham [72].

Yeh et al. [461, 462], who applied a measuring current passing the whole indium microbridge, observed that when a phase-slip center develops a voltage, this leads to an increased differential resistance of its nearest neighbour due to the ac quasiparticle effect while at high voltages all neighbouring phase-slip centers are affected by the dc quasiparticle diffusion current and they, thus, show a decrease of their voltage. Furthermore, Yeh et al. demonstrate that the critical current of the first part of their microbridge is enhanced by present phase-slip centers in other parts of the sample. For this purpose they also measured only the V-I characteristics of the first part of the sample by using a current which leaves the sample through the first potential probe.

Aponte and Tinkham [72] investigated the properties of two neighbouring parts of their tin microbridge with separately adjustable currents which leave the sample through the potential probes. While a fixed current is flowing through the first part they measure the V-I characteristic of the second part.

The result is that a phase-slip center in the first part enhances the critical current in the series case but depresses the critical current in the opposed case. In addition to this asymmetric interaction there is a symmetric interaction which acts to reduce the critical current. Both contributions could be separated. The asymmetric part, ΔI_c^a, grows proportional to the quasiparticle current in the influencing phase-slip center and is interpreted as quasiparticle diffusion current effect, while the symmetric part, ΔI_c^s, can be explained by Joule heating effects.

Recently, Kober, Clauss, and Huebener [473] investigated the interaction of two phase-slip centers in a long indium microbridge with an additional voltage probe attached to the center of the bridge. In the left and in the right part of the microbridge there is a notch. These two notches act as nucleation sites for the first two phase-slip centers. The central probe does not only serve to measure the V-I characteristics of each part of the microbridge, but also to feed a control current only flowing through one part of the bridge.

In a first type of experiment the transport current is flowing through the whole bridge. In this case the appearance of the second phase-slip center leads to a change of the differential resistance of the first phase-slip center. For temperatures very close to T_{c0} a voltage rise is observed while the opposite is the case at lower temperatures.

In a second type of experiment the transport current through the whole bridge is fixed to a value where one phase-slip center is present in one part of the microbridge. Then a control current is applied to the other part of the bridge, parallel to the measuring current. If the control current is enhanced, the voltage generated by the first phase-slip center changes. The change is especially strong, if the control current is so large that the other part changes into the dissipative phase-slip state. This experiment (and its results) is very similar to the 'other experiment' of Jillie et al. [33, 469] described above.

Parts of the phenomena observed by Kober et al. can be quantitatively understood as quasiparticle diffusion current effects. For this purpose results of the KSS model in the high voltage dc limit are considered. Moreover, there seem to be ac effects present. These effects cannot be understood by an overlap of the cores of the phase-slip centers, because the distance of the centers is much larger than the GL coherence length. They remain unexplained.

Thus, the experiments performed with microbridges show interaction effects which can be understood in terms of current induced order parameter depression, quasiparticle diffusion currents or ac quasiparticle current feedback effects. The decay length for current induced order parameter depression is the GL coherence length, $\xi(T)$, while quasiparticle diffusion currents decay with the charge imbalance conversion length Λ_{Q^*}. The ac quasiparticle currents decay with $\lambda_{d,KSS}$ ranging between Λ_{Q^*in} in the low frequency limit and $\Lambda_{Q^*in}(\tau_{OR}\tau_E)^{1/2}/[\frac{1}{2}(\tau_E+\tau_{OR})]$ in the high frequency limit (see section 5.8.4) for the case of charge imbalance relaxation due to inelastic electron phonon scattering only. Thus, also the decay length of the ac quasiparticle currents does not exceed the charge imbalance conversion length.

While the distance of the microbridges investigated (and, thus, the distance of the interacting phase-slip centers) is usually smaller than the charge imbalance decay length and comparable with the GL coherence length, this is not the case in the whisker of Fig. 33. In this sample the distance between the influencing and influenced phase-slip center is at least 300 μm. The GL coherence length $\xi(T)$ is only 5 μm at the measuring temperature. A value calculated for Λ_{Q^*in} (using Tinkham's estimate for τ_E) yields only 32 μm. The quasiparticle diffusion length Λ calculated according to the SBT model from the measured temperature independent differential resistance is 26 μm.

Assuming that the charge imbalance decay and, thus, the decay of the quasiparticle current is characterized by the quasiparticle diffusion length as determined from the experiment, we calculated the time averaged quasiparticle diffusion current which can act to enhance the critical current of part 1 of sample Sn 3 in the experimental situation of Fig. 33 [22]. For this purpose we assume the case of greatest possible stabilization, namely that the influenced phase-slip center in part 1 will appear near potential probe P 1, the influencing phase-slip centers in part 3 are localized near probe P 2, and their quasiparticle currents superimpose in an additive manner. Within the SBT model the time averaged quasiparticle current is given by eq. (136) of the present work (with $x_R - x$ and $x_R - X_{PSC}$ replaced by $x - x_L$ and $X_{PSC} - x_L$, respectively, where x_L is the right side of the left Wood metal contact and x increases from left to right). In the case of curve (a) of Fig. 33 with two influencing phase-slip centers it follows [22] that the time averaged quasiparticle current at the locus of the influenced phase-slip center is only $3 \cdot 10^{-3}$ μA which is several orders of magnitude smaller than the measured enhancement of the critical current of about 10 μA. Thus, the stabilizing effects observed in our whiskers cannot be explained by quasiparticle diffusion currents.

It is actually not clear which mechanism is responsible for the stabilization of superconductivity in our whiskers. Naturally stimulated superconductivity by HF radiation (photons, see section 5.5) has to be discussed in this context [22]. For such a mechanism the superconductivity would be stabilized by photons which are emitted by an active phase-slip center in one part of the whisker and then absorbed in another part. The frequency of these photons can be calculated from the voltage of the emitting phase-slip center by the Josephson relation. It is, however, below the lower frequency border of microwave radiation stimulated superconductivity (see section 7.3). Furthermore, we mentioned already in the beginning of this section, that there is no evidence for a systematic dependence of the stabilizing effect on the voltage developed by the influencing phase-slip center.

Thus, possible mechanisms for the explanation of the stabilizing effect observed can only be proposed in a speculative manner: Since phonons should be created in the relaxation processes of nonequilibrium quasiparticles, a phonon stimulated superconductivity could be possible. If nonequilibrium quasiparticles in higher places of the excitation spectrum would reach the locus of the influenced phase-slip center, they could lead to an enhanced Cooper-pair density due to their recombination rate which is higher compared to the equilibrium situation. Both phenomena have been discussed in section 5.5. Moreover, the influencing phase-slip center may excite a collective excitation in the sample, leading to local compressions of Cooper pairs and quasiparticles. If the collective mode is only damped by the motion of quasiparticles, which is dissipative due to scattering from impurities, its decay length, λ_d, may exceed the charge imbalance decay length Λ_{Q^*in} by several times for a sample in the clean limit (like a pure tin whisker) at temperatures some millikelvins below T_{c0} (see section 5.6). The problem is that one then has to state an (unknown) mechanism which switches off the damping of the mode due to charge imbalance relaxation by inelastic electron-phonon processes although τ_E is not much larger than τ_{OR}. In the other case the decay length of the mode, $\lambda_{d,KSS}$, cannot exceed Λ_{Q^*in} as discussed above.

In sample Sn 2 we did not only observe an enhancement but also a depression of the critical current. In this sample it cannot be excluded that the distance between influencing and influenced phase-slip center is rather close. Thus, it may be possible that quasiparticle overpopulation effects play a role which would lead to a depression of the energy gap (see section 5.5). Since the guaranteed distance between the phase-slip centers, on the other hand, is much larger than $\xi(T)$, current-induced depairing should be negligible. Also lattice heating should not be of any importance, because heating effects are very small in a pure sample close to T_{c0} (see chap. 11).

There also is no general explanation for the synchronization of phase-slip centers in our samples. As long as the effect is observed for two neighbouring parts of a whisker (as in sample Sn 1), it cannot be excluded that the synchronizing phase-slip centers appear close together and mechanisms invoked to explain the critical current pulling in closely spaced microbridges may be considered. The first problem for an explanation arises from the fact that in whiskers several phase-slip centers synchronize their critical currents. If one assumes that one pair of these is closely spaced, this cannot be the case for the others, because they avoid a close neighbourship to each other due to the stabilizing effect of their quasiparticle currents. Moreover, synchronous voltage steps are also observed in parts 1 and 3 of sample Sn 3 which shows more clearly, that a long range interaction is required for an explanation.

Finally, also the TDGL theory predicts a distance in current between the voltage steps in the V-I characteristics of a quasi-one-dimensional filament.

As we have discussed in detail in section 5.9, an isolated phase-slip center is predicted to exist only below the GL critical current, I_c, while periodic arrays of phase-slip centers can exist up to a maximum current which is larger than I_c. For a given γ value (that means for instance a fixed sample at a fixed temperature) this maximum current first increases and then decreases again, if the distance d_p between the phase-slip centers in the array is reduced. This can be seen from Fig. 1 of ref. 316, where the normalized maximum current density $\hat{j}_{max}$ is plotted as a function of $d_p/\xi_0(T)$.

The idea is that for a filament at a fixed temperature and for increasing current, the existence region of the dissipative phase-slip state can be extended to higher and higher measuring currents by the successive appearance of additional phase-slip centers because this leads to a reduction of their distance d_p. This is in qualitative agreement with the features of our measured V-I characteristics.

There are some problems if we want to do a quantitative comparison with the TDGL theory: The theory is only valid in the dirty limit, while our pure whiskers are clean superconductors. The theory deals with periodic arrays of phase-slip centers. These arrays do not seem to be present in the experiments, because we observe several voltage steps in the V-I characteristics of one part of a sample and the fully superconducting state in another part. To illustrate this, we refer to Fig. 32 b, demonstrating that in sample Sn1 there may be three phase-slip centers present in part 2, while the neighbouring part 3 is fully superconducting. As the length of both parts is very similar, the presence of a periodic array would also require about three phase-slip centers in part 3. (For a more detailed discussion in which sequence voltage steps appear in the characteristics of different whisker parts we refer to ref. 24, where we summarized this feature for several tin whiskers with multi-potential-probe contacts.) Finally, our experiments indicate that the onset of the first voltage step is related to the GL critical current (see section 6.1). This is contrary to the TDGL theory which predicts that the upper border of the current region which a single phase-slip center can exist in is limited by I_c.

In summary, the present section shows that a whisker in the region between the superconducting and the fully normal conducting state enters a dissipative state with a system of interacting phase-slip centers. There is a stabilizing effect of a present phase-slip center which enhances the critical current of another center, even if this center is situated a large distance away compared to the quasiparticle diffusion length. An active phase-slip center seems to influence the superconducting properties of the whole sample. There are voltage steps in the V-I characteristics of our samples which have a current difference caused by the stabilizing effect only. Thus, the large transition width of our samples can be explained, even if one assumes that the whisker consists of nearly homogeneous material. However, the mechanism of the long range interaction is not understood at present.

7.6. Electrochemical Potentials of Pairs and Quasiparticles - Evidence for Collective Excitations

In the nonequilibrium region around a phase-slip center the electrochemical potentials of Cooper pairs, μ_p, and quasiparticle excitations, μ, are different, because the charge imbalance, Q^*, is nonzero. Differences between μ and μ_p can be detected by an NC/SC potential probe pair, because a normal conducting probe measures μ and a superconducting probe measures μ_p (see section 5.3).

Dolan and Jackel [70] measured the time-averaged potentials $\langle\mu\rangle$ and $\langle\mu_p\rangle$ near a phase-slip center in long microbridges of tin and indium. They used opposed sets of normal and superconducting probes with oxide barrier tunnel junctions at the probe/sample interfaces. In their paper they show a measured spatial profile for both potentials in a tin strip which convincingly supports the behaviour predicted by SBT (Fig. 13 c): While $\langle\mu\rangle$ varies smoothly over several μm showing an exponential decay, $\langle\mu_p\rangle$ changes abruptly between two spatially constant values.

Aponte and Tinkham [72] measured the spatial dependence of $\langle\mu\rangle(x)$ near a phase-slip center in a long tin microbridge using normal tunnel junction probes. Stuivinga et al. [73] investigated $\langle\mu\rangle(x)$ in the vicinity of a phase-slip center in a long aluminium microbridge, using normal tunnel junction probes. Moreover, they performed a few measurements in which they simultaneously also measured $\langle\mu_p\rangle(x)$. All these experiments agree with the predictions of the SBT model recalled above.

Spatially resolved measurements of $\langle\mu\rangle$ and $\langle\mu_p\rangle$ similar to the Dolan and Jackel experiment have not yet been done for clean perfect single crystalline filaments with large electron mean free path such as our tin whiskers. There are only some first measurements of the potentials for very pure tin whiskers [21, 24, 367]. These measurements can, however, in most cases not be explained with a quasiparticle potential smoothly decaying with increasing distance from the phase-slip center. The results can only be understood by assuming that at a larger distance from the phase-slip center $\langle\mu\rangle$ differs from the proposal of SBT by showing a 'swinging over' and a 'swinging below' $\langle\mu_p\rangle$ on the negative polarity side and the positive polarity side of the phase-slip center, respectively.

There is no theoretical explanation resulting in the assumed slope of $\langle\mu\rangle$ except for a rough model calculation of the author [24, 474]. This calculation is based on the assumption that although the nonequilibrium quasiparticles diffuse away from the core region while they relax their charge imbalance (as proposed by SBT), there is, moreover, a collective excitation which modifies the resulting quasiparticle potential. The collective excitation is assumed only to be damped by the motion of the quasiparticles which is dissipative due to scattering from impurities. In the following we will briefly sketch some of the main results and we refer to the cited literature (especially refs. 367 and 474) for further information.

For the measurements we use a tin whisker with several potential probes (Fig. 1c), with one part of the sample in contact with an SC/NC potential-probe pair. We measure the V-I characteristics of this part and at the same time also of a different part which phase-slip centers are active in, before the first voltage step in the characteristics of the part with the SC/NC potential-probe pair occurs. If the nonequilibrium quasiparticles emitted by these phase-slip centers reach the NC probe, a voltage should be measured between the SC/NC potential-probe pair together with the appearance of a voltage step in the other part, although there is no phase-slip center present between the SC/NC potential-probe pair. This voltage is proportional to the difference between $\langle \mu_p \rangle$ and $\langle \mu \rangle$ at the NC potential probe. The reason is that the time averaged pair potential is constant between the SC/NC potential-probe pair as long as there is no phase-slip center present in the region between the probes and, thus, the SC probe measures the same $\langle \mu_p \rangle$ that is also present at the site of the NC probe.

In the temperature range of the measurements potential probes made of pure Sn whiskers can be superconducting, normal conducting, or change their state. To be sure to have normal conducting probes throughout the temperature range one may use Sn-In alloy whiskers. On the other hand, the Wood metal contacts are strongly superconducting all the time.

Two basic types of measurements were performed. In the first one the transport current flows through the whole sample. In this experimental situation nonequilibrium potentials were measured with one probe superconducting and the second changing its state from normal conducting to superconducting. Furthermore, there are experiments where one probe is superconducting and one normal conducting during the whole measurement. Finally, we also made investigations with two normal conducting probes. In the last case the potential-probe pair indicates the change of $\langle \mu \rangle$ from one probe to the other one.

For the second type of measurement, potentials are detected in whisker parts where no transport current is flowing. In these experiments the current only flows through one part of the whisker, leaving (or entering) the sample through a probe.

To describe current directions and voltage polarities we now need the full variety of the 'multi-potential-probe contact notation' introduced in Fig. 1c and explained in the beginning of section 7.4. In this notation I_{AB} is, for instance, a current flowing in the conventional current direction (from '+' to '-') which enters the sample through the Wood metal contact A and leaves the whisker through contact B, while $V_{AB} = \Phi_A - \Phi_B$ is the voltage between the Wood metal blocks A and B, where Φ_A and Φ_B are the conventional electrostatic potentials. For a normal conducting sample and a current $I_{AB} > 0$ we have $\Phi_A > \Phi_B$, and, thus, $V_{AB} > 0$. For the contribution of the electrostatic potential to the electrochemical potential of an electron with charge $-e$ we get $-e\,\Phi_A < -e\,\Phi_B$.

To give an example for the first type of measurement, we take some results for sample Sn 2, a tin whisker with three additional tin whiskers as potential probes. The sample geometry is just that one sketched in Fig. 1c. The lengths of the whisker parts are $L_1 = 0.338$ mm, $L_2 = 0.0375$ mm, $L_3 = 0.425$ mm, and $L_4 = 0.475$ mm. The state of the potential probes changes in the temperature range of the measurements. For higher temperatures probe P2 is superconducting and probes P1 and P3 are normal conducting. With decreasing temperature probe P3 and then probe P1 becomes superconducting.

We apply a current I_{EA} to the sample and measure the voltages V_{DB} and V_{CB} at the same run. As long as the potential probes P2 and P1 are an SC/NC pair we observe a voltage V_{CB} in part 2 occurring together with the first voltage step of the $V_{DC}(I_{EA})$ characteristics of part 3 (Fig. 34). This footlike structure ('voltage foot') develops before the onset of the V-I characteristic of part 2. It becomes smaller with decreasing temperature, but it does not disappear until probe P1 changes into the superconducting state. After the disappearance of the voltage foot the onset of voltage in part 3 still occurs before the onset of voltage in part 2.

We believe that this experiment measures the difference between $\langle \mu_p \rangle$ and $\langle \mu \rangle$ at the site of potential probe P1, caused by phase-slip centers in

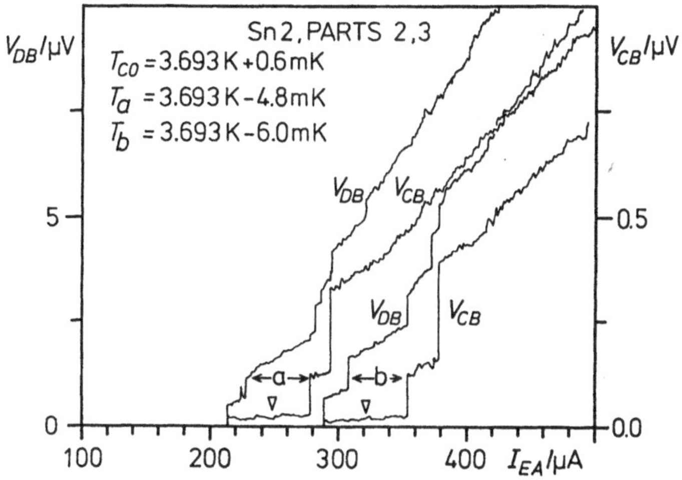

Fig. 34: *The V-I characteristics of a tin whisker with three additional potential probes (sample Sn 2) at increasing current [367].*

The $V_{DB}-I_{EA}$ characteristic of parts 3 plus 2 and the $V_{CB}-I_{EA}$ characteristic of part 2 were measured at the same run for two fixed temperatures $T_{(a)}$ and $T_{(b)}$. Together with the first voltage step in part 3 a foot-like structure ('voltage foot') appears (∇) as long as the potential probe P2 is superconducting and probe P1 is normal conducting.

Notice that the voltage axes for V_{DB} and V_{CB} differ by a factor of ten.

part 3 of the sample. Any other explanation for the voltage foot does not seem to be reasonable.

The measurements show that nonequilibrium quasiparticles are present in a certain distance from a phase-slip center. The exact magnitude of this distance is unknown, because we do not know, for instance, the site of the phase-slip center generating the first voltage step of part 3. The minimum distance possible between the potential probe P1 and the phase-slip center is $L_2 = 0.038$ mm, the maximum value of this distance is $L_2 + L_3 = 0.463$ mm.

There are other measurements, also performed with sample Sn 2, which show that nonequilibrium quasiparticles are present far away from a phase-slip center. In these experiments a large distance between the phase-slip center and the normal conducting probe is guaranteed: There is a temperature range where probe P2 is superconducting and probe P3 is normal conducting. Thus, an experiment has been performed demonstrating that a phase-slip center in part 1 leads to a voltage foot in the characteristics of part 3. Again this voltage foot appears together with the phase-slip center in part 1, the height of the foot decreases with decreasing temperature and no foot is observed if probe P3 has become superconducting.

We now turn to the second type of measurement, where there is not any transport current flowing through the whisker part with the SC/NC potential-probe pair. Again one example for such an experiment is given.

This experiment has been performed with sample Sn 4, a tin whisker with two additional potential probes. Thus, part 4 and contact D of Fig. 1c are not defined. Probe P1 is a superconducting tin whisker, while probe P2 is an Sn-In alloy whisker which is normal conducting throughout the measurements. The lengths of the sample parts are $L_1 = 0.313$ mm, $L_2 = 0.188$ mm, and $L_3 = 0.650$ mm.

We applied a current I_{AB} which flows through part 1 only. As a function of this current we measured the voltages V_{AC} and V_{CE} of parts 1 plus 2 and part 3 during the same run for several fixed temperatures. The voltage V_{CE} is measured with the normal conducting probe P2 and the superconducting Wood metal contact as potential probes.

The result of the measurement is shown in Fig. 35: As long as part 1 is fully normal conducting, no voltage is observed in part 3 (curves (a), $V_{CE} \equiv 0$). For temperatures below the critical temperature of the sample, so that part 1 can carry a superconducting current, a small voltage drop V_{CE} is found across part 3, with a sign which is opposite to that one of V_{AC} (curves (b)). The voltage V_{CE} is already present while part 1 carries the current in the superconducting state. The absolute value of V_{CE} first increases with increasing current and then decreases. The voltage vanishes at the current where part 1 becomes fully normal conducting. Qualitatively this behaviour is also observed at lower temperatures (curves (c) and (d)). Quantitatively the initial slope changes which $V_{CE}(I_{AB})$ developes with from $I_{AB} = 0$ in the range where I_{AB} can be carried in part 1 as a supercurrent. With decreasing

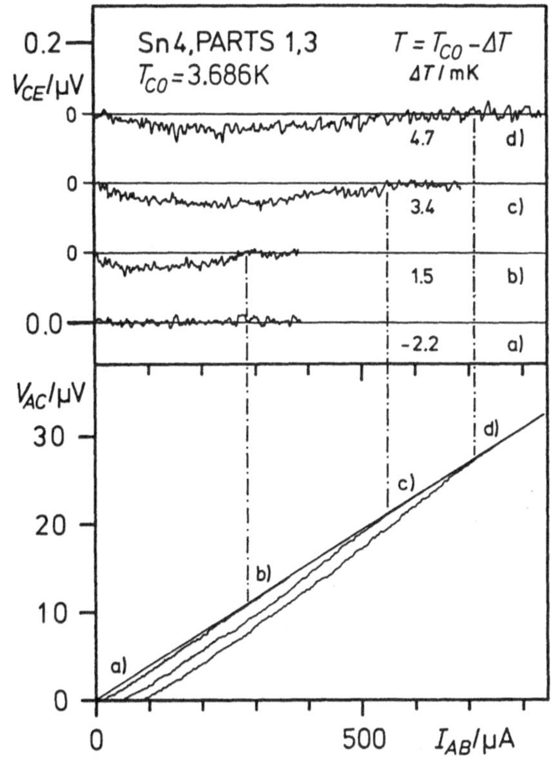

Fig. 35: *The V-I characteristics of parts 1 and 3 of sample Sn 4 (wiring as described in the text), measured with increasing current during the same run [367].*

Although the current I_{AB} flows through part 1 only, a small voltage is observed across part 3. This voltage, V_{CE}, has a sign opposite to that one of the voltage V_{AC}.

temperature the initial slope becomes smaller. Moreover, the maximum value of V_{CE} first increases with decreasing temperature and then decreases again. The broadening of the current range where $V_{CE} \neq 0$ is caused by the broadening of the $V_{AC}-I_{AB}$ characteristics with decreasing temperature.

Changing the polarity of the current leads to changes of the signs of both voltages, and so they have opposite signs again.

In this experiment we measure differences of $\langle\mu\rangle$ and $\langle\mu_p\rangle$ at the site of the normal conducting probe P2. Thus, phase-slip centers in part 1 cause nonequilibrium quasiparticles in part 3 of the sample, without any net transport current flowing from part 1 to part 3.

The measurements reported show that there must be a slope of the chemical potentials in our samples which is different from the SBT prediction at least at a larger distance from the phase-slip center. In the following we

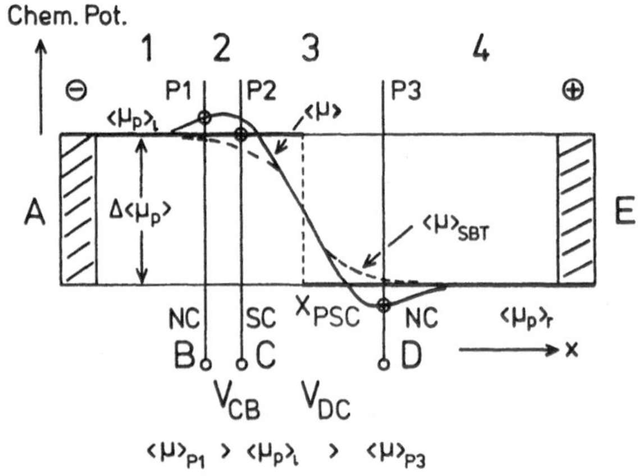

Fig. 36: *Sketch of sample Sn 2 for the discussion of the slope of the time-averaged electrochemical potential* $\langle\mu\rangle$ *for quasiparticles and* $\langle\mu_p\rangle$ *for Cooper pairs. The phase-slip center is situated at* X_{PSC}. *Here* $\langle\mu_p\rangle$ *changes in a step-like way from the value* $\langle\mu_p\rangle_l$ *on the left side of the phase-slip center to* $\langle\mu_p\rangle_r$ *on the right side. The dashed line shows* $\langle\mu\rangle_{SBT}$ *as proposed by SBT (see section 5.8.3). The solid line shows our proposal* $\langle\mu\rangle$, *which we can explain our experimental results with. The numerals 1 - 4 mark the parts of the sample and P1 - P3 the potential probes; NC, normal conducting; SC, superconducting. The polarity of the battery connected with the Wood metal contacts (shaded) is indicated by* ⊖, ⊕, *leading to a current* I_{EA}. *The circles at the crossover points of the electrochemical potentials with the potential probes indicate which potential is measured by the probe: Probe P1 measures* $\langle\mu\rangle(x_{P1}) =: \langle\mu\rangle_{P1}$; *probe P3 measures* $\langle\mu\rangle_{P3}$; *the superconducting probe P2 measures* $\langle\mu_p\rangle_l$.

discuss how the slope of $\langle\mu\rangle$ has to be modified to explain our experimental results. For this purpose we use qualitative sketches and only discuss the sign of the measured voltages.

We start with the discussion of the potential measurements in the presence of a transport current (Fig. 34). In Fig. 36 we sketch the experimental situation. The phase-slip center is situated in part 3 and the voltage foot is detected with the NC/SC potential probe pair P1/P2 applied to part 2. One has to explain that the voltage step generated by the phase-slip center in the characteristics of part 3 has the same sign as the voltage foot.

This can be achieved with our proposal for $\langle\mu\rangle$: The voltage foot has been interpreted as the difference between the electrochemical potentials of quasiparticles and Cooper pairs at the site of the normal conducting probe and is, thus, given by

$$V_{CBf} = (\langle\mu_p\rangle_l - \langle\mu\rangle_{P1})/(-e) \tag{201}$$

and the voltage step by

$$V_{DC} = (\langle\mu\rangle_{P3} - \langle\mu_p\rangle_l)/(-e) \qquad (202)$$

Since $\langle\mu\rangle_{P1} > \langle\mu_p\rangle_l > \langle\mu\rangle_{P3}$, we get the same sign for V_{CBf} and V_{DC}. This has been observed in the experiment.

Using $\langle\mu\rangle_{SBT}$ instead of $\langle\mu\rangle$, we have to replace the quasiparticle potentials at the probes P1 and P3 in eqs. (201) and (202) by $\langle\mu\rangle_{SBT,P1}$ and $\langle\mu\rangle_{SBT,P3}$. As $\langle\mu_p\rangle_l > \langle\mu\rangle_{SBT,P1} > \langle\mu\rangle_{SBT,P3}$ we would get V_{CBf} and V_{DC} with different signs which would be contrary to our experimental observation.

From Fig. 36 we also see that $V_{CBf}=0$ if probe P1 has become superconducting, because then P1 and P2 measure the spatially constant $\langle\mu_p\rangle_l$.

With the proposed slope of $\langle\mu\rangle$ we can also understand the voltage foot appearing in part 3 of sample Sn 2 if a phase-slip center generates a voltage step in part 1. Also in this case the signs of the voltage of the step and the foot are equal. To explain this, one may imagine that the phase-slip center is situated in part 1. Then probe P2 measures $\langle\mu_p\rangle_r$ and probe P3 measures the lower lying $\langle\mu\rangle_{P3}$.

Now we show that also the measurements in the absence of a transport current can be understood with the proposed slope of $\langle\mu\rangle$. In the experiments performed with sample Sn 4 (Fig. 35) the transport current leaves the sample through the superconducting probe P1 and does not flow through part 3 of the whisker where the differences between $\langle\mu\rangle$ and $\langle\mu_p\rangle$ are detected. We assume that a phase-slip center appears in the disturbed border region between the potential probe and the sample (Fig. 37 a), before phase-slip centers in the sample develop, leading to a slope of the electrochemical potentials as drawn in Fig. 37 b. (A similar slope would also be present with the phase-slip center localized at a point X_{PSC} situated in the

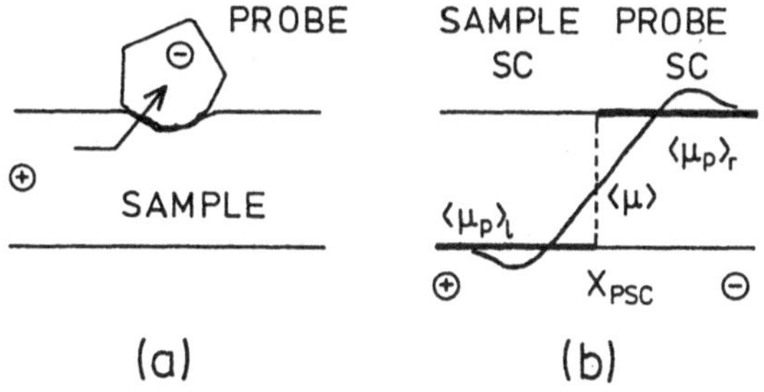

Fig. 37: *Slope of the time-averaged electrochemical quasiparticle potential $\langle\mu\rangle$ and pair potential $\langle\mu_p\rangle$ for a current flowing through one part of the sample only and then leaving the sample through a potential probe.*
(a) Sketch of the sample geometry. The arrows indicate conventional currents.
(b) Slope of the potentials if the potential probe and the sample are superconducting (SC) and a phase-slip center develops at X_{PSC} in the contact region.

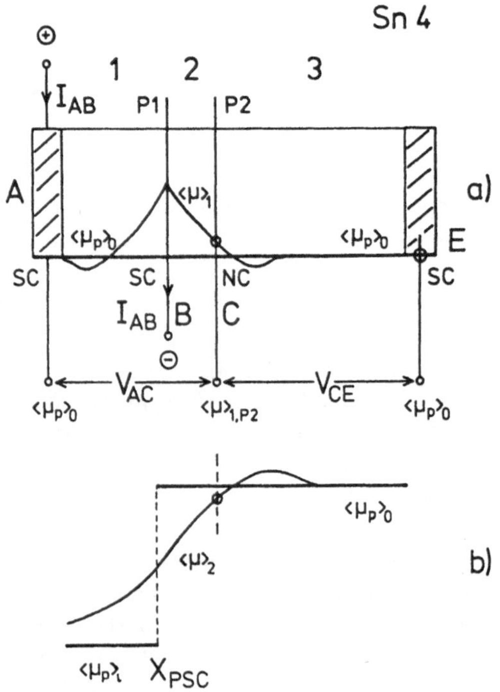

Fig. 38: *Slope of electrochemical quasiparticle and pair potentials in sample Sn 4 in the experimental situation of Fig. 35.*
(a) Before the appearance of phase-slip centers in part 1 of the sample.
(b) Change with the occurrance of a phase-slip center in part 1.

potential probe.) Then, following Fig. 37 b, we get the slopes of $\langle\mu_p\rangle_0$ and $\langle\mu\rangle_1$ as drawn in Fig. 38 a. Since $\langle\mu\rangle_{1,P2} > \langle\mu_p\rangle_0$, a small voltage $V_{CE} < 0$ can be measured. The small voltage V_{AC} developed for the same reason cannot be seen in the measurement because V_{AC} was only measured with a 160 times smaller resolution than V_{CE} (see Fig. 35). The development of phase-slip centers in part 1 for increasing current (no phase-slip centers can develop in part 2 because no transport current flows through this part) leads to $V_{AC} \neq 0$. These phase-slip centers can make V_{CE} larger or smaller. The first case has been drawn in Fig. 38 b, because it seems to dominate finally (see Fig. 35). Since $\langle\mu_p\rangle_1 < \langle\mu\rangle_{2,P2}$ but $\langle\mu\rangle_{1,P2} > \langle\mu_p\rangle_0$, the voltages V_{AC} and V_{CE} have different signs.

In the case of sample Sn 4 also $\langle\mu\rangle_{SBT}$ could be used to explain the experimental observations. However, similar experiments have been performed with sample Sn 2 [367]. In these experiments the transport current flows through part 1 only, while the SC/NC potential-probe pair P 2/P 3 indicates a difference between $\langle\mu_p\rangle$ and $\langle\mu\rangle$ at the site of probe P 3. In this sample the NC probe is situated in a much larger distance compared to sample Sn 4. As expected the results can only be understood assuming that probe P 3 measures the quasiparticle potential after its swinging over (or below) the pair potential. It is remarked that the transport current enters the sample through the normal conducting probe P1 and that the slope of $\langle\mu\rangle$ at this NC/SC boundary has been assumed to be similar to that near a phase-slip center (see Fig. 14 of ref. 367).

Now we come to a discussion concerning a theoretical explanation. No systematically derived theory is known to the author that results in a slope of $\langle\mu\rangle$ as proposed in the present work. The TDGL theory predicts a continuously decaying quasiparticle potential. Also the results of the KSS model do not show any evidence for the 'swinging-over' (or 'swinging-below') effect. Thus, we briefly sketch the main ideas of our own model calculation [474].

As already mentioned above, the idea is that a phase-slip center excites a collective excitation (Carlsson-Goldman mode) in addition to a diffusive behaviour of the quasiparticles. The features of the Carlsson-Goldman mode are described in detail in section 5.6, where we also pointed out that the mode leads to differences of the electrochemical quasiparticle potential and pair potential at a given locus. The reason is that the mode generates regions of enhanced and lowered Cooper-pair density. To maintain overall charge neutrality more electron-like excitations are removed from regions of enhanced Cooper-pair density into regions of lowered Cooper-pair density and more hole-like excitations are moving vice versa. Thus, in regions of enhanced Cooper-pair density there is an excess of more hole-like over more electron-like excitations ($Q^* > 0$, $\mu < \mu_p$), while there is an excess of more electron-like over more hole-like excitations in regions of lowered Cooper-pair density ($Q^* < 0$, $\mu > \mu_p$).

It is assumed that the core region of a phase-slip center is the source of the collective excitation, because this region periodically emits more electron-like (or more hole-like) quasiparticle excitations into the bordering superconductor representing a space charge causing Cooper pairs to leave the injection region (or pulling Cooper pairs into the injection region). Thus, on the side of the phase-slip center with an excess of more electron-like quasiparticles a kind of Cooper-pair compression wave is generated, while on the side with an excess of more hole-like quasiparticles a kind of Cooper-pair dilution wave builds up. To simplify the description we assume that both waves are generated at the core of the phase-slip center and we thus neglect the extension of the injection region.

To formulate this idea in a quantitative form, the phase-slip center may be located at $x=0$ and the polarity of the battery may be chosen as in Fig. 12c (the negative pole to the left where $x<0$ and the positive pole to the right where $x>0$). Since the problem is symmetric in space, only the one half of the slope of the electrochemical potential is calculated which $x \geq 0$ for. The total time-averaged electrochemical quasiparticle potential, $\langle\mu\rangle$, is assumed to be the sum of two terms. The first term is the quasiparticle potential proposed by SBT, $\langle\mu\rangle_{SBT}$. The second term, $\langle\mu\rangle_K$, is generated by the Carlsson-Goldman mode. Thus,

$$\langle\mu\rangle = \langle\mu\rangle_{SBT} + \langle\mu\rangle_{K} \tag{203}$$

The 'SBT potential' is given by eq. (135) to be

$$\langle\mu\rangle_{SBT} = \langle\mu_p\rangle_r + \frac{e\,V_{RL}}{2}\,\frac{\sinh[(x_R - x)/\Lambda]}{\sinh(x_R/\Lambda)} \tag{204}$$

Since V_{RL} can be obtained from experiment, this equation describes the slope of $\langle\mu\rangle_{SBT}(x)$ for fixed but arbitrary value of V_{RL} if the point x_R where $\langle\mu\rangle_{SBT} = \langle\mu_p\rangle_r$ is fixed in a reasonable way.

The calculation of the 'collective-oscillation potential', $\langle\mu\rangle_K$, is much more involved: Assuming that $\langle\mu_p\rangle_r$ remains constant in space, we make the Ansatz

$$(\langle\mu\rangle_K - \langle\mu_p\rangle_r)(x) = -C_K\, f_1(x)\, \mathrm{Re}\langle f_2(x)\, u_s(x,t)\rangle \tag{205}$$

and

$$(\langle\mu\rangle_{SBT} - \langle\mu_p\rangle_r)_{x=0} = C_K\, \mathrm{Re}\langle u_s(0,t)\rangle \tag{206}$$

Here C_K is a constant and $u_s(x,t)$ is the space and time dependent displacement amplitude of the superfluid part (in one dimension) as introduced in eq. (94) of section 5.6. to describe the Carlsson-Goldman mode as a propagating longitudinal wave in which the density of the superfluid part (Cooper pairs) oscillates.

This Ansatz can be justified by the following argument: Equation (206) expresses that the collective excitation is excited at $x=0$ by an excess of quasiparticles which $(\langle\mu\rangle_{SBT} - \langle\mu_p\rangle_r)_{x=0}$ is a measure for. This excess of quasiparticles generates a proportional enhancement of the Cooper-pair density compared to the equilibrium situation. With regard to eq. (205), the change of the Cooper-pair density caused by the excess of quasiparticles at $x=0$ leads to a propagating density fluctuation of the Cooper pairs, resulting in a difference between quasiparticle and pair potentials. This density fluctuation is influenced by the diffusion of the quasiparticles. Thus, in the vicinity of the phase-slip center the density fluctuation does not completely build up, because there is an excess of more electron-like quasiparticles and the Cooper pairs escape from this region. We consider the fact by introducing the function $f_2(x)$. Furthermore, the escape-motion of more electron-like quasiparticles from regions of enhanced Cooper-pair density is hindered by the diffusing more electron-like excess quasiparticles. This fact is considered by inserting the function $f_1(x)$. Following SBT, the diffusing excess quasiparticle population is spatially reduced in proportion to $\sinh[(x_R - x)/\Lambda]/\sinh(x_R/\Lambda)$. Thus, we assume

$$f_1(x) = f_2(x) = 1 - \frac{\sinh[(x_R - x)/\Lambda]}{\sinh(x_R/\Lambda)} \tag{207}$$

Finally, the negative sign in eq. (205) has been inserted because for the case of enhanced Cooper-pair density (that means $\langle u_s(x,t)\rangle > 0$) it yields an excess of more hole-like quasiparticles over more electron-like quasiparticles, which should lead to $\langle\mu\rangle_K < \langle\mu_p\rangle_r$.

To calculate $\langle u_s(x,t)\rangle$ we consider $u_s(x,t)$ at an arbitrary but fixed site x as a function of time. We assume that together with each phase-slip process a new excitation of the collective mode takes place and only take into account contributions from the just excited wave. In other words, we take the following time average:

$$\langle u_s(x,t)\rangle = \frac{1}{\tau_{PSC}} \int_{\Delta t}^{\Delta t+\tau_{PSC}} u_s(x,t)\,dt \tag{208}$$

Here τ_{PSC} is the period of the phase-slip process and $\Delta t = x/c_{CG}$, where c_{CG} is the velocity of propagation of the mode into the superconductor (see section 5.6 for a discussion of c_{CG}). Thus, we average over one period of the phase-slip process, starting at the moment at which the density fluctuation begins to build up at the site x.

The function $u_s(x,t)$ is obtained from eq. (94) by a linear combination of two terms, one term with the positive sign and the other one with the negative sign in the expression for ω_{CG} (see eq.(95)). It is assumed that the amplitude of both terms is equal. We furthermore assume that the collective mode is forced to have the same frequency, $2\pi/\tau_{PSC} = h/2eV_{RL}$, as the phase-slip process, by setting $2\pi/\tau_{PSC} = \mathrm{Re}\ \omega_{CG} = (G - F^2)^{1/2}$. Due to eq.(97) it is $G = c_{CG}^2 q_{CG}^2$ leading to $q_{CG} = (1/c_{CG})[(2\pi/\tau_{PSC})^2 + F^2]^{1/2}$. The further evaluation is straightforward and results in the expression [474],

$$\langle\mu\rangle_K = \langle\mu_p\rangle_r - \frac{eV_{RL}}{2}\left\{1 - \frac{\sinh[(x_R - x)/\Lambda]}{\sinh(x_R/\Lambda)}\right\}^2 \cdot [\cos(q_{CG}x)]\left[-F\cos\left(\frac{2\pi}{\tau_{PSC}}\Delta t\right) + \frac{2\pi}{\tau_{PSC}}\sin\left(\frac{2\pi}{\tau_{PSC}}\Delta t\right)\right]$$

$$\cdot\frac{\exp[-F(\Delta t + \tau_{PSC})] - \exp(-F\Delta t)}{-F[\exp(-F\tau_{PSC}) - 1]} \tag{209}$$

To verify that this result can at least qualitatively describe the experiments, we calculated $\langle\mu\rangle_K$ using eq. (209), $\langle\mu\rangle_{SBT}$ using eq.(204), and then $\langle\mu\rangle$ using eq. (203) for sample Sn 2 in the experimental situation of Fig. 34 where a transport current passes the whole sample. For this calculation we insert $x_R = 650\,\mu m$, $\Lambda = 65.5\,\mu m$ and, furthermore, the clean limit expressions for F, c_{CG}, and G with $\ell = 2.803\,\mu m$ and plot the spatial dependence of the electrochemical potentials for different values of the temperature and V_{RL}. To compare the calculated results with the measurements, temperatures are considered for which experiments were carried out and values for $V_{RL} = V_{DC} \approx V_{DB}$ are taken from the measurement [474].

An example for such a calculation is shown in Fig. 39. Only the right side of the phase-slip center has been plotted and one may think of a symmetric branch to get the sketched slope of $\langle\mu\rangle$ in Fig. 36. The calculated

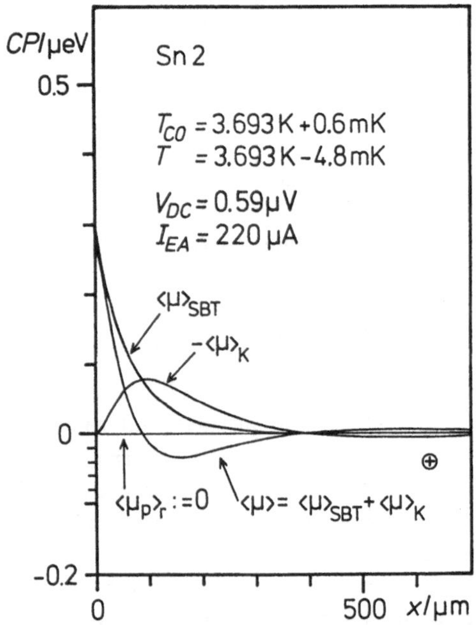

Fig. 39: *Slope of the electrochemical potentials (CP), namely $\langle\mu_p\rangle_r$, $\langle\mu\rangle_{SBT}$, $\langle\mu\rangle_K$, and $\langle\mu\rangle$ as a function of the site on the right side of a phase-slip center situated in part 3 of sample Sn 2. The phase-slip center is located at x = 0. We set $\langle\mu_p\rangle_r = 0$. The potentials are measured in $\mu eV = 10^{-6}$ electron-volt. The voltage $V_{DC}(I_{EA})$ is taken from the experiment at the given temperature. The polarity of I_{EA} is indicated by the polarity (⊕) of the battery [474].*

potential $\langle\mu\rangle$ shows the postulated space dependence: Near the phase-slip center its slope is similar to $\langle\mu\rangle_{SBT}$ and at a certain distance it 'swings' below $\langle\mu_p\rangle_r$. At larger distance from the phase-slip center it approaches $\langle\mu_p\rangle_r$ in a kind of asymptotic behaviour, with a slight swinging-over at large x-values. The distance of the potential $\langle\mu\rangle$ from $\langle\mu_p\rangle_r$ in the 'swinging-below-region' is a measure of the voltage $V_{CBf}(I_{EA})$ of the voltage foot. It is found that this distance decreases with decreasing temperature, although the total value of V_{DC} increases (Fig. 2 of ref. 474). This behaviour is in qualitative agreement with the experimental observation that the voltage foot becomes smaller for decreasing temperature.

The quantitative agreement of our model calculation and experiment is good. The calculated voltages $V_{CBf}(I_{EA})$ have a magnitude of order 10 nV in agreement with the experiment (see Fig. 34). Assuming that the normal conducting voltage probe P1 has a distance of 150 µm from the phase-slip center in part 3 of sample Sn 2 it follows, for instance, $V_{CBf}(I_{EA} = 220\,\mu A) = 34$ nV at T = 3.693 K - 4.8 mK and $V_{CBf}(I_{EA} = 300\,\mu A) = 14$ nV at T = 3.693 K - 6.0 mK. In the first case the experiment yields a foot voltage of 19 nV and in the second case of 13 nV.

The fact that there may be a factor two between calculated and measured values is not worrisome, since one should not expect too much from our 'rough' model. More important than the equality of numerical values is the fact that, given the possibilities of the sample geometry (the phase-slip center has to be localized in part 3), a distance between the normal conducting potential probe and the phase-slip center can be found that leads to an agreement of calculated and measured voltages in order of magnitude.

Naturally, our model calculation may be criticized. The main problem is that we assume the collective mode to be damped by the quasiparticle motion only (which is dissipative due to scattering from impurities), although the inelastic electron-phonon scattering time τ_E is not much larger than the supercurrent response time τ_{OR}. Therefore, one would expect that a damping of the mode due to inelastic electron-phonon processes should be of importance (compare sections 5.6, 5.7, and 5.8.4) and an unknown mechanism has to be assumed which switches off these processes. Another problem is that we simply combine the solutions of two limiting cases, namely the limit of exponential diffusive charge imbalance decay and the limit of the quasiparticle-motion damped Carlsson-Goldman mode. It is not clear whether this is allowed to come to a description of the general case. Thus, the calculation can only be the first step toward an understanding of the phenomena observed.

The model needs an active phase-slip center and, therefore, does not describe the experimental situation where a current leaves or enters the sample through a normal conducting probe ('dc injection'). However, also in this case Cooper pairs should leave a region with an overpopulation of more electron-like excitations and more electron-like excitations should leave regions of enhanced Cooper-pair density. Moreover, it is possible that more hole-like quasiparticles from regions lying deeper in the superconductor migrate toward a region of more electron-like quasiparticle excess.

Our results bring to mind the screening problem of an excess charge in a solid. It also turns out in this case that the smoothly decaying Thomas-Fermi solution does not seem to be the complete description of the problem, but that long-range order charge-density oscillations appear, so-called 'Friedel oscillations' [475].

The final question is, why measurements performed for long microbridges do not show any evidence for the 'swinging-over' effect we have observed. The reason may be that the electron mean free path is much smaller in microbridges and, therefore, the Carlsson-Goldman mode is damped much stronger than in our 'clean' samples.

8. The Limit of Long Quasiparticle Relaxation Times

8.1. Introductory Remarks

The characteristic relaxation times of nonequilibrium quasiparticles in a superconductor, such as the charge imbalance relaxation time, τ_{Q^*}, the order parameter relaxation time, τ_Δ, the recombination time, τ_r, and the thermalization time, τ_{th}, all scale with the inelastic electron-phonon collision time, τ_E. These times become large if τ_E grows (see sections 5.4 and 5.5 and ref. 103, where $\tau_0 = 8.4\,\tau_E$). Furthermore, the dynamics of charge imbalance depends on τ_E, because the damping of the collective excitations in a superconductor decreases with increasing τ_E (see sections 5.6, 5.7, and 5.8.4).

We decided to make an experimental study with whiskers made of two different materials with extremely different inelastic electron-phonon collision times. Suitable materials are Zn with large τ_E and Pb with small τ_E (see Tab. A 1). If we use Tinkham's estimate for τ_E of the pure bulk materials, the τ_E values of Zn and Pb should differ by a factor of about $3.3 \cdot 10^4$. If we take the result of Kaplan et al. the values are predicted to differ by a factor of about $4 \cdot 10^3$.

Thus, the inelastic collision times τ_E of these materials differ by several orders of magnitude. At the same time Zn and Pb represent the limit of long and short τ_E in our experiments. While τ_E is of order 10^{-11} s for Pb, it is of oder 10^{-10} s for Sn and In and of order 10^{-7} s for Zn. It is remarked that τ_E for Al is of order 10^{-8} s. (Al whiskers have not been investigated, but experiments with long microbridges are reported in the literature.)

Experiments on Zn and Pb require enhanced cryogenic efforts, because the critical temperature T_{c0} of Zn is below 1 K, while T_{c0} of Pb is above 4.2 K. Therefore, these materials cannot be investigated in a bath cryostat and special cryostats were developed which are described in chap. 3 of the present work.

While experiments on Pb whiskers are discussed in chap. 9, results for Zn whiskers are given in the present chapter. Whiskers of Zn are expected to have much larger charge imbalance decay lengths (quasiparticle diffusion lengths) than all samples investigated before, because not only τ_E and thus τ_{Q^*} are large but, due to the ideal crystalline structure, these specimens also have a large electron mean free path ℓ.

8.2. Experiments on Zn and Zn-Ag Whiskers

Qualitatively, whiskers of Zn and Zn-Ag alloys behave similar to tin whiskers [20, 41] *1: For small measuring currents of about 1 μA or less, the V-T transition curves of pure Zn whiskers have a very small transition width, often only a fraction of one millikelvin. For Zn-Ag alloys this transition width is usually some millikelvins. This indicates that the samples consist of very homogeneous material. In both cases the transition curves show a shape typical for a fluctuation-governed phase transition. For larger currents the transition width of the curves rises and voltage steps build up.

These voltage steps are also observed in the V-I characteristics which we have studied in more detail. The number of voltage steps and their properties strongly depend on the length, L, of the sample. Short samples only show a single voltage step (Figs. 40 and 41). The shorter a sample is, the lower the temperature has to be chosen until a sharp voltage jump can be observed. For very short samples, down to several millikelvins below T_{c0}, the voltage step is rounded, although fluctuations should not dominate the transition (Fig. 40). As long as no sharp voltage jump occurs, no hysteretic behaviour is observed and for increasing and decreasing current the same characteristic is obtained.

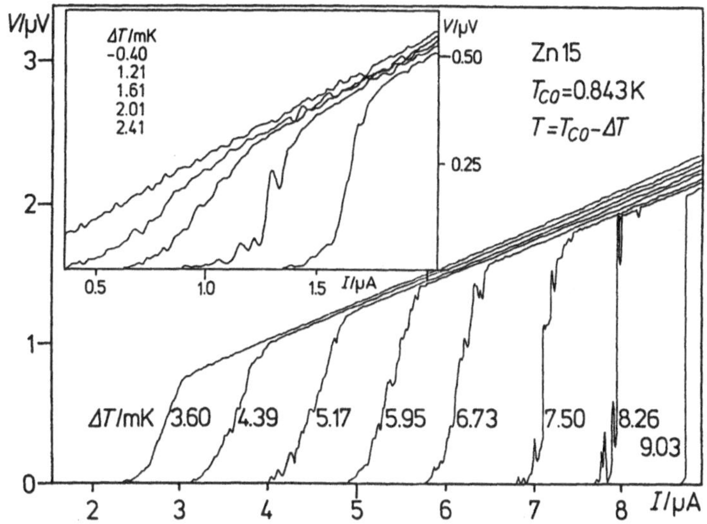

Fig. 40: *The V-I characteristics of a very short zinc whisker (Zn 15, L = 50 μm) for several fixed temperatures, T, below the critical temperature, T_{c0}, for increasing current [20]. The inset shows the low current region at higher magnification.*

*1 There is an error in the right hand scale of Fig. 4 in ref. 41. The numbers given have to be multiplied by a factor of 1.144. Moreover, it is remarked that sample Zn 2 of ref. 41 is called Zn 5 in ref. 20.

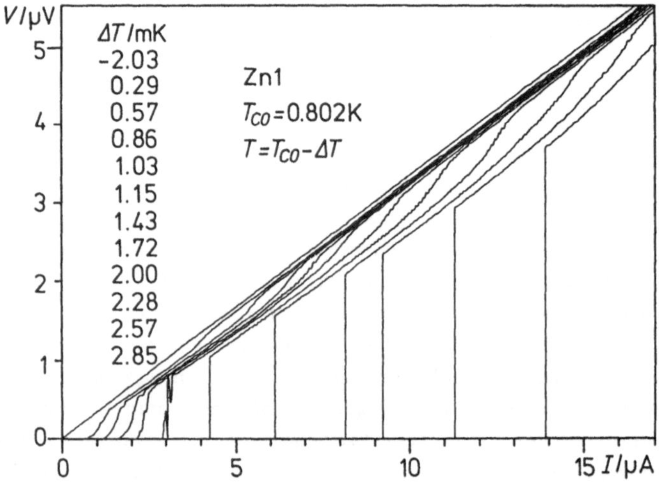

Fig. 41: *The V-I characteristics of a short zinc whisker (Zn 1, L=226 μm) for several fixed temperatures, T, below the critical temperature, T_{c0}, for increasing current [41]. (The value of L given here is obtained from SEM pictures [20] and slightly differs from that one of ref. 41 which was measured with a light microscope.)*

With increasing length also the number of voltage steps increases. A regular voltage step structure with a series of steps, similar to whiskers of Sn or In, is in the case of Zn only observed for samples with lengths above ~1200 μm (Fig. 42). Whiskers of Zn with Ag impurities qualitatively behave like samples of pure zinc.

Generally, the number of voltage steps in the characteristics of Zn and Zn-Ag whiskers is much lower than for whiskers of Sn or In with comparable lengths. Naturally, long samples also show hysteresis as soon as the voltage steps become sharp.

As we discussed in chap. 6, for Zn and Zn-Ag whiskers the critical current obeys the GL theory and the behaviour of the critical temperature, T_{c0}, is understood within the theory of Markowitz and Kadanoff.

Now we turn to the properties of an isolated phase-slip center represented by the first voltage step in the V-I characteristics. As discussed in detail in section 7.2, the first voltage step is characterized by its differential resistance (yielding a normal-like length, L_{An1}, which may be related to the quasiparticle diffusion length, or charge imbalance relaxation length, within the SBT model) and the zero voltage intercept, I_0 (which may be interpreted within the SBT model as a time averaged supercurrent). For all Zn and Zn-Ag whiskers both quantities are independent of temperature. As an example we show the results for a very short and a long pure Zn whisker in Fig. 43.

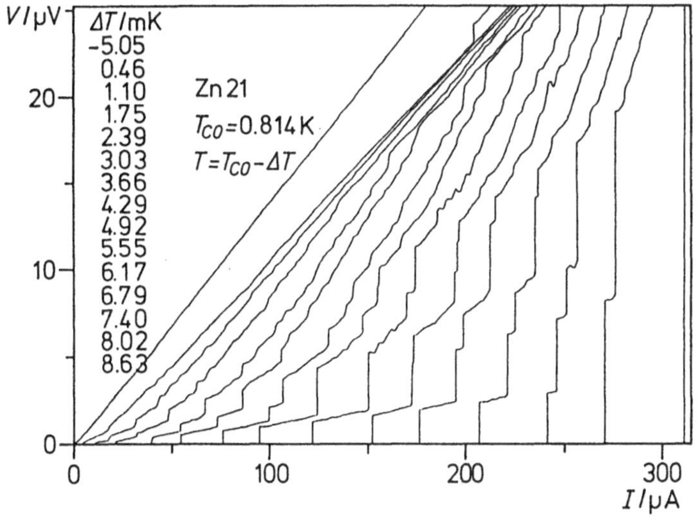

Fig. 42: *The V-I characteristics of a long zinc whisker (Zn 21, L=1375 μm) for several fixed temperatures T (decreasing from left to right) and increasing current [20].*

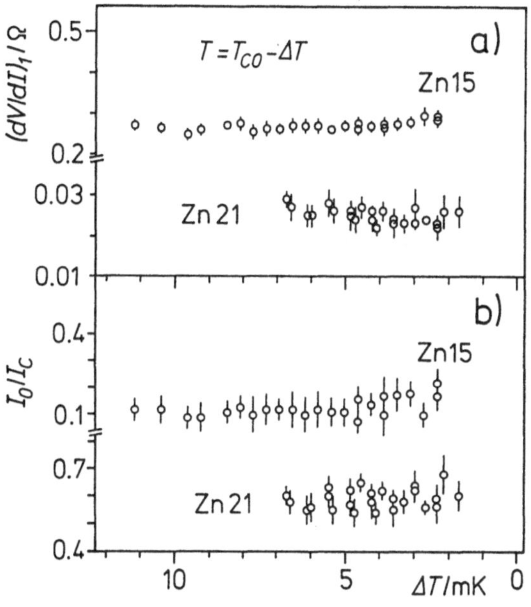

Fig. 43: *Differential resistance (dV/dI), and ratio I_0/I_c of the first voltage step in the V-I characteristics of a very short zinc whisker (Zn 15, T_{c0}=0.843 K) and a long zinc whisker (Zn 21, T_{c0}= 0.814 K) as a function of the temperature difference $\Delta T = T_{c0} - T$, with T_{c0} the critical temperature of the mentioned sample [20].*

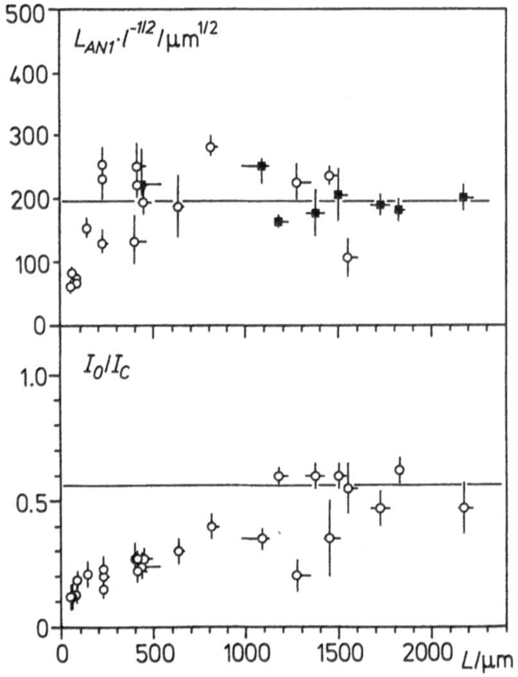

Fig. 44: *Normalized normal-like length $L_{An1}\ \ell^{-1/2}$ and ratio I_0/I_c as a function of the sample length L. The solid lines indicate values for a phase-slip center appearing unaffected by the contacts. Full squares in (a) indicate results manifested by the investigation of higher voltage steps.*

Since $(dV/dI)_1$ and I_0/I_c are temperature independent, averaged values can be calculated for each sample. From the resulting average of the differential resistance the normal-like length of the sample, L_{An1}, can be calculated. In Fig. 44 we plot $L_{An1}/\ell^{1/2}$ and the averaged I_0/I_c as a function of the length of the sample, L. Both quantities increase with increasing L. As we will discuss in the following section, this length dependence is caused by an influence of the strongly superconducting Wood metal contacts. While $L_{An1}/\ell^{1/2}$ rapidly increases toward a length-independent value, the increase to a length-independent behaviour is more slow for I_0/I_c. For $L \gtrsim 1200\,\mu m$ both quantities are length independent. We conclude that above this length phase-slip centers occur in our whiskers without being affected by the contacts. For these unaffected phase-slip centers mean values are $L_{An1}/\ell^{1/2} = 196 \pm 23\,\mu m^{1/2}$ and $I_0/I_c = 0.56 \pm 0.06$.

It is remarked that samples marked in Fig. 44 a by a full square have well developed higher voltage steps with a differential resistance which is a multiple of $(dV/dI)_1$. This behaviour is expected for weakly interacting phase-slip centers. However, in section 7.5 we also found that active phase-slip centers may not change the differential resistance of an influenced center although its critical current is enhanced.

In Fig. 44 a we normalized L_{An1} by $\ell^{1/2}$, because we assume that L_{An1} changes proportional to $\ell^{1/2}$. This assumption has been verified by a log-log plot of $L_{An1}(\ell)$. In this plot (Fig. 10 of ref. 20) L_{An1} follows a straight line with slope 1/2 for phase-slip centers appearing unaffected from the contacts in long whiskers. Moreover, there is a group of results from very short whiskers ($L \lesssim 100\,\mu m$) joining a different lower lying straight line, also with slope 1/2. For I_0/I_c no dependence on the electron mean free path, ℓ, has been observed.

For all clean whiskers ℓ as calculated from the residual resistance is much larger than the sample diameter d_W. From the behaviour of T_{c0} we know that surface scattering affects the superconducting properties of the sample. To include the influence of the surface on the relaxation behaviour of quasiparticle excitations, we assumed that the electron mean free path is limited by the diameter, thus replacing ℓ by d_W for all samples with $\ell \gg d_W$ in Fig. 44 a and in the further discussion of L_{An1}.

We did not undertake a detailed investigation of the interaction of phase-slip centers as in the case of tin whiskers (section 7.5). However, the manner in which the V-I characteristics of our samples develop with decreasing temperature indicates an interaction of phase-slip centers also in the case of Zn whiskers: Once the first voltage step has developed, the distance in current to the second step increases and then decreases until finally both steps occur at the same current (Fig. 42). For a quantitative investigation we plotted the onset currents, $I_1^{2/3} = I_c^{2/3}$ and $I_2^{2/3}$, for the first and the second step, respectively, as a function of temperature (Fig. 45). Very close to T_{c0} and for low temperatures both steps appear together at I_c. In the intermediate region I_2 is larger than I_c, indicating an enhancement of I_2 due to the influence of the first phase-slip center.

8.3. Interpretation by Comparison with Theoretical Work

First we will discuss the dependence of L_{An1} and I_0/I_c on the sample length L within the framework of the SBT model: The relation $L_{An1} \approx 2\Lambda$ is only valid for a phase-slip center with an undisturbed quasiparticle relaxation, where the distances between the site of the phase-slip center, X_{PSC}, and the points x_R and x_L, where $\mu = \mu_p$, are large compared to Λ. In the more general case it follows from eq. (137)

$$(dV/dI)_1 = (2\rho_n \Lambda / A) \tanh(L/2\Lambda) \tag{210}$$

and, thus,

$$L_{An1} = 2\Lambda \tanh(L/2\Lambda) \tag{211}$$

These results are valid for a phase-slip center situated in the middle of the whisker and the assumption that the time-averaged electrochemical potentials

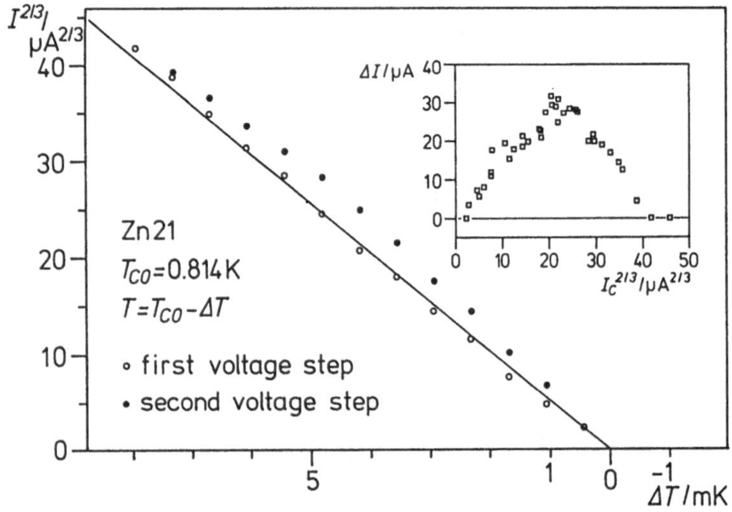

Fig. 45: *Current at the onset of the first and second voltage steps of a zinc whisker (sample Zn 21) as a function of $\Delta T = T_{c0} - T$. Data taken from Fig. 42. The solid line represents $I_c^{2/3} \sim \Delta T$. The inset shows the difference between the onset current of both steps as a function of $I_c^{2/3}$. For the inset data were taken from Fig. 42 and from further measurements.*

of pairs and quasiparticles are equal at the contact edges. This equality of the electrochemical potentials of pairs and quasiparticles is the consequence of an Andreev reflection problem at the whisker/contact interface as will be discussed within the interpretation of the behaviour of I_0/I_c.

Thus, the relation $L_{An1} = 2\Lambda$, where $\Lambda = ((1/3)\ell v_F \tau_2)^{1/2}$, only can be applied in the region where L_{An1} is independent of the sample length and the phase-slip centers are not affected by the contacts. With $v_F = 9.1 \cdot 10^5$ m/s one has $\tau_2 = (3.2 \pm 0.75) \cdot 10^{-8}$ s which may be called τ_{0ZnW} according to the systematic notation introduced in section 7.2.

With this value for τ_2 we calculated Λ for each sample and then invoked eq. (211) for a prediction of L_{An1} for the sample. The ratio of measured and thus calculated values of L_{An1} is close to unity. This indicates that the length dependence of L_{An1} is described by the SBT model and supports our evaluation of τ_2.

Let us now discuss the behaviour of I_0/I_c which for an unaffected phase-slip center has a magnitude similar to that one measured for whiskers of Sn and In (not too close to T_{c0}) and to the SBT prediction, but which decreases for decreasing sample length. In our opinion this reduction of I_0/I_c is the consequence of an enhanced overpopulation of the excitation spectrum with nonequilibrium quasiparticles: For short samples a large number of nonequilibrium quasiparticles generated by the phase-slip process reach the superconducting contacts. At the interface between the whisker and the contact the energy gap rises rapidly (typical length $\xi(T)$) from the small gap of the sample toward the large gap of the Wood metal. (The temperature is close to the critical temperature of the Zn whisker, but very low compared to the critical temperature of Wood's metal of about 8.2 K [19].) In the contact region we, thus, have an Andreev reflection problem [275]. A quasiparticle can only penetrate into the contact region until its energy is equal to the lowest energy of the spatially varying excitation spectrum. At this point there has to be a direct conversion from the quasiparticle current into a supercurrent. Essentially all of the quasiparticle excitations have to undergo this conversion process. The conversion process is called Andreev reflection, because in this process the quasiparticle which tries to penetrate into the strong superconductor vanishes and a quasiparticle with opposite effective charge is emitted out of the strong superconductor into the opposite direction. The quasiparticles seem to be reflected at the interface with their effective charge changed to the opposite [94, 275]. Thus, if a current of electron-like quasiparticles penetrates into the interface, a current of hole-like excitations leaves the interface in the opposite direction. The consequence is a reduction of the charge imbalance to zero in the contact region, leading to an equality between the electrochemical potentials of quasiparticles and Cooper pairs. In the whisker the charge imbalance is reduced by the reflected excitations, but the density of nonequilibrium quasiparticles is enhanced compared to an undisturbed relaxation in a long sample. Therefore, an overpopulation of the excitation spectrum is created at the core of the phase-slip center which we assume to be characterized by an effective temperature $T^* > T$ in the sense of Parker's model (see section 5.5). In this case the whisker behaves as if it had the temperature T^*, so that its critical current is reduced to $I_c(T^*) < I_c(T)$.

This effect has to be considered in the SBT model (Fig. 13 c), because the amplitude of the oscillating supercurrent in the core now is the critical current $I_c(T^*)$ and not $I_c(T)$. The consequence is a reduction of the time averaged supercurrent I_0, leading to a smaller ratio I_0/I_c.

The T^* effect also contributes to the hysteresis and leads to the obvious broadening of the hysteresis width for short zinc whiskers (chap. 11).

Next, we discuss the properties of an isolated phase-slip center in a zinc whisker which is not affected by the contacts.

Such a phase-slip center is related to the first voltage step in a long whisker. As found for all samples, this step is characterized by a temperature independent ratio I_0/I_c and a temperature independent differential resistance. This differential resistance leads to a temperature independent normal-like length and, thus, within the SBT model to a temperature independent quasiparticle relaxation time τ_2. The time has to be identified with the charge imbalance relaxation time τ_{Q^*}. As already pointed out in section 7.2 no temperature independent steady state charge imbalance relaxation time is known.

For Sn and In (not too close to T_{c0}) we also obtained a temperature independent quasiparticle relaxation time (section 7.2). In that case we found an empirical rule which connects the value of τ_2 with the scaling time τ_0 of Kaplan et al. [103]. This rule is not valid for Zn whiskers, because it would require a value of τ_2 for Zn whiskers which is about two orders of magnitude larger than the measured result [20, 41].

Also Kulik's electron-bogolon approach, widely discussed in section 7.2, does not explain the observed phenomena. The resulting normal-like length $2\Lambda_{Ku}(0)$ is two orders of magnitude too large.

Next, we are going to discuss the predictions of the TDGL theory (section 5.9) as given by Kramer and Rangel (KR): The calculations of KR are based on the generalized TDGL equations (eqs. (172) and (173)) and are valid for dirty filaments ($\ell \ll \xi_0$) and in local equilibrium ($\Lambda_E < \xi_D(T)$). The first condition is not the most serious restriction. Although $\ell \ll \xi_0$ is only strictly fulfilled for our Zn-Ag alloy whiskers, also for our pure samples it is $\ell \approx \xi_0$. The local equilibrium condition, however, restricts the applicability of the theory to a very small temperature interval of only some one-hundredth parts of one millikelvin below the critical temperature [20]. Thus, we cannot expect a quantitative description of our measurements.

KR give results as a function of the pair-breaking parameter γ. The problem is that for our samples it is $\gamma > 1000$. For such high γ values KR do not give a prediction for the normal-like length and the zero-voltage intercept. For $\gamma \approx 1000$, KR obtain that L_{An1} should be close to $2\Lambda_{Q^*in} \sim \Delta T^{-1/4}$. Furthermore, I_0 should be close to zero. Both predictions do not agree with our experiments.

Baratoff calculated the behaviour of a phase-slip center (in a gapless quasi-one-dimensional superconductor) in a situation, where the local equilibrium approximation does not hold. This approach has been widely discussed in sections 5.9 and 7.2. It predicts a temperature independent normal-like length (of a given sample) as soon as the temperature is not too close to T_{c0}, so that $\tau_{GL} \leq \tau_E$ (that means $\Lambda_E \leq \xi_D(T)$). In this temperature-independent regime τ_2 may be identified with τ_E which does not depend on the temperature (for a given sample).

From the condition $\tau_{GL} \leq \tau_E$ it follows that the temperature-independent regime for zinc should already be present if the temperature is lowered to only some one-hundredth parts of one millikelvin below T_{c0} [20]. Moreover, for pure polycrystalline zinc we get $\tau_E = 1.8 \cdot 10^{-7}$ s using Tinkham's formula and $\tau_E = 9.3 \cdot 10^{-8}$ s after Kaplan et al. (compare section 5.4 and Tab. A 1). Our experimental result for τ_2 is only a factor of 3 smaller than Kaplan's value for τ_E. Baratoff's theory, therefore, gives a possible explanation for our measurements.

Since Zn is a material with a large inelastic electron-phonon scattering time, τ_E, we also calculated the electron-electron scattering time, τ_{ee}, from theoretical results given in section 5.4. For pure three-dimensional polycrystalline zinc at its critical temperature, τ_{ee} is of order 10^{-6} s. In the three-dimensional case with finite ℓ at T=0 K it is τ_{ee} of order 10^{-5} s for an electron with energy $\varepsilon_K = \Delta(T_{c0} = 0.8\,\mathrm{K}, \Delta T = 10\,\mathrm{mK})$, if it is assumed that $\ell = 1\,\mu\mathrm{m}$. Here, $\Delta(T_{c0} = 0.8\,\mathrm{K}, \Delta T = 10\,\mathrm{mK})$ is the BCS result for the energy gap of a weak-coupling superconductor with $T_{c0} = 0.8$ K at a temperature which is 10 mK below T_{c0}, as may be reasonable for a Zn whisker.

Furthermore, one may evaluate the expression for thicker two-dimensional disordered systems. (For polycrystalline zinc with $\ell \approx 1\,\mu\mathrm{m}$ and a film thickness of $d \approx 1\,\mu\mathrm{m}$ at 0.875 K it is $d^2 \lesssim \hbar v_F \ell / 3kT$ and $k_F d \gg 1$.) In this case τ_{ee} is of the order of 10^{-6} s.

Thus, in the three-dimensional case and also in the two-dimensional case (assuming values for ℓ and d which are similar to the properties in our whiskers) it follows a value for τ_{ee} which is larger than τ_E and much larger than our measured τ_2. We conclude that inelastic electron-electron scattering should not play an important role in our experiments.

Finally, also in the case of Zn whiskers there is evidence for the enhancement of the critical current of a phase-slip center by an active phase-slip center. Considering the large quasiparticle relaxation length in our zinc whiskers one may speculate [22], that the enhancement effect is caused by quasiparticle currents generated by the active phase-slip center. This mechanism has been widely discussed in sections 5.8.3 and 7.5. The reduction of the enhancement may then be caused by a weakening process due to an overpopulation of the excitation spectrum with nonequilibrium quasiparticles (section 5.5). A proof of this explanation is not possible, because the distance between the involved phase-slip centers is unknown. Nevertheless, also in the case of zinc whiskers a stabilizing effect explains the large current region between the first onset of voltage and the fully normal conducting state.

9. The Limit of Short Quasiparticle Relaxation Times

9.1. Introductory Remarks

In section 8.1 we discussed that the quasiparticle relaxation times scale with the inelastic electron-phonon scattering time, τ_E, and that also the dynamics of charge imbalance depends on this characteristic time. Moreover, we remarked that Pb whiskers were chosen for the experimental study of the limit of short quasiparticle relaxation times, because the value of τ_E is very short in this material.

For the measurements of Pb whiskers a special ^{4}He overpressure cryostat had been developed because highly stabilized temperatures around 7.2 K, that means above the normal boiling point of ^{4}He, were needed (chap. 3).

The astonishing result of these experiments is that there are no voltage steps visible in the transition characteristics. At a first view we thought that there was not any structure in the characteristics. However, more detailed investigations with a higher resolution showed, that, although there are no voltage jumps in the V-I characteristics, one can clearly recognize portions where the voltage depends linearly on the current.

In our opinion also these linear portions are generated by phase-slip centers: There is a reasonable behaviour of the normal-like length assigned to their differential resistance. Moreover, the characteristics exhibit 'current steps' (section 7.3) in a HF-radiation field.

Nevertheless, the absence of voltage jumps and a ratio $I_0/I_c \approx 1$ seemed so unusual to us that we started a systematic investigation of whiskers made of materials with different electron-phonon coupling strengths. For this purpose we investigated numerous whiskers made of In-Pb alloys with lead concentrations spread over the whole alloy system. As In is a weak-coupling superconductor and Pb is a strong-coupling superconductor, it is, thus, possible to investigate the continuous change of properties toward the strong coupling behaviour of Pb.

Special attention has been paid to the investigation of In-rich alloys and Pb with small In concentrations. To prove the universality of the results of Pb with small In impurity content, we also investigated Pb whiskers with Bi impurities.

The results of In-rich alloys may be partly interpreted within the SBT model and contain interesting information for a comparison with the TDGL results of KR. The results obtained for whiskers of pure Pb and Pb-In or Pb-Bi whiskers may be interpreted within the SBT and KSS model. The KSS model leads to a qualitative understanding of the results. It turns out that it is indeed the strong-coupling nature of the material with its small inelastic electron-phonon scattering time, τ_E, which makes visible the observed, somewhat unusual phenomena over a large range of measuring temperatures. The small value of τ_E acts to spread the temperature range where there are no voltage jumps but, nevertheless, linear portions. In principle this behaviour should also be present in the transition behaviour of filaments of other materials. However, usually it cannot be investigated, because the related temperature range is only small and very close to T_{c0} and the phenomena seem to be disturbed by fluctuation effects.

In the following the basic results of our investigations will be summarized. For further information see refs. 40, 359, and 372. All experiments were performed in the mentioned overpressure cryostat.

9.2. Change from Weak to Strong Coupling Behaviour: Experiments on Whiskers of the In-Pb Alloy System

9.2.1. In-Rich Alloys

All whiskers from In-rich alloys (that means from the In crystalline solid solution which extends up to a lead concentration $c_{Pb} = 12.7$ at%) show characteristics very similar to those of samples from pure indium [16]. Thus, their shape is similar to the pure tin case (Figs. 5 and 6) except that the differential resistance $(dV/dI)_1$ and the ratio I_0/I_c of the first voltage step depend on temperature (section 7.2).

The differential resistance and, thus, also the normal-like length, L_{An1}, show the whole variety of temperature dependences known from the investigations of In whiskers (section 7.2): The normal-like length may be independent of temperature or change proportional to $\Delta T^{-1/4}$, $\Delta T^{-1/2}$, and ΔT^{-1}, respectively. Some examples are shown in Figs. 46 and 47.

For larger values of ΔT the normal-like length does not show any temperature dependence. These temperature independent results may be averaged for each sample (the average is called $(L_{An1})_{0\underline{In}PbW}$) and plotted as a function of the electron mean free path ℓ of the sample together with the results for pure In whiskers (called $(L_{An1})_{0InW}$). These 'low temperature values' show the familiar $\ell^{1/2}$ dependence (Fig. 48).

The ratio I_0/I_c exhibits the behaviour expected from the fact that $V_1(I_c)$ is a straight line, requiring that $1 - I_0/I_c$ compensates the temperature dependence of $(dV/dI)_1$ (section 7.2). Thus, I_0/I_c is temperature independent for large values of ΔT and shows an increasing tendency if the temperature approaches T_{c0} (Fig. 49). Also in the case of I_0/I_c the temperature

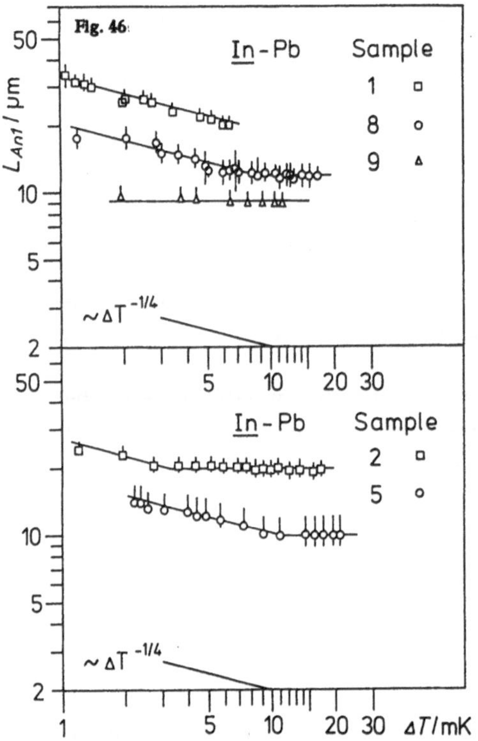

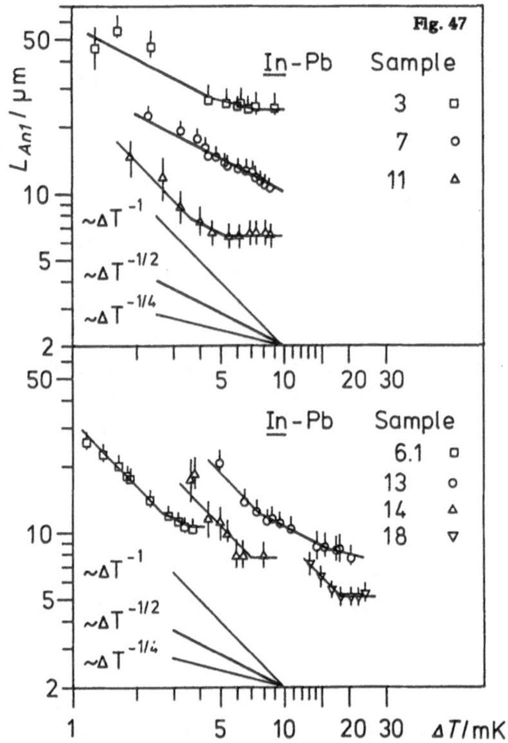

Fig. 46: *Normal-like length, L_{An1}, as a function of the temperature difference $\Delta T = T_{c0} - T$, using log-log coordinates, for several In-Pb whiskers with $L_{An1} \sim \Delta T^{-1/4}$, represented by a straight line with slope $-1/4$ [40]. Horizontal lines indicate that L_{An1} is independent of temperature. Sample 9 shows this behaviour only. For T_{c0} of each sample see refs. 14 and 40.*

Fig. 47: *Normal-like length, L_{An1}, as a function of the temperature difference $\Delta T = T_{c0} - T$, using log-log coordinates, for several In-Pb whiskers with other temperature behaviour besides the $\Delta T^{-1/4}$ law [40]. Horizontal lines indicate that L_{An1} is independent of temperature. For T_{c0} of each sample see refs. 14 and 40.*

independent values may be averaged for each sample. In Fig. 50 we plotted these 'low temperature values' (called $(I_0/I_c)_{0InPbW}$) as a function of ℓ, together with the results of $(I_0/I_c)_{0InW}$ of pure indium whiskers. This plot shows that indeed there is a slight decrease of the pure indium results with decreasing ℓ, as mentioned in section 7.2. In the In-Pb whiskers ℓ is further decreased, leading to a further decrease and then to an increase of the ratio. A similar behaviour was observed for Sn-In whiskers (Fig. 27).

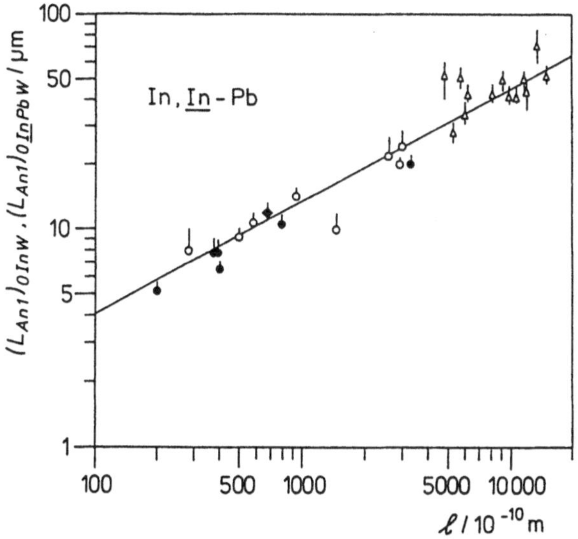

Fig. 48: *Length $(L_{An1})_{O\underline{In}PbW}$ as a function of the electron mean free path, ℓ, for several In-Pb whiskers, using log-log coordinates [40]. Results for $(L_{An1})_{OInW}$ of pure In whiskers are added.*
(Δ): Pure In whiskers from ref. 17
(○), (●), (◆): In-Pb whiskers from ref. 40. In that work we introduced different symbols to distinguish between samples where the first voltage step is well developed (light circle) or occurs together with the second step (full circle) and the second and third step (full diamond), respectively.
The full line represents an $\ell^{1/2}$ dependence.

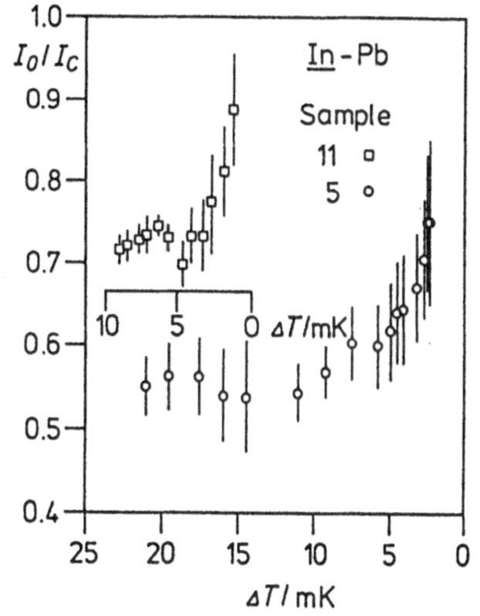

Fig. 49: *Ratio I_0/I_c as a function of the temperature difference $\Delta T = T_{c0} - T$ for two In-Pb alloy whiskers [40]. For T_{c0} of each sample see ref. 14.*

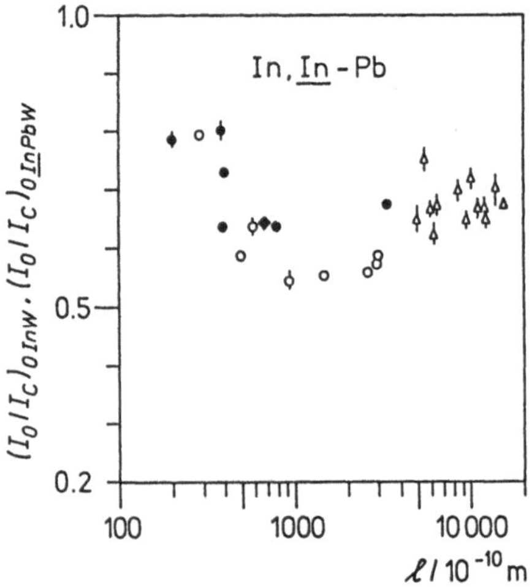

Fig. 50: *Ratio $(I_0/I_c)_{0\underline{In}PbW}$ as a function of the electron mean free path, ℓ, for several In-Pb whiskers [40]. Results for $(I_0/I_c)_{0InW}$ of pure In whiskers are added. The symbols are the same as those of Fig. 48.*

9.2.2. The Range of Substantial Alloy Contents

The In-Pb alloy system shows a nearly complete solubility over the whole range of concentrations. There is a small range between the In crystalline solid solution and the Pb crystalline solid solution which a α_1 mixed crystal exists in. The two-phase regions are located between lead concentrations of 12.7 and 13.7 at% and between 28.5 and 35.5 at%, respectively [14, 476].

We investigated several samples with a lead content between 13.5 and 97.5 at%. It appears to be impossible to evaluate the characteristic properties of voltage steps generated by phase-slip centers. The reason is that 'phase-slip structures' disappear within the rise of big voltage steps (which we call 'resistive steps'). In the α_1 mixed crystal region evidence for the residue of a phase-slip structure may be detected [40]. This was not possible in the characteristics of whiskers from the Pb crystalline solid solution.

The behaviour of the resistive steps is very different from the behaviour of a phase-slip center. We assume that the observed structures are generated by so called 'self-heating hot spots' [40].

Thus, although c_{Pb} was enhanced up to 97.5 at% in the investigations of ref. 40, the In content was still too large to observe voltage steps generated by phase-slip centers. We recently performed experiments with small In impurity concentrations which allow the study of phase-slip structures [359]. The results will be reported in section 9.2.4. It is more comfortable first to discuss the results of pure Pb whiskers, because small impurity concentrations of In (or Bi) act to modify the behaviour of pure samples.

9.2.3. Pure Pb

Also in the case of pure Pb whiskers, the V-T transition curves for small measuring currents show a sharp, fluctuation governed transition from the superconducting to the normal state. If the measuring current is adjusted to higher fixed currents, the transition width (in temperature) is broadened. However, no voltage steps are observed [40].

Also the V-I characteristics do not show any voltage steps. Nevertheless, in the low voltage region of the characteristics, portions of constant differential resistance are observed, where the V-I characteristic is straight. In Figs. 51 - 53 the V-I characteristics of sample Pb 7* are shown in three different solutions [40]. There is no hysteresis so that the characteristics follow the same line for increasing and decreasing current. For larger currents there is a separation of the characteristics for increasing and decreasing current in the medium range, while close to the critical current both curves still join together (Fig. 51). This behaviour is generated by Joule heating effects.

It is clear from Fig. 53 that the slope of the first linear portion is well described by $V(I)=(dV/dI)_1(I-I_0)$ with $I_0 \approx I_c$ for $I \geq I_c$, leading to $V(I_c)=0$. The result $I_0/I_c \approx 1$ is typical for pure Pb whiskers.

The figures also show that the first linear portion is followed by others and that there is no separation of the linear portions by voltage steps. In sample Pb 7* the first linear portion is generated by a single phase slip center while the second linear portion is generated by three phase-slip centers. In Fig. 54 we plotted the corresponding normal-like lengths L_{An1} and L_{An3} as a function of the temperature. We added results obtained from another lead whisker (sample Pb 12*) in which already the first linear portion is generated by three phase-slip centers.

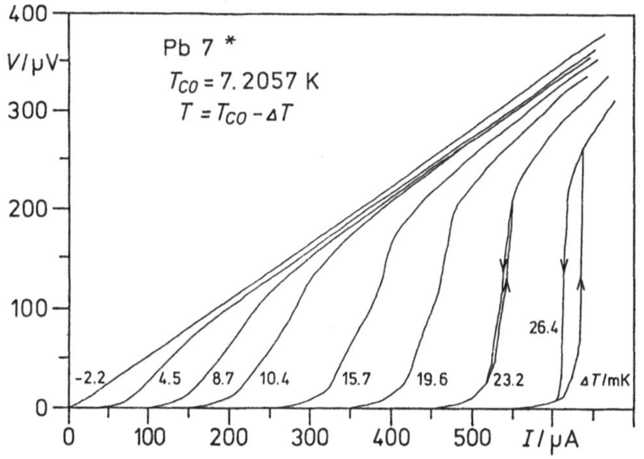

Fig. 51: *The V-I characteristics (increasing and decreasing current) of the lead whisker 7* for several fixed temperatures T [40]*

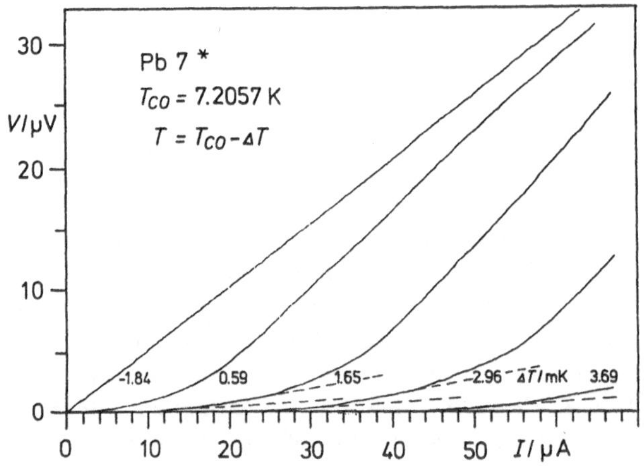

Fig. 52: *Lower part of the V-I characteristics (increasing and decreasing current) of lead whisker 7*. The dashed lines represent the linear portions with constant differential resistance [40].*

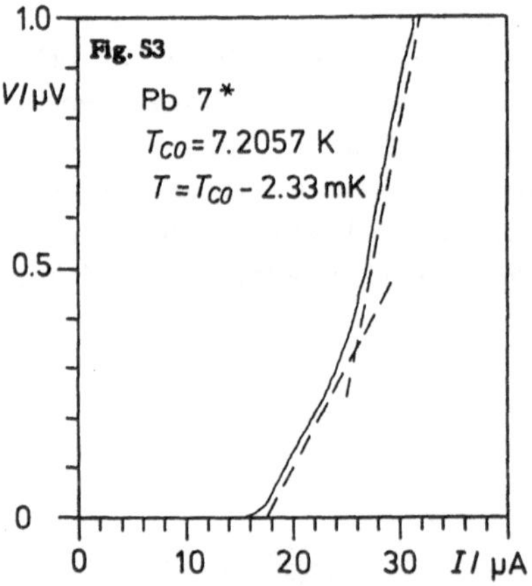

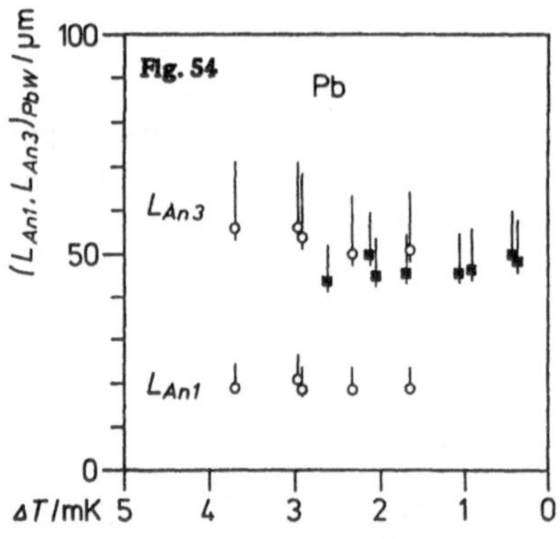

Fig. 53: *Onset of a V-I characteristic (increasing current) of lead whisker 7*. Dashed lines are parallel to the linear portions with constant differential resistance [40].*

Fig. 54: *(○): Normal-like lenghts L_{An1} and L_{An3} for sample Pb 7* as a function of temperature [40].*

For this sample L_{An1} has been calculated with the differential resistance of the first linear portion, while L_{An3} is the result for the second portion. The index '3' indicates that three phase-slip centers generate this part of the characteristics.

(■) : Results for sample Pb 12 as calculated from the first linear portion [40].*

Here, $\Delta T = T_{c0} - T$, with T_{c0} the critical temperature of each sample as given in refs. 14 and 40 (or also in Fig. 51 for Pb 7).*

It is chracteristic for lead whiskers that the normal-like length L_{An1} does not depend on temperature. Thus, again an averaged normal-like length L_{An1} may be calculated for each sample and plotted as a function of the electron mean free path ℓ, showing the familiar $\ell^{1/2}$dependence [40]. The results obtained for pure Pb whiskers are summarized in a common plot together with normal-like lengths of Pb-In and Pb-Bi whiskers, shown in the following section. There are several samples where the first linear portion is not generated by a single phase-slip center. We refer to refs. 40 and 359 for a discussion how to handle this problem and how to evaluate a suitable value for L_{An1} in this case.

Further detailed investigations of the linear portion phenomenon in the V-I characteristics of pure Pb whiskers show that a curvature may appear in the beginning of the portion. To demonstrate this we refer to Fig. 55: This figure shows results of sample Pb 1 from ref. 359. The first linear portion is generated by a single phase-slip center in the whisker. As the differential resistance of the second linear portion is twice that of the first one, we conclude that the second linear portion is generated by two phase-slip centers. Close to T_{c0} the linear portions are only very slightly bended. (This is probably the reason why no curvature can be recognized in Fig. 54.) However, the curvature becomes stronger for decreasing temperature until the slope becomes infinite at the critical current. Below this temperature the transition becomes hysteretic.

Due to the bending of the linear portion, the ratio I_0/I_c slightly falls below unity. At the temperature where the slope of the characteristics at the critical current has become infinite, I_0/I_c is still about 0.94 and seems to saturate.

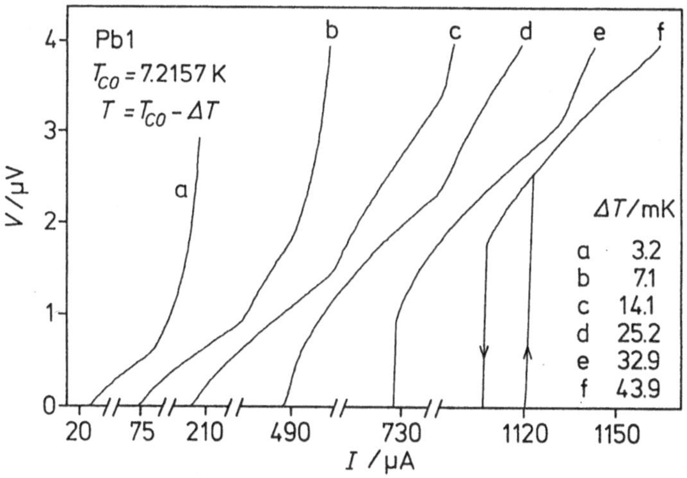

Fig. 55: *V-I characteristics of the pure lead whisker Pb 1 (for increasing and decreasing current) traced down to lower measuring temperatures to demonstrate the bending of linear portions [359].*

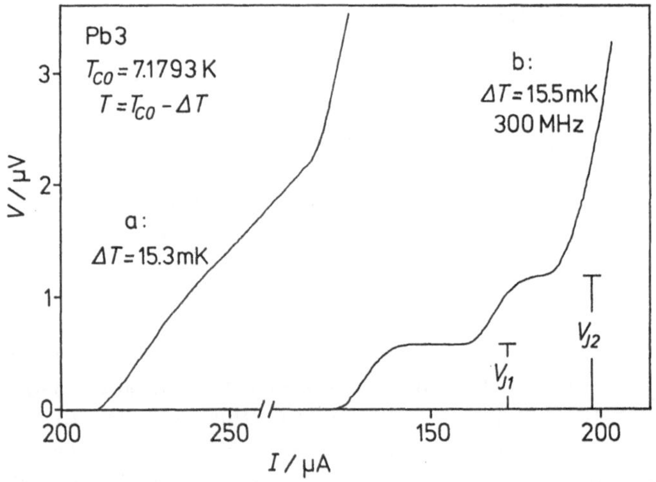

Fig. 56: *V-I characteristics of the pure lead whisker Pb 3 for increasing current [372].*
(a) Without HF radiation
(b) With applied HF radiation of $\nu = 300$ MHz. Current steps appear at $V_{J\tilde{n}} = \tilde{n}(h/2e)\nu$, with $\tilde{n} = 1$ and 2.

We also applied a HF-radiation field to a pure lead whisker (sample Pb 3 of ref. 372), using a loop turned around the whisker to couple the radiation to the sample (see section 7.3 for details about this technique). The result of such a measurement is shown in Fig. 56. Without HF radiation the V-I characteristics exhibit a linear portion generated by a single phase-slip center in the whisker. In the HF field 'current steps' are observed. As well the main step at $V_{J1}=(h/2e)\nu$ as also the first harmonic step at $V_{J2}=2(h/2e)\nu$ occur, where ν is the frequency of the HF field.

In Fig. 57 a sequence of V-I characteristics of the same sample in a HF field is shown, where the frequency of the radiation is adjusted to different values from 220 to 600 MHz. The characteristics exhibit the main current step at V_{J1}. We have widely discussed in section 7.3 that the observation of these 'inverse ac Josephson-like effects' establish the existence of phase-slip processes at Josephson frequency in our samples.

Finally, also in pure Pb whiskers there is evidence for an interaction of the phase-slip centers. Similar to the case of Zn whiskers (section 8.2) we did not undertake a detailed investigation of the interaction of phase-slip centers as in the case of tin whiskers (section 7.5), but we draw this conclusion from the behaviour of the current range ΔI over which the first linear portion exists until the next portion appears [359].

In Fig. 58 we plotted the range ΔI as a function of $I_c^{2/3}$ for samples Pb 1 and Pb 3, where the first linear portion is generated by a single phase-slip center in the sample. Also for Pb whiskers $I_c^{2/3}$ is proportional to $\Delta T = T_{c0} - T$. For both samples ΔI first increases with increasing critical

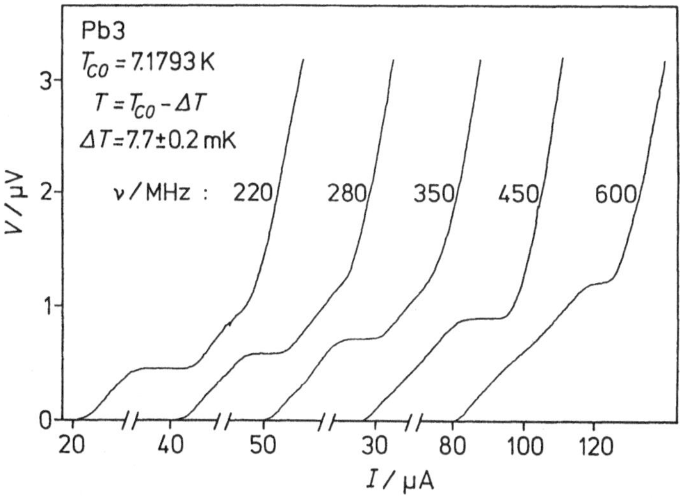

Fig. 57: *V-I characteristics of the pure lead whisker Pb 3 for increasing current in a HF radiation field at several different frequencies ν between 220 and 600 MHz [372].*

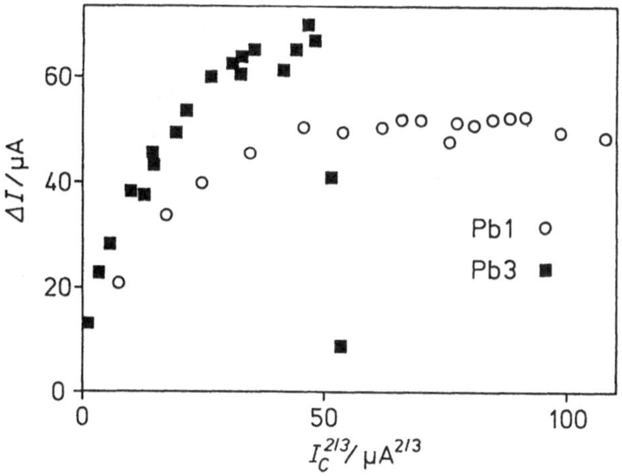

Fig. 58: *Current range ΔI which the first linear portion (or for lower temperatures the first voltage step) exists in, as a function of $I_c^{2/3}$ for two lead whiskers, samples Pb 1 and Pb 3 [359].*

current. While ΔI then decreases for Pb 3 (similar to the investigated Zn whisker in section 8.2), it seems to saturate for sample Pb 1.

Thus, also in the case of Pb whiskers the distance in current between the appearance of the first phase-slip center and the next phase-slip center (or phase-slip centers, if they appear together at the same current) is small for small critical currents but grows if the critical current becomes larger at lower temperatures. From this behaviour we conclude that the presence of a phase-slip center enhances the critical current of the following centers. Thus, also in the case of Pb whiskers a stabilizing effect generates the large current region between the first onset of voltage and the normal state. However, also in this case the mechanism is still unknown.

9.2.4. Pb with Small Impurity Concentrations of In (or Bi)

The V-I characteristics of Pb whiskers with small impurity concentrations of In or Bi are very similar to those of pure Pb whiskers. In Fig. 59 the characteristics for a Pb whisker with 0.4 at% In (sample Pb In 5) are shown [359]. Results for a Pb whisker with 1.45 at% Bi (sample Pb Bi 3) are given in ref. 359. The main difference to the case of pure Pb is that In and Bi impurities act to shift the temperature at which the transition becomes unstable at I_c toward the critical temperature and that the characteristic is much more straight after the voltage jump. For lower temperatures a distinct voltage step structure similar to that one of Sn or In is observed.

Evaluating the differential resistance of the first linear portion (or at lower temperature the first voltage step) one finds that the differential resistance is temperature independent. There may be a slight increase of the differential resistance close to T_{c0} and at low temperatures.

Close to T_{c0} where the characteristics show linear portions I_0/I_c is close to unity. According to the development of voltage steps in the characteristics, the ratio I_0/I_c then decreases with decreasing temperature. For lower temperatures, I_0/I_c becomes independent of temperature. As well for sample Pb In 5 as for Pb Bi 3, I_0/I_c roughly equals 0.85 in the low temperature range [359].

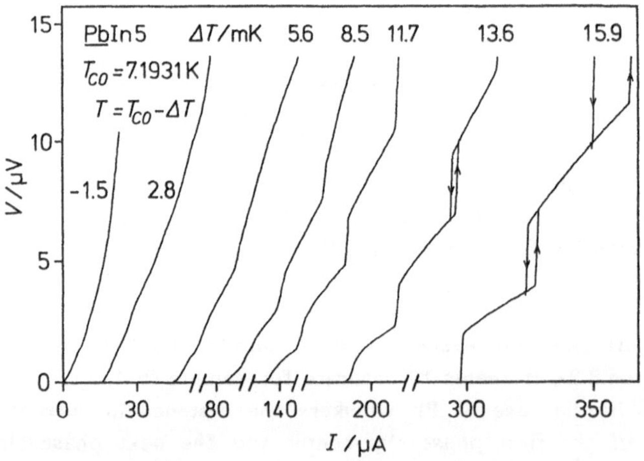

Fig. 59: *V-I characteristics of a Pb whisker with 0.4 at% In impurities (sample Pb In 5) for increasing and decreasing current [359].*

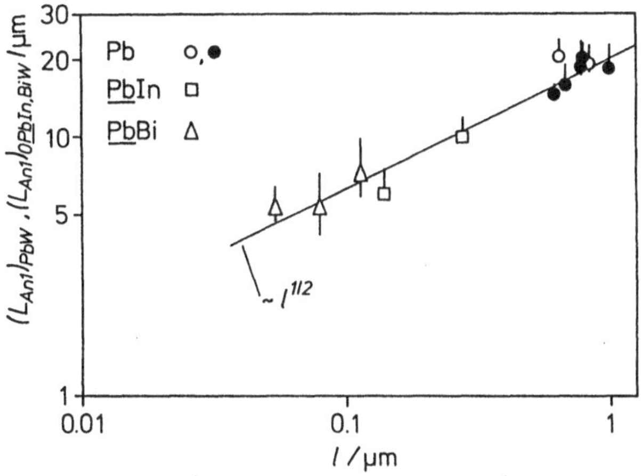

Fig. 60: *Normal-like length, L_{An1}, as a function of the electron mean free path, ℓ, using log-log coordinates.*

The plot shows results for whiskers of Pb-In, denoted '$(L_{An1})_{OPbInW}$', Pb-Bi, denoted '$(L_{An1})_{OPbBiW}$', and pure Pb, denoted '$(L_{An1})_{OPbW}$'.

The results for the alloys are from ref. 359 as well as the light symbol marked Pb values, while the full symbol marked Pb results are from ref. 40.

In Fig. 60 we plotted the normal-like lengths for several Pb-In and Pb-Bi whiskers as a function of the electron mean free path, adding results of pure Pb whiskers. For the alloys, we calculated L_{An1} by averaging the differential resistance in the temperature region where it is temperature independent. The results show the familiar $\ell^{1/2}$ dependence.

We also investigated the mean free path dependence of the averaged low temperature values $(I_0/I_c)_0$ for Pb, Pb-In, and Pb-Bi whiskers [359]. With decreasing ℓ the results decrease from about unity to about 0.85 for short electron mean free paths.

9.3. Interpretation within Phase-Slip Models and the TDGL Theory

First, we will interpret our experimental results in the framework of the SBT model.

For In-rich alloys the normal-like length L_{An1} exhibits a variety of temperature dependences. For several samples we found a certain temperature range where $L_{An1} \sim \Delta T^{-1/4}$. As we pointed out in section 7.2, L_{An1} may be identified with $2\Lambda_{Q^*in}$ in this case (where Λ_{Q^*in} is the charge imbalance relaxation length due to inelastic electron–phonon processes) and an experimental value for $\tau_{Q^*in}(0) \sim \tau_E$ may then be evaluated. It is found that the experimental values for $\tau_{Q^*in}(0)$ as evaluated from the experiment and as calculated theoretically (using Tinkham's estimate of eq. (72) for τ_E) have the same order of magnitude (see Tab. I of ref. 40). Moreover, the experimental results obtained for τ_E, using $\tau_{Q*in}(0) = 0.42\,\tau_E$, systematically fit into Fig. A 2 concerning the dependence of τ_E on the electron mean free path. Therefore, we conclude that in the region where the normal-like length is proportional to $\Delta T^{-1/4}$ the healing of quasiparticles is governed by the relaxation of charge imbalance due to inelastic phonon processes.

In this context it is remarked that in the case of In-rich alloys we did not strictly follow Tinkham's formula to calculate τ_E in ref. 40. Nevertheless, the results obtained seem to be 'closer to the truth' and, thus, will also be used in the discussion of the experimental results given in the present work. The problem is explained in detail in appendix 5 where we also give the values of τ_E which would follow from a strict application of Tinkham's formula.

For sufficiently low temperatures the normal-like length of In-rich alloys becomes independent of temperature and shows the $\ell^{1/2}$ dependence predicted by the SBT model. As described in section 7.2 a temperature independent quasiparticle relaxation time τ_2 can be evaluated from the straight line of Fig. 48, representing a graphically averaged value for all $\underline{\text{In}}$-Pb whiskers considering the results of pure indium whiskers. Assuming that the value of v_F for indium as given in Tab. A 2 also holds for In-rich alloys, it follows $\tau_2 = 1.90 \cdot 10^{-9}$ s [40], subsequently referred to as $\tau_{O\underline{In}PbW}$.

For whiskers of Pb and Pb with small concentrations of In or Bi impurities, the normal-like length is independent of temperature and shows the $\ell^{1/2}$ dependence predicted by SBT. Also in this case a temperature independent quasiparticle relaxation time can be evaluated. For Pb whiskers alone it follows from Fig. 15 of ref. 40 $\tau_2 = 5.42 \cdot 10^{-10}$ s (called τ_{OPbW} in the following). For $\underline{\text{Pb}}$-In and $\underline{\text{Pb}}$-Bi whiskers, considering also samples of pure Pb and assuming that the value for v_F of Pb (Tab. A 2) also holds for Pb-rich alloys, it follows from Fig. 60, that $\tau_2 = 4.93 \cdot 10^{-10}$ s [359] (called $\tau_{O\underline{Pb}In,BiW}$ in the following).

In section 7.2 we discussed that a temperature independent quasiparticle relaxation time which may be identified with τ_E is expected from theory for sufficiently low temperatures where $\tau_{GL} < \tau_E$. This interpretation is not successful for the samples discussed in the present chapter:

In the case of In-rich alloys samples 2, 5, and 8 show a distinct region where $L_{An1} \sim \Delta T^{-1/4}$ besides the temperature independent behaviour (Fig. 46). As mentioned above, an experimental value of τ_E can be evaluated in this case which may be used to calculate the temperature border below which a change to a temperature independent normal-like length should occur. Using $\tau_{Q*in}(0) = 0.42\,\tau_E$, we get from Tab. 1 of ref. 40 $\tau_E = 1.09 \cdot 10^{-10}$ s, $0.95 \cdot 10^{-10}$ s, and $2.57 \cdot 10^{-10}$ s for samples 2, 5, and 8, leading to temperature borders of $\Delta T = 27.6$ mK, 31.5 mK, and 11.7 mK, respectively. While the result is reasonable for sample 8, the change to a temperature independent behaviour occurs much closer to T_{c0} for the other specimens (Fig. 46). Moreover, in all cases the values of τ_E are an order of magnitude smaller than $\tau_{O\underline{In}PbW}$.

For whiskers of Pb and Pb with small concentrations of In and Bi, only a temperature independent behaviour was observed. Therefore, values for τ_E of Pb cannot be obtained from our experiments but results from Tab. A1 may be used to estimate the temperature border. Using $\tau_E = 0.0555 \cdot 10^{-10}$ s, as calculated by applying Tinkham's estimate, we get $\Delta T = 540$ mK. For $\tau_E = 0.38 \cdot 10^{-10}$ s, as measured, we get $\Delta T = 79$ mK. In both cases the change to a temperature independent behaviour is predicted for temperatures far below the range of our measurements. Furthermore, $\tau_{0Pb\underline{In},BiW} \approx \tau_{0PbW}$ is at least an order of magnitude larger than all known values for τ_E of lead (Tab. A1).

Next, also identifying L_{An1} with two times the temperature independent length $\Lambda_{Ku}(0)$, as discussed in section 7.2, does not lead to an explanation of the temperature independent regime. Already in section 7.2 we discussed that $\Lambda_{Ku}(0)$ is much too large in the case of pure In whiskers and that its proportionality to ℓ contradicts the observed $\ell^{1/2}$dependence of the normal-like length. Since pure In whiskers and $\underline{In}$-Pb whiskers join the same line $L_{An1} \sim \ell^{1/2}$ (Fig. 48), the length $\Lambda_{Ku}(0)$ is also not appropriate to explain the experiments performed with whiskers of In-rich alloys. The same conclusion is drawn in the case of Pb and Pb with small concentrations of In and Bi. Using the material parameters of pure bulk Pb and $\ell \approx 1\,\mu m$ as typical for a pure whisker, we get $\Lambda_{Ku}(0) = 312\,\mu m$ which is much larger than $L_{An1}/2$ in this case (Fig. 60). Moreover, the predicted ℓ dependence contradicts the observed $\ell^{1/2}$ law. Thus, also in this case electron-bogolon scattering is no appropriate explanation for the temperature independence of the normal-like length.

It is remarkable that the temperature independent quasiparticle relaxation times for In-rich and Pb-rich alloys join the empirical rule which connects the quasiparticle relaxation times with the scaling times τ_0 introduced by Kaplan et al. [103]. For a detailed discussion see section 7.2. Together with the result of that section it follows

$$\tau_{0SnW}/\tau_0(Sn) = (\tau_{0InW} \approx \tau_{0\underline{In}PbW})/\tau_0(In) = (\tau_{0PbW} \approx \tau_{0\underline{Pb}In,BiW})/\tau_0(Pb) \quad (212)$$

where $\tau_0(Pb) = 0.196 \cdot 10^{-9}$ s [103].

It is somewhat astonishing that results obtained from strong-coupling and weak-coupling materials join a common rule. As we have discussed in detail in ref. 40, this result indicates that the charge imbalance relaxation in lead is caused by scattering processes in which low energy phonons are involved for which the electron-phonon coupling function is approximately proportional to the quadrat of the phonon frequency as in the case of weak-coupling superconductors.

Thus, only the $\Delta T^{-1/4}$ law observed for L_{An1} of several $\underline{In}$-Pb alloy whiskers in a certain temperature range can be understood within the SBT model. In this temperature region inelastic electron-phonon scattering governs the charge imbalance relaxation. The temperature independent results of the normal-like length lead to experimental values of a temperature independent quasiparticle relaxation time. However, these times cannot be interpreted. Also the stronger temperature laws observed for several In-rich alloys cannot be understood within the SBT model. One runs into the same problems as discussed in detail in section 7.2 for the case of pure In whiskers.

Also the temperature dependence observed for I_0/I_c for In-rich alloys cannot be understood within the SBT model. In this model it is $I_0/I_c \approx 0.65$. This value is in agreement with the temperature independent value of I_0/I_c for pure In whiskers (Fig. 50). However, the electron mean free path dependence of these values (Fig. 50) also cannot be understood in the SBT model. Moreover, the result that I_0/I_c is close to unity as observed for whiskers of pure Pb and Pb-In or Pb-Bi whiskers close to T_{c0} is not predicted by SBT.

The shape of the V-I characteristics of Pb, Pb-In, and Pb-Bi whiskers can, however, qualitatively be understood within the KSS model widely discussed in section 5.8.4:

In Fig. 16 we plotted results of the analytical approximation of the KSS model. The shape of the I-V characteristics depends strongly on the ratio τ_E/τ_{OR}. For $\tau_E/\tau_{OR} \ll 1$ (T close to T_{c0}), the KSS model predicts that the characteristic is a straight line, developing continuously at I_c, so that $I_0 = I_c$. For increasing τ_E/τ_{OR} (decreasing T) the characteristic bends up until for $\tau_E/\tau_{OR} = 1$ it rises vertically at the critical current and then bends over to a quasi-ohmic behaviour for $I \gg I_c$. By a further decrease of the temperature ($\tau_E/\tau_{OR} > 1$) the KSS characteristic bends back so that a voltage exists for currents less than I_c. For measurements with an impressed current a voltage jump would now appear at I_c and the transition would be hysteretic.

The development of the characteristics with decreasing temperature is in qualitative agreement with the experimental results given in this chapter. Especially the straight-line behaviour of our V-I characteristics close to T_{c0} and, thus, $I_0 = I_c$ would be predicted by the KSS model. The transition to a hysteretic behaviour, however, occurs already for temperatures much closer to T_{c0} than predicted by KSS. In ref. 359 we showed that this observation may be caused by heating effects and, therefore, does not necessarily contradict the KSS model.

In whiskers made of Pb the linear-portion phenomenon is observed over a wide temperature range. Small impurity concentrations of In or Bi act to shift the temperature below which voltage jumps appear toward T_{c0}. For whiskers of other materials (Sn, In, and Zn) the development of the V-I characteristic with decreasing temperature seems to be qualitatively similar. However, the temperature range which linear portions or rounded voltage steps are observed in is very small and very close to T_{c0}, where fluctuation effects may influence the characteristics, and the effect is only observed in a few experiments. There may be evidence for the phenomenon in the V-I characteristics closest to T_{c0} given in refs. 16, 36, 37, and 20 Fig. 2 (equal to

Fig. 42 of the present work) and in the characteristics of some of our unpublished experiments.

Also these features can qualitatively be understood within the analytical approximation of the KSS model: The temperature range of the linear-portion phenomenon is wide in the strong-coupling materials because τ_E is so small. From Tab. A 1 one can see that τ_E of Pb is at least one order of magnitude smaller than the values for Sn or In. Thus, there is a much wider temperature region in which τ_E/τ_{0R} is much smaller than unity.

If the electron mean free path, ℓ, of a Pb whisker is reduced by impurities, the ratio τ_E/τ_{0R} is larger than in the clean case, for a given temperature difference $\Delta T = T_{c0} - T$, and the temperature range shrinks which linear portion phenomena are predicted in by KSS. The reason for the increase of τ_E/τ_{0R} is its proportionality to χ/ℓ which increases with decreasing ℓ as long as the sample is not in the dirty limit. The ratio τ_E/τ_{0R}, furthermore, depends on T_{c0} and τ_E. The critical temperature of our Pb whiskers with small concentrations of In or Bi is similar to that one of pure Pb whiskers so that its influence on the change of τ_E/τ_{0R} is of no importance. A decrease of τ_E of Pb with decreasing ℓ (Fig. A 1), if there really is any (compare Tab. A 1) would weaken the increase of τ_E/τ_{0R}.

The influence of impurities on the shape of the V-I characteristics of Pb whiskers, thus, is in qualitative agreement with the KSS model. We should, however, remark that the onset of a sharp hysteretic transition into the voltage-carrying state and its dependence on the electron mean free path is quantitatively better described by Joule heating effects [359].

Thus, the general shape of the V-I characteristic of Pb whiskers (and Pb whiskers with In and Bi impurities) is qualitatively predicted by the analytical approximation of the KSS model. The temperature independence of the differential resistance, however, cannot be understood within the KSS model either.

We do not only state the mentioned qualitative agreement, but also speculate somewhat about a physical interpretation of the result $I_0/I_c \approx 1$:

In the KSS model the expression for the dependence of the time averaged voltage on the total current is formally the same as the SBT result. Therefore, the measured intercept I_0 also in the KSS model may be interpreted as time averaged supercurrent in the core of the phase-slip center. However, as discussed in detail in section 5.8.4, the SBT result is actually only recovered in the high voltage dc limit. The reason is that in the SBT model it is assumed that the phase angle difference between two points of the superconductor develops linearly in time. In this case the time average of the supercurrent in the core can be calculated by averaging its dependence on the phase angle within one period. In general (especially for low voltages) it is not allowed to replace the time averaged supercurrent in the core, $\langle I_s(X_{PSC}, t)\rangle$, by its phase angle average. The reason is that the voltage across the core region and thus (via the Josephson relation) the phase angle difference is a complicated function of time due to the excitation of charge imbalance waves by the phase-slip cycle of the core. Therefore, in general $\langle I_s(X_{PSC}, t)\rangle$ is not independent of the time-averaged voltage (as in the SBT model) but changes along the voltage-current characteristics. This leads to a significant deviation from the high voltage dc straight line behaviour for low voltages.

The experimental result $I_0 \approx I_c$ can for instance be understood if the supercurrent in the core increases very rapidly in time to a value close to I_c, then creeping toward I_c which is reached at the end of the phase-slip cycle. This behaviour must be a consequence of the time development of the phase difference across the core region (related to the time development of the voltage across the core region by the Josephson relation) in connection with the current-phase relation connecting the phase angle difference with the supercurrent in the core. In other words, if the current-phase relation has a simple form (for instance sinusoidal), the voltage across the core region must be strongly time dependent. (One may consult the results of the RSM paper [265] in this context.) On the other hand, if the voltage is nearly time independent, the current-phase relation must be complicated.

From the experiments performed with Pb whiskers in a HF radiation field one would expect a sinusoidal current-phase relation, because no subharmonic current steps are observed in the V-I characteristics (see Figs. 56 and 57, and compare the discussion in section 7.3). This conclusion was drawn considering our calculation of the time-averaged supercurrent in the core which assumes a time independent voltage across the core region (see section 7.3). If the present argumentation concerning $I_0/I_c \approx 1$ should hold, a complicated strongly time dependent behaviour of the voltage across the core would be required. Let us assume that our argument concerning the I_0/I_c problem given above is correct. Then this discrepancy seems to indicate that the current steps occurring in the V-I characteristics in a HF field can be described by the calculation of section 7.3 although the properties in the core are much more involved in detail than considered in that calculation.

Finally, the experimental results of the present chapter may be compared with the predictions of the TDGL theory as elaborated by KR and widely discussed in section 5.9. The conclusions of this comparison will be summarized briefly and refs. 40 and 359 may be consulted for further details*¹. Moreover, the following chapter deals with a comparison of measured properties of an isolated phase-slip center with the predictions of the TDGL theory.

The central result elaborated by KR from the TDGL theory is a prediction Λ_{KR} for the behaviour of the normal-like length L_{An1} and a prediction $\beta_{KR} I_c$ for the zero voltage intercept I_0. (One may compare eqs. (174) and (187) to see that $L_{An1} = \Lambda_{KR}$ and $I_0 = \beta_{KR} I_c$.) The results for $\Lambda_{KR}/\xi_D(T)$ as a function of the pair-breaking parameter, γ, are plotted in Fig. 20, while the

*¹ Note that the quantity $L_{An1,KR}$ introduced in ref. 40 is equal to Λ_{KR}.

ratio $I_0/I_c=\beta_{KR}$ is tabulated in the work of KR [316]. The quantities are discussed in detail in section 5.9 and at the end of section 7.2.

For an isolated phase slip center in a homogeneous filament it is $L_{Anl}=2\Lambda_{Q^*in}\sim\Delta T^{-1/4}$ for very large pair-breaking parameters ($\gamma\gtrsim 1000$) only. However, also for somewhat smaller γ values the $\Delta T^{-1/4}$ temperature law is approximately obtained, because L_{Anl} only changes to about $2.6\Lambda_{Q^*in}$ if γ is lowered to about 100. If γ is further decreased a continuous change to a stronger temperature law is predicted. For $\gamma\approx 10$ it is $L_{Anl}\approx 7\xi_D(T)\sim\Delta T^{-1/2}$. For $\gamma<10$ the temperature dependence of the normal-like length becomes even stronger. Between $\gamma\approx 5.5$ and 6 the results seem to approach a line which is given by $L_{Anl}\approx 47.6\,\xi_D(T)/\gamma\sim\Delta T^{-1}$.

Since $\gamma\sim\tau_E\,\Delta T$, the properties of a phase-slip center in a filament can be changed by changing τ_E or ΔT. For a fixed sample (i.e. τ_E fixed), γ increases for increasing $\Delta T=T_{c0}-T$, that means decreasing temperature, and there should be a change in the temperature dependence of the normal-like length from $L_{Anl}\sim\Delta T^{-1}$ close to T_{c0} over $L_{Anl}\sim\Delta T^{-1/2}$ somewhat further away from T_{c0} to $L_{Anl}\sim\Delta T^{-1/4}$ at even lower temperatures.

The ratio $\beta_{KR}=I_0/I_c$ is predicted to decrease with increasing γ. For a fixed sample I_0/I_c should, therefore, decrease for decreasing temperature.

At a first view one may think that changing ΔT allows the adjustment of any desired γ value. This is not true in a real experiment. If ΔT becomes too small fluctuations may influence the measurement, while heating effects may lead to disturbances if ΔT becomes too large. Therefore, several samples with different τ_E values have to be used to extend the investigated γ range.

The TDGL theory is valid for dirty superconductors ($\ell\ll\xi_0$) as long as the local equilibrium approximation ($\Lambda_E\ll\xi_D(T)$) holds. The breakdown of the local equilibrium approximation occurs at least if $\Lambda_E=\xi_D(T)$. This condition defines a low temperature border ΔT_E in the sense that the TDGL theory is only valid if $\Delta T<\Delta T_E$ *1. For weak-coupling superconductors (such as our In-rich alloys) it follows $\Delta T_E=\pi\hbar/8k\tau_E$, as given in section 5.9, leading to $\Delta T_E=\tau_E^{-1}\cdot 3\cdot 10^{-12}$ K s. To get an expression for ΔT_E for strong-coupling superconductors (such as Pb or Pb with In or Bi impurities) we consider that the BCS coherence length, ξ_0, in $\xi_D(T)$ may change in the strong-coupling case. From eqs. (7) and (70) it follows for a weak-coupling superconductor that $\xi_0=\hbar v_F/\pi\mathbf{\Delta}(0)$ where $\mathbf{\Delta}(0)=1.76\,k\,T_{c0}$. According to the discussion in section 5.10 we replace the gap at zero temperature by $\mathbf{\Delta}(0)=2.15\,k\,T_{c0}$ as valid for Pb [14]. Then it follows from the condition $\Lambda_E=\xi_D(T)$, that $\Delta T_E=\pi\hbar/9.72\,k\tau_E$ or $\Delta T_E=\tau_E^{-1}\cdot 2.5\cdot 10^{-12}$ s for our strong-coupling samples.

If we formally apply the expression for ΔT_E for dirty and also for samples which are not in the dirty limit (using Tinkham's estimate for τ_E), it turns out that the local equilibrium approximation holds for all In, <u>In</u>-Pb, Pb, and <u>Pb</u>-In whiskers explicitly mentioned in section 7.2 and chapter 9 of the present work (see Figs. 28 - 30, 46, 47, 49, 51 - 59) at all measuring temperatures mentioned [40, 359]. Thus, the TDGL theory could be applied, if

*1 Note that ΔT_E is called ΔT_V in ref. 40.

the samples would be dirty. The last requirement is not fulfilled for all samples, but it may be the case for the In-Pb whiskers 6.1, 7, 8, 11, 13, 14, and 18.

The In whiskers and the In-Pb whiskers show a variety of temperature laws for the normal-like length. For a fixed sample the temperature dependence of L_{An1} becomes weaker for lower measuring temperature. If the temperature is sufficiently far away from T_{c0}, the normal-like length becomes temperature independent. All temperature laws (except the temperature independence) experimentally observed are also predicted by KR. Thus, there is a qualitative agreement between the TDGL theory and our experiments, except that the temperature independence of L_{An1} cannot be understood within the TDGL results of KR.

Usually a sample does not show all temperature laws ($L_{An1} \sim \Delta T^{-1}$, $\Delta T^{-1/2}$, $\Delta T^{-1/4}$, and temperature independent), if the measuring temperature is lowered (starting from T_{c0}). There are whiskers with $L_{An1} \sim \Delta T^{-1/4}$ and L_{An1} temperature independent only. Others directly change from exhibiting $L_{An1} \sim \Delta T^{-1}$ to a temperature independent normal-like length.

For a more quantitative comparison of the TDGL theory with our measurements we tabulated the temperature borders and γ value borders within which a given temperature law for L_{An1} is experimentally observed for the different samples [40]. It turns out that $L_{An1} \sim \Delta T^{-1}$ is usually observed if γ is between 5 and 11 and $L_{An1} \sim \Delta T^{-1/2}$ for γ between 9 and 22. This may be regarded as an agreement with the TDGL theory *1 which predicts a change to a ΔT^{-1} behaviour if γ falls below 10 and a $\Delta T^{-1/2}$ law for γ between 10 and 20. ($\Lambda_{KR}/\xi_D(T)$ only slightly changes if γ is enhanced from 10 to 20.) A normal-like length $L_{An1} \sim \Delta T^{-1/4}$ is observed for γ values between 7 and 39. This observation is not in agreement with the TDGL theory which expects a $\Delta T^{-1/4}$ law for $\gamma \gtrsim 100$. In the mentioned γ range a $\Delta T^{-1/2}$ law or something between a $\Delta T^{-1/2}$ and $\Delta T^{-1/4}$ law is predicted by the TDGL theory.

In the region of the $\Delta T^{-1/2}$ law one may compare the absolute values of the measured normal-like lengths with the prediction of the TDGL theory. The calculated values are larger than the measured ones (by up to a factor of 2.6).

It may be interesting to summarize the results for dirty whiskers which the TDGL theory can be strictly applied for: These samples usually show $L_{An1} \sim \Delta T^{-1}$ for γ between 5.4 and 11.2 and $L_{An1} \sim \Delta T^{-1/2}$ for γ between 8.9 and 17.9. Furthermore, there is one sample with $L_{An1} \sim \Delta T^{-1/4}$, however, at values of γ between 6.5 and 18.1. The calculated values for the normal-like length in the $\Delta T^{-1/2}$ temperature law region are larger than the measured ones (by up to a factor of 2.6).

---

*1 Concerning the ΔT^{-1} dependence this conclusion could not be drawn in ref. 40, because the results of the TDGL theory for $\gamma < 10$ were not known at that time.

Now we will discuss the results of Pb, Pb-In, and Pb-Bi whiskers. The value of τ_E for these materials is so small that all experiments are performed for γ values below $\gamma=3$ [40, 359]. For sample 7* and 12*, the temperature is, moreover, so close to T_{c0} that $0.48<\gamma<0.74$ and $0.21<\gamma<0.62$, respectively. In this γ range no prediction for the normal-like length of a homogeneous filament is given by KR. Only results of the static approximation are available, but also only down to $\gamma=1$. Therefore, no comparison between experiment and theory can be done if the specimens are regarded as homogeneous filaments. However, also the static approximation would not yield the temperature independence of the normal-like length observed experimentally.

Finally, the ratio $\beta_{KR}=I_0/I_c$ is predicted to grow with decreasing γ (Tab. 1 of ref. 316). For a fixed sample I_0/I_c thus should increase if approaching T_{c0}. This is in qualitative agreement with our experimental observations for In and In-Pb whiskers. Moreover, Pb, Pb-In, and Pb-Bi whiskers show this tendency.

One may also argue that the increasing tendency of the temperature independent values of I_0/I_c of whiskers from the In-Pb alloy system with increasing lead content (Fig. 50) is a consequence of the decrease of γ with decreasing τ_E. The problem is that the temperature independence of I_0/I_c cannot be understood within the TDGL theory.

10. Universal Behaviour of an Isolated Phase-Slip Center: Experiment and TDGL Theory

One of the basic problems in the scope of the current induced dissipative state in a quasi-one-dimensional superconductor is to come to an understanding of the behaviour of an isolated phase-slip center. Within the TDGL theory an isolated phase-slip center is a phase-slip state with infinite periodic length. In other words, an isolated phase-slip center is a single phase-slip center in a filament which develops without being influenced by other centers or affected by the contacts.

The first voltage step in the characteristics of a whisker or a long microbridge may be regarded as generated by an isolated phase-slip center. However, if the quasiparticle relaxation length is very large, as in Zn whiskers, the sample has to be sufficiently long to exclude influences of the superconducting contacts.

There are two basic characteristic properties of the first voltage step in a V-I characteristic which have been widely discussed in the preceding chapters. These are the normal-like length, L_{An1}, and the ratio of the extrapolated zero voltage intercept and the critical current, I_0/I_c. The experiments show that the behaviour of both quantities may be very different for samples of different materials. Both quantities may be temperature independent over the whole temperature range investigated. In other cases a temperature independent behaviour is only observed at somewhat lower temperatures, T, and there is a strong temperature dependence if one approaches the critical temperature, T_{c0}. In the case of the normal-like length, a variety of temperature laws is observed. Besides being temperature independent, L_{An1} may be proportional to ΔT^{-1}, $\Delta T^{-1/2}$ or $\Delta T^{-1/4}$, where $\Delta T = T_{c0} - T$. For a given sample the normal-like length usually runs through different temperature dependences with increasing strength if T becomes closer to T_{c0}.

The situation seems to be rather envolved. Therefore, we looked for a suitable parameter that governs the behaviour of both characteristic properties. Theoretical results for L_{An1}/ξ_D and I_0/I_c elaborated by Kramer and Rangel (KR) from the TDGL equations for dirty superconductors are plotted or tabulated as a function of the pair-breaking parameter, γ (see section 5.9). Here ξ_D is the GL coherence length in the dirty case. In the preceding chapter we found agreement with several predictions of KR. Although not all phenomena observed in our experiments could be explained, we assume that γ is a suitable parameter. Therefore, we plotted experimental values for L_{An1}/ξ and I_0/I_c as a function of γ for whiskers of different materials [147, 477]. Here, ξ is the GL coherence length for arbitrary electron mean free path.

We summarized experimental results from whiskers made of Pb, In, Sn, and Zn and of In-Pb, Sn-In, and Zn-Ag alloys. Due to the very different values of the inelastic electron-phonon collision time, τ_E, in these materials the investigated γ range extends from 0.1 in the case of Pb whiskers up to 20000 for Zn whiskers. We found that L_{An1}/ξ and I_0/I_c systematically develop with growing γ.

Our plots may be regarded as an experimentally obtained universal prediction for the behaviour of an isolated phase-slip center, only depending on the parameter γ. Moreover, it turns out that our experiments in a certain γ range are in qualitative and reasonable quantitative agreement with the results of the TDGL theory.

First we are going to discuss the range of small and medium pair-breaking parameters, γ, [147]. The corresponding universal plots are shown in Figs. 61, 62, and 63. In Figs. 61 and 63 experimental results for 24 different samples are shown together with the predictions of the TDGL theory [*1]. Some representative results of the single sample behaviour of L_{An1}/ξ are redrawn in Fig. 62. The theoretical prediction of L_{An1}/ξ_D is just that one given in Fig. 20 of the present work for the case of an isolated phase-slip center ($d_p=\infty$). The quantity I_0/I_c was calculated by KR for a periodic array of phase-slip centers only. As already pointed out in section 5.9, I_0/I_c is nearly independent of the periodic length, d_p, for not too small periodic lengths. Therefore, the results for $d_p=12\,\xi_D$ as given in Tab. 1 of ref. 316 may be applied for an isolated phase-slip center and, thus, are plotted in Fig. 62.

In Figs. 61 and 63 we do not use different symbols for the different samples. Therefore, it is remarked that due to the small value of τ_E, whiskers of Pb and Pb-In alloys have values for γ between 0.22 and 2.7. The In-Pb alloys have γ values between 5 and 43. For pure In whiskers γ is between

---

*1 All values for γ belonging to the experimental results in these figures are calculated using Tinkham's estimate for τ_E given in section 5.4. As mentioned in section 9.3 and discussed in detail in appendix 5, we did not strictly follow Tinkham's formula in the case of In-Pb whiskers, but inserted the total resistivity at room temperature instead of the phonon-induced part. The same was done for Sn and Sn-In whiskers. In this case the total resistivity is calculated from the phonon-induced part (which we identify with the bulk material resistivity of pure tin) and the measured residual resistance ratio, applying Mathiessen's rule (section 5.4). In the case of Sn whiskers the results nearly do not deviate from those strictly evaluated, because the residual resistance ratios of these samples are very small.

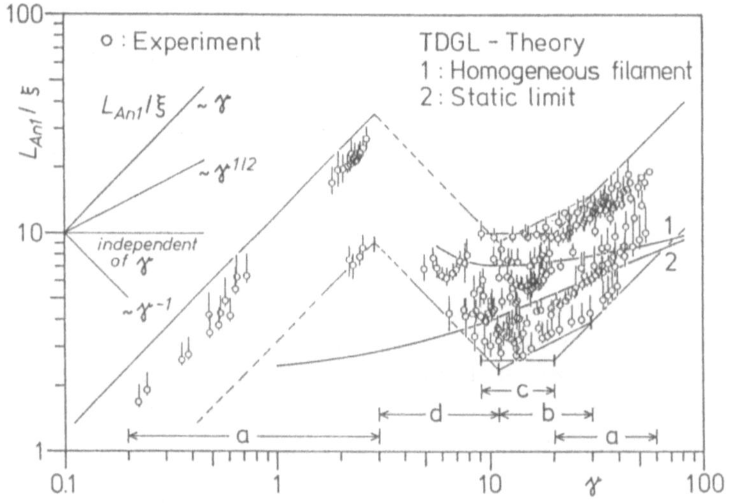

Fig. 61: *Normal-like length, L_{An1}, normalized by the GL coherence length, ξ, as a function of the pair-breaking parameter, γ, for small and medium values of this parameter [147]. Experimental values for an isolated phase-slip center in whiskers of different materials (Pb, In, Sn, Pb-In, In-Pb, Sn-In) and at different temperatures, T, close to T_{c0}, together with the prediction of the TDGL theory.*

The straight solid lines (dashed if estimated) indicate the borders of experimental values and at the same time indicate the dependence experimentally found for L_{An1}/ξ on γ. We observed for the regions 'a'-'d', respectively, that $L_{An1}/\xi \sim \gamma$, $\sim \gamma^{1/2}$, independent of γ, and $\sim \gamma^{-1}$, corresponding to a temperature dependence of L_{An1} for a sample in these regions given by L_{An1} independent of temperature, $\sim \Delta T^{-1/4}$, $\sim \Delta T^{-1/2}$, and $\sim \Delta T^{-1}$.

To see the correspondence of the dependence of L_{An1}/ξ on γ and of L_{An1} on ΔT, note that $\gamma \sim \Delta T^{1/2}$ while $\xi \sim \Delta T^{-1/2}$, where $\Delta T = T_{c0} - T$.

As we did not introduce different symbols for the different samples, the behaviour of each single sample is not directly visible.

10.5 and 39, while γ ranges between 19 and 56 for pure Sn whiskers and between 25 and 54 for Sn-In alloy whiskers.

First let us discuss the behaviour of L_{An1}/ξ given in Fig. 61. The experimental results of all samples are lying within a certain region of the (L_{An1}/ξ)-γ plane. The straight lines mark the borders of this region and at the same time indicate which dependence of L_{An1}/ξ on γ (and, thus, which temperature dependence of L_{An1}) has been observed in the experiments. For γ between 9 and 30 the indicated dependences of L_{An1}/ξ on γ overlap. This means the observation of several different dependences in this γ range.

The behaviour of the single samples cannot be distinguished from each other in Fig. 61. However, the dependence of L_{An1} on the temperature was published before for most samples. A complete list of references is given in

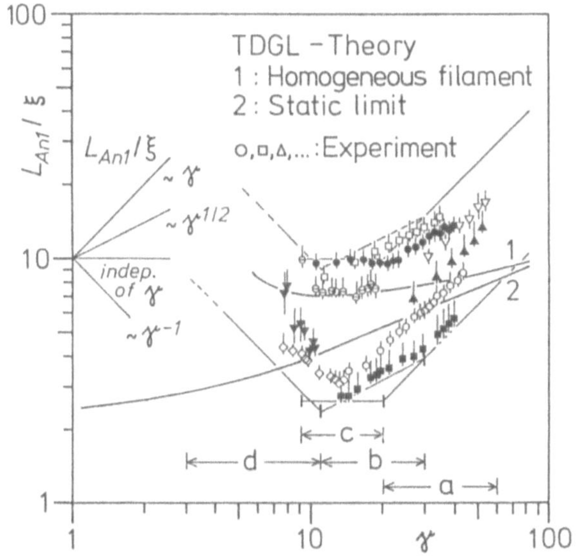

Fig. 62: *Normal-like length, L_{An1}, normalized by the GL coherence length, ξ, as a function of the pair-breaking parameter, γ. Experimental values for an isolated phase-slip center in whiskers of different materials and at different temperatures close to T_{c0}, together with the prediction of the TDGL theory.*

This figure shows some selected samples appearing in the medium γ range of Fig. 61. To demonstrate the single-sample behaviour we used different symbols for the different whiskers: (●) In 17, (□) In 20, (○) In-Pb 2, (■) In-Pb 5, (◇) In-Pb 6.1, (⊖) In-Pb 13, (▼) In-Pb 14, (▽) Sn 3 part 3, (▲) Sn-In 7.

The meaning of the full lines and the regions 'a'-'d' is the same as explained in the caption of Fig. 61.

the appendix of ref. 147. Several results are also given in the present work, such as for two of the Pb whiskers (Fig. 54), all In-Pb whiskers (Figs. 46 and 47), and all In whiskers (Fig. 30). The results for a single sample do not scatter very much, so that the temperature dependence of L_{An1} can clearly be observed.

For some representative samples we plotted L_{An1}/ξ as a function of γ in Fig. 62, making the behaviour of the individual sample visible by using different symbols for each specimen. The systematic development of L_{An1} on the temperature stated above leads to the systematic change of L_{An1}/ξ on γ.

It appears, that for different specimens of the same type of material the magnitude of the quantity L_{An1}/ξ at the same value of γ differs by a factor of up to four. We usually found that whiskers of very homogeneous pure material show values of L_{An1}/ξ close to the upper border marked in Fig. 61. Alloy whiskers have lower values, down to the lower border. Samples which are less homogeneous generally have low values of L_{An1}/ξ.

A very important information which may be obtained from Fig. 60 concerns the temperature dependence of L_{An1} for a given sample. This temperature dependence can be concluded from a known dependence of L_{An1}/ξ on γ, considering that for a fixed sample it is $\gamma \sim \Delta T^{1/2}$, while $\xi \sim \Delta T^{-1/2}$. In the range where it is $L_{An1}/\xi \sim \gamma$, it is L_{An1} independent of the temperature. If $L_{An1}/\xi \sim \gamma^{1/2}$ it follows $L_{An1} \sim \Delta T^{-1/4}$, while L_{An1}/ξ independent of γ leads to $L_{An1} \sim \Delta T^{-1/2}$ and $L_{An1}/\xi \sim \gamma^{-1}$ yields $L_{An1} \sim \Delta T^{-1}$.

Thus, Fig. 61 is also a map for the temperature laws of the normal-like length, giving information about the observed temperature law of a fixed sample in a certain γ range. While L_{An1} is found to be temperature independent for small and large values of γ (regions 'a'), there is an intermediate range where L_{An1} shows a $\Delta T^{-1/4}$ law (region 'b') or a $\Delta T^{-1/2}$ law (region 'c') or even a ΔT^{-1} law (region 'd'), depending on the value of γ. If there is a certain γ range where the mentioned regions overlap, different temperature laws were observed for different specimens.

Whiskers made of Pb and Pb-In alloys have very small values of γ whereas samples of Sn and Sn-In alloys have large values of γ. Therefore, these specimens show a length L_{An1} that is independent of temperature. On the other hand, samples made of pure In and In-Pb alloys have γ values in the intermediate range and, thus, their normal-like length may show different temperature laws.

The prediction of the TDGL theory is within the range of our experimental data, except the result for the static limit for $\gamma \lesssim 7$. The static limit is the lower border down to which inhomogeneities can depress the value of L_{An1}/ξ below the result for the homogeneous filament. The main difference between experiment and theory is that the theory does not predict $L_{An1}/\xi \sim \gamma$ that means L_{An1} independent of temperature. Furthermore, the increase of L_{An1}/ξ with growing γ is not as strong as observed experimentally. Therefore, the change from L_{An1}/ξ independent of γ to $L_{An1}/\xi \sim \gamma^{1/2}$ (related to a change of $L_{An1} \sim \Delta T^{-1/2}$ to $L_{An1} \sim \Delta T^{-1/4}$) is predicted for much larger values of γ than observed in our experiments. At the same time the intermediate region between both dependences is predicted to be much broader than experimentally observed.

Now let us discuss the behaviour of I_0/I_c plotted in Fig. 63: The experimental results decrease from $I_0/I_c = 1$ at very small values of γ to $I_0/I_c \approx 0.5$ at $\gamma \approx 50$. The prediction of the TDGL theory is below the experimental results. Nevertheless, for γ between 2 and 20 the ratio I_0/I_c decreases with the same slope as the theoretical prediction for γ between 10 and 80. Then the dependence of our experimental results becomes much weaker. Indeed, the experiments performed with Zn whiskers discussed below indicate that there seems to be no continuous decrease of I_0/I_c with decreasing γ.

Next we will discuss how far long microbridges of Sn and In join our universal plots. For the most part we only will consider the behaviour of L_{An1}/ξ on γ because data for I_0/I_c are too rare. The history of experimental

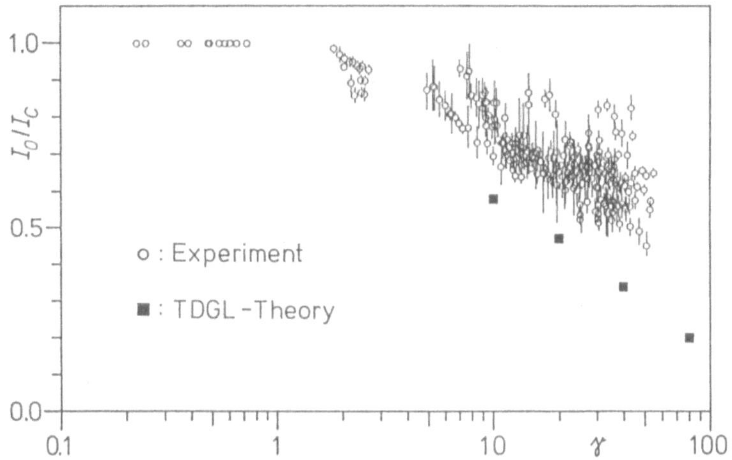

Fig. 63: *Ratio I_0/I_c as a function of the pair-breaking parameter, γ, for small and medium values of γ [147]. Experimental values for an isolated phase-slip center in whiskers of different materials (Pb, In, Sn, Pb-In, In-Pb, Sn-In) and for different temperatures close to T_{c0} , together with the prediction of the TDGL theory. We did not use different symbols for the different samples in this figure.*

investigations of phase-slip centers in long microbridges was reported in section 7.2, containing informations concerning the different experiments. If not explicitly given, detailed information about the evaluation procedure of the values for L_{An1}/ξ and γ of microbridges given below may be obtained from ref. 147.

First, we are going to discuss a group of measurements where the normal-like length has been evaluated from the differential resistance of the V-I characteristics.

The Sn microbridges of SBT showed a temperature independent normal-like length [32]. A typical result for their bridge would be $\gamma \approx 51$ and $L_{An1}/\xi \approx 5$. The γ value lies in region 'a' of Fig. 61. The ratio L_{An1}/ξ is somewhat smaller than we observed in our experiments. In this range of γ values our samples also show a temperature independent L_{An1}. Since the bridge mentioned does not consist of homogeneous material (see Fig. 2 of ref. 32) we should not worry about the small value of L_{An1}/ξ.

On the other hand, Kadin, Skocpol, and Tinkham measured an Sn microbridge showing a differential resistance changing proportional to $\Delta T^{-1/4}$ between $\Delta T = 28.6$ and 72.4 mK (that means for γ ranging from 77.8 to 123.8) *1 and becoming temperature independent for even lower temperatures

*1 For the determination of γ we calculated τ_E after Tinkham's estimate given in section 5.4, with $T_{c0} = 3.998$ K as obtained from Fig. 9b of ref. 84. The other material parameters are the same as those used for the other Sn microbridges discussed (see appendix A 2 of ref. 147).

(see Fig. 9b of ref. 84). Since the temperature dependence of the related normal-like length is the same as that one of the differential resistance, it follows $L_{An1} \sim \Delta T^{-1/4}$ or $L_{An1}/\xi \sim \gamma^{1/2}$ in the mentioned γ range. (The absolute magnitude of L_{An1}/ξ cannot be calculated, because the sample parameters needed are not given in ref. 84.) The measurements of Kadin, Smith, and Tinkham are performed in a γ range which lies above the region of experimental values of Sn and Sn-In whiskers in Fig. 61 or 62. In this γ range no experimental results from whiskers are available.

A temperature dependent normal-like length $L_{An1} \sim \Delta T^{-1/4}$ (that means $L_{An1}/\xi \sim \gamma^{1/2}$) also was obtained from measurements of the differential resistance of a long indium microbridge by Jillie [33], with L_{An1}/ξ increasing from 3.3 to 5.7 for γ growing from 11.4 to 33.8. Since these γ values range within region 'b' of Fig. 61 we would expect the observed temperature law. The ratios L_{An1}/ξ are within the borders of our experimental results, close to the lower border.

Also Weissbrod et al. [369] investigated phase-slip centers in superconducting indium microbridges. In Fig. 7 *1 of their paper they give results for L_{An1}/ξ as a function of γ for two different samples, called WInIV (100 μm long) and WInXb (130 μm long). To characterize the behaviour of the samples one may fit straight lines through the plotted results. In the case of sample WInIV it is $L_{An1}/\xi \sim \gamma^{1/2}$ (that means $L_{An1} \sim \Delta T^{-1/4}$) for γ between 36.6 and 78.1 and $L_{An1}/\xi \sim \gamma$ (that means L_{An1} independent of temperature) for γ between 78.1 and 128.1. The absolute values for L_{An1}/ξ are 15.5, 23.1, and 37.8 for $\gamma = 36.6$, 78.1, and 128.1, respectively. In the case of sample WInXb it is $L_{An1}/\xi \sim \gamma^{1/2}$ over the whole γ range from 25.6 to 70.3, with $L_{An1}/\xi = 5.6$ and 9.1 for $\gamma = 25.6$ and 70.3, respectively. In both cases temperature dependent normal-like lengths are observed in regions where our whiskers already show a temperature independent behaviour of this quantity. The change to a temperature independent behaviour observed in one of the samples occurs at a γ value which is outside the range of our measurements. The absolute values for both samples are lying within the borders (or extrapolated borders) of our experimental results (close to the upper border in the case of sample WInIV and close to the lower border for WInXb).

Weissbrod et al. also evaluated the ratio I_0/I_c (for two In microbridges of 30 μm length) and plotted the results in Fig. 5 of ref. 369 as a function of

---

*1 The results for L_{An1}/ξ given in that figure have to be multiplied by a factor of $2\xi_0/\xi$. If one uses $\xi_0 = 440$ nm as taken in that work, this factor yields 4.94 and 3.74 for sample WInIV and WInXb, respectively. For our discussion we prefer to take $\xi_0 = 0.2944$ μm [40] so that the factor is 5.91 for WInIV and 4.33 for WInXb.

The authors of ref. 369 use $\tau_E = 80$ ps. For our discussion we prefer to calculate τ_E after Tinkham's estimate given in section 5.4., leading to $\tau_E = 250$ ps for both samples, if we use their measured critical temperature and the material parameters for pure In as given in ref. 40. Thus, we multiplied the γ values obtained from Fig. 7 of ref. 369 by 250 ps / 80 ps = 3.125.

the temperature. For decreasing temperature the experimental values first decrease and then increase again. We will not discuss the temperature dependent behaviour in detail, but only state that I_0/I_c of their samples shows values between 0.7 and 0.5 in the temperature range between 3.3954 K and 3.3354 K, that means for γ between 30.5 and 96.7 (if we again calculate τ_E as described in the last footnote). This result is in agreement with our measurements given in Fig. 63, as far as there is an overlap of the γ ranges, and establishes the saturating tendency of I_0/I_c.

Now, we are going to discuss a group of experiments where the quasiparticle diffusion length, Λ, was obtained from spatially resolved measurements of the electrochemical quasiparticle potential near a phase-slip center and where the normal-like length may be calculated as $L_{Anl} \approx 2\Lambda$ according to SBT: In all cases these measurements yield a quasiparticle diffusion length Λ (and, thus, also a normal-like length L_{Anl}) changing proportional to $\Delta T^{-1/4}$ (that means proportional to $\gamma^{1/2}$). From the measurements of Dolan and Jackel [70] for a tin bridge we get L_{Anl}/ξ increasing from 5.7 to 7.9 for γ growing from 63.8 to 121.3. Measurements of Aponte and Tinkham [72] lead to L_{Anl}/ξ increasing from 3.4 to 6.9 for γ growing from 30.1 to 123.5. These results are below the border of our experimental values in Fig. 61. As far as there is an overlap of the γ ranges with our experiments, the γ values are so large already, that our measurements show a temperature independent normal-like length.

Although in some cases the behaviour of Sn and In microbridges are in agreement with our measurements (SBT and Jillie), the characteristic difference between whiskers and Sn and In microbridges seems to be that microbridges show a temperature dependent normal-like length in a γ range where L_{Anl} for whiskers has already become temperature independent (Weissbrod et al., Aponte and Tinkham). In microbridges this temperature dependence was found up to a γ value of about 120 (Kadin et al., Dolan and Jackel, Aponte and Tinkham).

The reason for the different behaviour of microbridges and whiskers is finally not clear. For whiskers there should be no masking of a $\Delta T^{-1/4}$ law by heating effects which are usually invoked to explain the observation of a temperature independent normal-like length in the case of microbridges. We have already discussed this point somewhat in detail in section 7.2. May be that the different grade of homogeneity plays a role. Already in section 7.2 we discussed that compared to whiskers, microbridges are usually much more inhomogeneous samples with strong variations of the critical temperature along the bridge and that pinning centers, such as notches, are widely used to fix the phase-slip center in a bridge.

The influence of inhomogeneities may be indeed a probable explanation. As already mentioned in section 7.2, the experiments of Liengme et al. discussed below seem to indicate this. Moreover, very recent experiments on tunable weak links seem to establish this statement (chap. 12). In these experiments the properties of a phase-slip center were studied which appeared in a weakly superconducting section induced into a whisker. The temperature dependence of the differential resistance of the V-I characteristics generated by this phase-slip center shows a dependence on the strength of superconductivity in the weakly superconducting section.

Also from the TDGL theory one would expect a strong influence of inhomogeneities on the properties of a phase-slip center developing in an inhomogeneous section of a specimen (see the discussion in the present section and section 5.9). The problem is that the theory predicts a weaker temperature dependence of the normal-like length if inhomogeneities are considered.

Naturally we should also discuss why microbridges only yield normal-like lengths with $L_{An1} \sim \Delta T^{-1/4}$, but why there is no experiment reported where L_{An1} changes proportional to $\Delta T^{-1/2}$ or ΔT^{-1} for T approaching T_{c0}. The reason is probably that the values of γ for microbridges were not small enough to observe a normal-like length with one of the stronger temperature laws.

Concerning the TDGL theory a very important question is, how far the change to a temperature independent normal-like length may have something to do with going beyond the validity range of the local equilibrium approximation. The local equilibrium range is at least left if $\xi_D(T)$ becomes smaller than Λ_E which means that ΔT exceeds ΔT_E, or γ exceeds γ_{max}, or τ_{GL} becomes smaller than τ_E (see section 5.9).

For all Pb and Pb-In whiskers mentioned in Fig. 61 it is $\gamma_{max} = 8.8$. The γ values of these samples are much smaller than this limit.

For all In and In-Pb whiskers of this figure (and for sample In) we have already discussed in section 9.3, that $\Delta T < \Delta T_E$ for all measuring temperatures. The values of ΔT_E for these samples are summarized in Tab. 8 of ref. 40. It turns out that there are some specimens where a change to a temperature independent behaviour is observed for ΔT close to ΔT_E. For the pure In whiskers In, In 17, and In 20 it is $\Delta T_E = 11.4$ mK, 11.5 mK, and 11.5 mK, respectively, which is close to the value of ΔT, where the normal-like lengths of these specimens become temperature independent (compare Figs. 29 and 30). The temperature borders ΔT_E for our pure In whiskers given here are calculated using Tinkham's estimate for τ_E of each whisker. They are similar to the value for pure polycrystalline indium and also close to the results obtained with τ_E evaluated from the region where $L_{An1} \sim \Delta T^{-1/4}$ for these specimens (compare the discussion in section 7.2).

In the case of In-Pb whiskers it is $\Delta T_E = 13.7$ mK and 15.9 mK for samples 3 and 5, respectively, not being far away from the temperature where L_{An1} becomes temperature independent (see Figs. 46 and 47). We should remark that this result is not obtained for sample 5 if the measured value of τ_E is employed. Using the measured value of τ_E leads, however, to $\Delta T_E = 11.7$ mK in the case of sample 8, close to the temperature where L_{An1} changes to a temperature independent behaviour for that sample (see Fig. 46 and compare the discussion given in section 9.3). For all other specimens the change to a temperature independent L_{An1} occurs much closer to T_{c0} than given by the border ΔT_E.

Next, we are going to discuss the Sn and Sn-In specimens for which results are also plotted in Fig. 61. Using again τ_E as calculated after Tinkham's estimate (see appendix A1 of ref. 147 for the material parameters needed) it is $\Delta T_E = 8.1$ mK (or $\gamma_{max} = 51.4$) for samples Sn [19] and Sn 3 [22], while $\Delta T_E = 11.1$ mK (or $\gamma_{max} = 43.9$) for sample Sn 2 [22] and tin whisker 1 of ref. 23. Furthermore, it is $\Delta T_E = 10.8$ mK, 11.4 mK, and 11.9 mK (or $\gamma_{max} = 43.9$, 42.9, and 42.0) for samples Sn-In 7 [15], Sn-In 10 [15], and Sn-In 17 [15], respectively. As already discussed in section 7.2, it is $\Delta T_E = 8.4$ mK (or $\gamma_{max} = 50.6$) for polycrystalline pure tin. It turns out that for Sn and Sn-In whiskers a temperature independent behaviour of L_{An1} (or the dependence $L_{An1}/\xi \sim \gamma$) is found up to temperatures which are much closer to T_{c0} than the border given by ΔT_E (or for γ much smaller than γ_{max}). For experimental data which demonstrate this statement see Fig. 10 of ref. 16 (samples Sn, Sn-In 7 (1.1 at% In), Sn-In 17 (3.2 at% In)), the intrinsic data in Fig. 9 of ref. 23 (tin whisker 1), and Fig. 62 of the present work (samples Sn 3, Sn-In 7).

Thus, only the In whiskers mentioned and some of the In-Pb whiskers show a change of their normal-like length from a temperature dependent to a temperature independent behaviour at the limits of the local equilibrium approximation given above. For the other samples L_{An1} already becomes independent of temperature (In-Pb whiskers) or is already temperature independent (Pb, Pb-In, Sn, Sn-In whiskers) at temperatures closer to T_{c0}.

Before we shall try to draw a conclusion, we recall some results of other materials: For Zn and Zn-Ag whiskers it is ΔT_E only some hundredth parts of a millikelvin (section 8.3), so that all experiments are performed outside the local equilibrium approximation. For all measuring temperatures these specimens deliver a temperature independent normal-like length. The resulting quasiparticle relaxation time τ_2 is not too different from τ_E. The identification of τ_2 with τ_E was proposed by Baratoff, who calculated the properties of a phase-slip center outside the local equilibrium approximation (see sections 5.9, 7.2, and 8.3).

While a cross-over from a temperature dependent to a temperature independent behaviour of L_{An1} is not observed for Zn and Zn-Ag whiskers, this transition occurs in the Al microbridges of Liengme et al. [136] (see the discussion of their experiments given in section 7.2). The quasiparticle relaxation time τ_2 determined from the results of the temperature independent regime is in agreement with τ_E as calculated for Al after Tinkham's estimate (section 7.2). From the temperature dependent regime where $L_{An1} \sim \Delta T^{-1/4}$ they evaluated a time τ_E which is much smaller. Setting τ_{GL} equal to the latter result for τ_E, they found that the cross-over from the temperature dependent to the temperature independent regime occurs at the border of the local equilibrium approximation range given by this criterion.

Naturally, it is not easy to draw a conclusion from these results. There are, however, several experiments which give evidence that the sample is outside the local equilibrium approximation in the temperature range where the normal-like length is temperature independent. Therefore, with some caution, we state that the observation of a temperature independent normal-like length indicates that the sample has left the temperature range (or γ range) of the local equilibrium approximation and, thus, cannot be described by the TDGL equations.

There are several arguments which may explain that there are a lot of samples for which a change to the temperature independent behaviour occurs at a temperature (or γ value) which differs from the upper border of the local equilibrium range given by the condition $\Lambda_E = \xi_D(T)$ which means that $\Delta T = \Delta T_E$, $\gamma = \gamma_{max}$, or $\tau_{GL} = \tau_E$. One problem is that this condition gives the border where the break-down of the local equilibrium approximation at least occurs. The more rigorous condition for the validity of the local equilibrium approximation is $\Lambda_E \ll \xi_D(T)$. Therefore, it may be possible that a sample leaves the local equilibrium range at a temperature which is closer to T_{c0} than ΔT_E (or at a γ value which is smaller than γ_{max}). Moreover, only a few of our samples are in the dirty limit. Besides this there is a substantial uncertainty in the knowledge of the material parameter τ_E (see section 5.4 together with appendix 3 of the present work and refs. 147 and 477 for a critical discussion of this quantity).

In the remainder of this chapter we will discuss the universal behaviour of an isolated phase-slip center in a material with large pair-breaking parameters γ, such as Zn, Zn-Ag, and Al [477]. The universal plots for L_{An1}/ξ and I_0/I_c as a function of γ are shown in Figs. 64 and 65. (Again γ was calculated using Tinkham's estimate for τ_E. For details and material parameters see ref. 477.)

In Fig. 64 we plot L_{An1}/ξ versus γ for our Zn and Zn-Ag whiskers (see chap. 8) together with results for Al microbridges of Liengme et al. [136] and Stuivinga et al. [73]. The curve of the TDGL theory (after KR [316]) for high γ values plotted in this figure is valid for an isolated phase-slip center in a homogeneous filament as well as in the static limit. The curve is also given in Fig. 20 of the present work. Calculated values are only available up to $\gamma \approx 1000$. The prediction of the TDGL theory is close to $L_{An1}/\xi \sim \gamma^{1/2}$ which is also expected for even higher γ values [316].

For the Al microbridges of Liengme et al. [136] in a certain γ range a $\Delta T^{-1/4}$ temperature dependence for L_{An1} has been observed leading to $L_{An1}/\xi \sim \gamma^{1/2}$, in agreement with the theory. The γ range of the $\Delta T^{-1/4}$ temperature law depends on the homogeneity of the sample. For very homogeneous samples the range is only small and an early cross-over to a temperature-independent behaviour ($L_{An1}/\xi \sim \gamma$) is observed. The microbridges of Stuivinga et al. [73] only deliver a $\Delta T^{-1/4}$ law. These samples contain a notch to pin the phase-slip center. It may be assumed that the cross-over to a temperature independent behaviour takes place at higher γ values which no measurements were carried out for.

For whiskers of Zn and Zn-Ag alloys the smallest γ values are beyond the cross-over region of the Al microbridges. The temperature independence

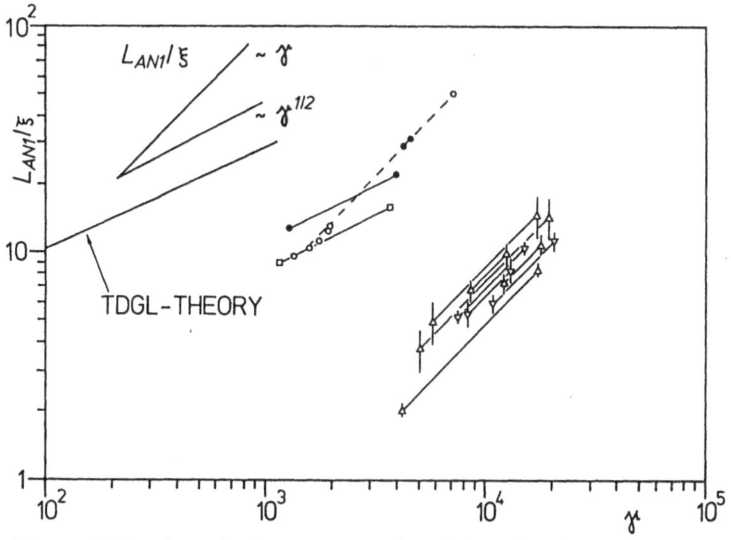

Fig. 64: *Normal-like length, L_{An1}, normalized by the GL coherence length, ξ, as a function of the pair breaking-parameter, γ, for large γ [477]. Experimental values for an isolated phase-slip center in whiskers of Zn and Zn-Ag and in Al microbridges for different temperatures, T, close to T_{c0}, together with the prediction of the TDGL theory:*

(Δ) Zn whiskers, (∇) Zn-Ag whiskers, (□) Al microbridge (sample 3 from ref. 73), (O) homogeneous Al microbridges from ref. 136 (the result with the largest γ value is from sample 4, the further points from sample 11), (●) inhomogeneous Al microbridge (sample 9 from ref. 136).

Symbols with the same error bars mark the range of a measurement. The experimental values ly on the connecting lines between the symbols. For $L_{An1}/\xi \sim \gamma$ and $\sim \gamma^{1/2}$ it is L_{An1} temperature independent and $\sim \Delta T^{-1/4}$, respectively, where $\Delta T = T_{c0} - T$.

The dashed line indicates the estimated slope for the temperature independent normal-like length (normalized by ξ) of Al microbridges.

of L_{An1} for whiskers supports the measurements of microbridges and extends the γ range of our universal behaviour study to values of about 20000. The absolute values of L_{An1}/ξ are smaller than those for Al microbridges. Furthermore, we found that for a fixed γ the value of L_{An1}/ξ decreases with decreasing grade of homogeneity of a whisker.

In Fig. 65 we plot the ratio I_0/I_c as a function of γ for our Zn and Zn-Ag whiskers and for Al microbridges. Since for high γ values ($\gamma > 100$) the static approximation can be applied, the TDGL theory expects I_0/I_c close to zero [316]. The measurements do not confirm this prediction. For an isolated phase-slip center, I_0/I_c scatters around 0.5. While I_0/I_c does not depend on temperature for our whiskers, a temperature dependence of this quantity is observed for the microbridge of Stuivinga et al. [73]. For this sample I_0/I_c increases from 0.3 close to T_{c0} to 0.7 at lower temperatures. Two of our

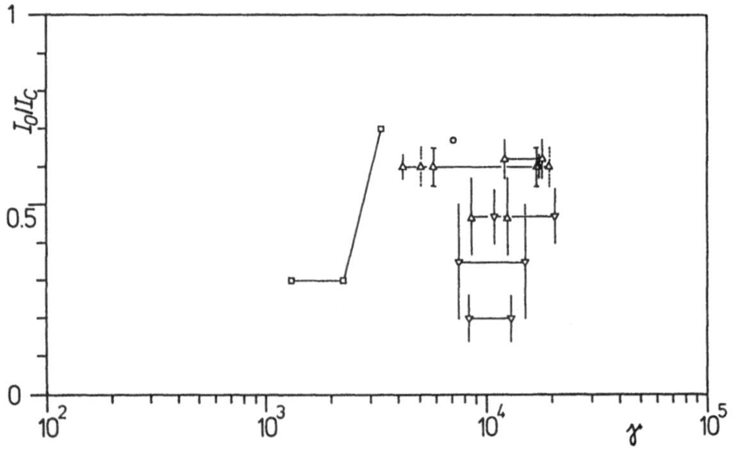

Fig. 65: *Ratio I_0/I_c as a function of the pair-breaking parameter, γ, for large γ. Experimental values for a single phase-slip center in whiskers of Zn and Zn-Ag and in Al microbridges for different temperatures close to T_{c0}. Same notation as used in Fig. 64. The single light circle is from sample 4 of ref. 136.*

samples also show values of about 0.3. This result is not typical for an isolated phase-slip center in a whisker. It can be explained by a nucleation of the phase-slip center close to one of the superconducting contacts (see the discussion in section 8.3). With some caution an analogous explanation may be applied concerning the low values for the Al microbridges of Stuivinga et al. [73] (see ref. 20 for a detailed discussion). It is remarked that in the two whiskers with a depressed ratio I_0/I_c the distance between the phase-slip center and the contact is large enough for an undisturbed quasiparticle relaxation. Thus, they show a value for L_{An1} which is typical for an isolated phase-slip center.

The investigation of the universal behaviour of an isolated phase-slip center presented in the present chapter gives us an overview over a γ range of about six decades. There are three γ ranges for which the normal-like length L_{An1} does not depend on temperature, separated by intermediate regions with temperature dependent L_{An1}. The ratio I_0/I_c decreases with increasing γ from unity to a saturating value of about 0.55. The behaviour of the temperature dependent regime of L_{An1} may be understood within the TDGL theory. Also a decrease of I_0/I_c with increasing γ is predicted by the TDGL theory, but not the saturating tendency. In the temperature independent regime of L_{An1} the samples seem to be outside of local equilibrium. In this case other theoretical methods have to be applied. However, only some first approaches of this kind exist.

11. Hysteresis of the Critical Current

11.1. Introductory Remarks

One of the characteristic features of the current-induced breakdown of superconductivity in a quasi-one-dimensional superconductor is its hysteretic behaviour. Our whiskers show a hysteresis as well in their V-T transition curves as also in their V-I characteristics.

In the case of the V-T characteristics the current has a fixed value and the voltage is measured as a function of the temperature, T. For increasing temperature the sample enters the voltage-carrying state at the transition temperature, T_c. (Note that T_c is the transition temperature in the current-carrying state and, therefore, more or less lower than the critical temperature, T_{c0}. The difference between T_c and T_{c0} depends on the magnitude of the transport current.) A further increase of the temperature leads to the normal conducting state. If then the temperature is lowered, the sample recovers the superconducting state at a bath temperature T_R which is smaller than T_c.

In the case of a V-I characteristic the temperature is fixed and the voltage is measured as a function of the current. As sketched in Fig. 25, for increasing current the dissipative phase-slip state is entered at the critical current I_c, while for decreasing current the superconducting state is recovered at the jump-back current, I_R, which is smaller than the critical current.

The problem is to find out which mechanisms generate the observed hysteresis. Since energy is dissipated in the filament a very important question is to estimate how far Joule heating effects can warm up the lattice of the sample and, thus, may lead to hysteretic behaviour. For this purpose the thermal boundary resistance ('Kapitza resistance') between the whisker and the surrounding helium has to be determined. If the contribution of the thermal hysteresis is known, the remaining intrinsic nonthermal hysteretic behaviour of a phase-slip center can be investigated. It turns out that this part cannot be described by a single mechanism but seems to be generated by an overpopulation of the excitation spectrum with nonequilibrium quasiparticle excitations and by the influence of charge imbalance waves on the phase-slip process.

Before we go into details, we should briefly sketch the history of the investigations of hysteresis effects of phase-slip centers and then add some remarks on the organization of the present chapter.

Already J.D. Meyer [19] observed hysteresis effects in his investigations of pure Sn whiskers (see Fig. 6 of the present work). In the case of V-I characteristics he found that the hysteresis width, $I_c - I_R$, increases if the temperature is lowered. Moreover, he reportet that the jump-back voltage, $V_R(I_R)$, does not show the linear increase as observed for the height of the first voltage jump for which $V_1(I_c)$ is a straight line.

The same behaviour was found for whiskers of Sn-In alloys [15, 28] and pure In [478]. For these specimens $I_c - I_R$ grows linearly with increasing critical current (after an initial nonlinear increase). Thus, we characterized the development of the hysteresis width in the V-I characteristics of a sample by $d(I_c - I_R)/dI_c$ and plotted the results as a function of the electron mean free path ℓ of the samples. Moreover, the saturating value, V_{RS}, of the jump-back voltage, V_R, was determined for different samples and plotted as a function ℓ. In the case of Sn-In alloys, $d(I_c - I_R)/dI_c$ is roughly 0.5 for ℓ between 0.045 and 0.2 µm and then smoothly decreases to about a value of 0.2 at $\ell = 3$ µm. The results for V_{RS} first increase with increasing ℓ (from $V_{RS} \approx 6$ µV at $\ell \approx 0.045$ µm to $V_{RS} \approx 19$ µV at $\ell \approx 1$ µm) and then decrease again (to $V_{RS} \approx 0.7$ µV at $\ell \approx 3$ µm). In the case of In whiskers, $d(I_c - I_R)/dI_c$ scatters between 0.1 and 0.3 for ℓ between about 0.15 and 1.6 µm, while V_{RS} continuously decreases with decreasing ℓ from about 4.6 to 1.2 µV in the same range of mean free paths.

An interesting question is whether a re-entrant part of the V-I characteristics with negative differential resistance which cannot be passed for impressed current leads to the sharp voltage jumps at I_c and I_R, respectively. J.D. Meyer [479] found evidence for such a re-entrant part by shunting a tin whisker with a small parallel normal conducting resistor, thus, working with impressed voltage instead of impressed current.

Also the Sn microbridges of SBT [32] showed hysteresis effects in their V-I characteristics. SBT argue that these effects may be partly understood as heating effects, but that in addition nonthermal intrinsic hysteresis must be present. Subsequently, Kadin, Skocpol, and Tinkham [84] also investigated the hysteresis of Sn microbridges and conclude that for their specimens thermal effects alone are sufficient to understand the whole hysteresis width.

The TDGL theory [316] predicts an intrinsic nonthermal hysteresis (see section 5.9). Also the KSS model [158] leads to an intrinsic hysteretic behaviour (as discussed in detail in section 5.8.4). It is remarkable that the KSS model predicts a saturating behaviour of the jump-back voltage [158]. These theoretical results will be discussed in detail in the present chapter.

The characteristic saturating behaviour of the jump-back voltage was also observed for whiskers of In-Pb alloys [40]. For these samples we, furthermore, compared the measured onset temperature, T_{Hy}^{exp} or $\Delta T_{Hy}^{exp} = T_{c0} - T_{Hy}^{exp}$, of the hysteretic behaviour in the V-I characteristics with theoretical predictions. In the KSS model the hysteretic regime is predicted for $\tau_E/\tau_{OR} > 1$, while the criterion is $\gamma > \gamma_c \approx 5.5$ in the TDGL theory. Thus, the criterion for the onset of hysteresis is given by $\tau_E/\tau_{OR} = 1$ and $\gamma = \gamma_c$, respectively. We calculated $\tau_E/\tau_{OR}(\Delta T_{Hy}^{exp})$ and $\gamma_{Hy}^{exp} = \gamma(\Delta T_{Hy}^{exp})$ for several In-Pb whiskers [40] and found that the KSS criterion is fulfilled within a factor of 2. Here, the experimental results $\tau_E/\tau_{OR}(\Delta T_{Hy}^{exp})$ show values between 0.5 and 1.9. In the case of the TDGL criterion we observed that γ_{Hy}^{exp} is always larger than γ_c (except for one sample). This means that the measured onset of hysteresis occurs at lower temperatures than predicted by the TDGL theory.

For Pb, Pb-In, and Pb-Bi whiskers no intrinsic hysteresis was observed. If there is any hysteretic behaviour present, it seems to be caused by thermal effects [40, 359]. This observation is consistent with the KSS criterion $\tau_E/\tau_{OR} = 1$ which leads to a prediction ΔT_{Hy}^{KSS} for the onset of hysteresis which is larger than 800 mK for pure Pb whiskers. In the case of Pb-In and Pb-Bi whiskers ΔT_{Hy}^{KSS} ranges between 140 and 370 mK, at least an order of magnitude larger than the measured ΔT_{Hy}^{exp}. The observation is, however, also consistent with the TDGL criterion, because for Pb, Pb-In, and Pb-Bi whiskers the measuring temperatures are so close to T_{c0} that the related γ values are much smaller than γ_c [40, 147, 359]. It is remarked that Tinkham's estimate for τ_E (see section 5.4) is used throughout.

We also investigated the hysteretic behaviour of V-I characteristics of our whiskers under the influence of HF radiation for whiskers made of Sn [371], In [16, 372], and In-Pb alloys [40]*¹. In all cases HF-radiation field induced 'current steps' are also observed in the hysteretic part of the characteristics so that the occurrance of phase-slip processes at Josephson frequency is established also in the hysteretic regime.

Recently, Weissbrod et al. [369] investigated the hysteretic behaviour of the V-I characteristics of long indium microbridges. They found the development of I_R/I_c with decreasing temperature to be consistent with the

*¹ See the long version of that work (denoted by ref. 9 in that paper).

KSS model which predicts a saturating behaviour of this quantity. The onset of hysteresis in their samples occurs much closer to T_{c0} than predicted by KSS *1.

For a systematic investigation of hysteresis effects in whiskers we first studied the heat transfer properties of whiskers from the In-Pb alloy system to the surrounding helium [480]. For this purpose we evaluated the width of the temperature hysteresis, δT_H, at fixed current I_F. If the currents are sufficiently large, self-heating hot spots appear in the sample and the hysteresis is generated by thermal effects. Since in these experiments the heat removal occurs mainly through the surface of the sample, the heat transfer coefficient and thus the Kapitza resistance can be easily evaluated from these experiments. Similar investigations have been performed with Zn and Zn-Ag whiskers to obtain the Kapitza resistance of zinc below 1 K [481].

Then we started detailed investigations of the hysteretic behaviour for lower measuring currents, where the appearance of phase-slip centers governs the dissipative state. Also in this case the temperature hysteresis, δT_H, was chosen to characterize the hysteretic behaviour of the dissipative phase-slip state. The reason is that the behaviour of this quantity in the case of a thermally dominated hysteresis is well known from our investigations of the heat transfer properties of our filaments. Therefore, deviations from that behaviour are expected to be caused by intrinsic, nonthermal effects.

Already in the experiments concerning the heat transfer properties of In-Pb whiskers we observed a 'footlike' structure in the $\delta T_H(I_F^2)$ plot at small currents and we suggested that this structure is caused by contributions of a nonthermal hysteresis [480]. Therefore, we performed detailed measurements of δT_H and plotted the quantity as a function of I_F^2. This can be done by a direct observation of $\delta T_H(I_F) = T_c(I_F) - T_R(I_F)$ in a V-T curve at fixed current I_F. Another possibility is to measure V-I characteristics at several fixed bath temperatures and to plot $I_c^{2/3}$ and $I_R^{2/3}$ as a function of the bath temperature (see for instance Fig. 21). Then the quantity $\delta T_H(I_F)$ is obtained as the distance in temperature between the $I_c^{2/3}$ and $I_R^{2/3}$ branch of such a plot at an arbitrary but constant value $I_F^{2/3}$. Naturally, also the directly measured $T_c(I_F)$ and $T_R(I_F)$ values join this plot. Thus, our usual procedure is to summarize both kinds of measurements in a common plot, then to draw

*1 We do not agree with the calculation of τ_{0R} as performed in ref. 369. Since the microbridges are not 'dirty', the expression for arbitrary ℓ as given in eq. (112) of the present work should be applied and not the dirty limit result.

Using the material parameters given in ref. 369 and v_F for pure In [40], ΔT_{Hy}^{KSS} for the microbridges investigated ranges between 28 and 104 mK for the choice $\tau_E = 80$ ps and between 20 and 75 mK for the choice $\tau_E = 110$ ps.

We also estimated τ_E after Tinkham using the measured critical temperatures of ref. 369 and the other parameters for pure In as given in ref. 40, resulting in $\tau_E \approx 2.5 \cdot 10^{-10}$ s for all microbridges. With $\xi_0 = 0.2944\,\mu$m and the measured ℓ and T_{c0} it is ΔT_{Hy}^{KSS} between 9 and 33 mK.

estimated slopes through the experimental points of the $I_c^{2/3}$ and $I_R^{2/3}$ branch, and then to determine $\delta T_H(I_F)$ and, thus, $\delta T_H(I_F^2)$ from such a plot. We will illustrate the method somewhat in detail in section 11.3.

Using this method we first studied the hysteretic behaviour of pure Sn whiskers [332]. Indeed a footlike structure appeared in the $\delta T_H(I_F^2)$ plot for small values of I_F^2. Since this structure is assumed to be generated by intrinsic hysteretic effects of the phase-slip process, we compared the experiments with the theoretical predictions of KSS [158] and the TDGL results of KR [316]. It turns out that the measurements are not described by one of the existing theories [332].

Therefore, we developed a phenomenological hysteresis model [332], considering self-heating effects of the phase-slip center, hysteresis effects generated by charge imbalance waves ('KSS mechanism'), and an overpopulation of the excitation spectrum by nonequilibrium quasiparticles ('T^* mechanism'). It turns out that none of the mechanisms alone is able to describe the observed hysteresis. A combination of the three mechanisms is needed for a successful explanation of the measured hysteresis effects in tin whiskers.

We subsequently investigated the intrinsic hysteresis of phase-slip centers in Zn whiskers [482]. Also in this case our phenomenological hysteresis model is able to describe the observed effects and to give us information how strong the different mechanisms contribute to the total hysteresis width.

As a next step we somewhat improved the phenomenological model and then again successfully applied the model to the materials listed above [483]. Moreover, we measured the hysteresis of whiskers from In, <u>In</u>-Pb, and Pb and again successfully applied the phenomenological model [483].

In this version, our phenomenological model still contained two fitting parameters. These parameters are needed for the calculation of the T^* contribution to the hysteresis. To get a phenomenological model without any fitting parameter, we developed a more detailed description of the generation and healing processes of nonequilibrium quasiparticles during the phase-slip cycle. The calculations result in an expression for the T^* contribution to the hysteresis in which one of the former fitting parameters is equal to unity, while the other parameter is not simply a constant, but a function of the voltage developed by the phase-slip center [483].

A comparison of the fitting parameter free version of the phenomenological hysteresis model with experimental hysteresis widths of Sn, In, <u>In</u>-Pb, and Zn whiskers results in a satisfactory agreement between measurement and model calculation [483]. This agreement indicates that our model seems to contain the basic mechanisms which generate the hysteresis of a phase-slip center in our quasi-one-dimensional filaments.

In the following sections the ideas of our systematic investigation of hysteresis effects will be discussed somewhat in detail. We shall start with a description of the heat transfer or Kapitza resistance problem in section 11.2. Then measurements of the hysteretic behaviour of a tin whisker will be shown and the evaluation procedure of the temperature hysteresis will be described (section 11.3). In section 11.4 our phenomenological hysteresis model is explained. We start with the version containing two fitting parameters which we compare with experiments, before the development of the model without fitting parameters will be described. Then, also this model will be compared with experiments. Finally, there is a separate section which we discuss the predictions of the TDGL theory in and compare the calculated hysteresis with our experiments.

11.2. Heat Transfer from a Metallic Filament into Helium – The Kapitza Resistance

While the current-induced breakdown of superconductivity in thin filaments very close to the critical temperature, T_{c0}, is characterized by a region of nonequilibrium superconductivity between the superconducting and the normal conducting state, the transition is governed by the formation of self-heating hot spots for temperatures far below T_{c0}. There are detailed theoretical and experimental investigations of the hot-spot phenomenon reported in the literature [260, 484 – 498]. In the hot-spot regime there is a large thermal hysteresis, δT_H, which may be measured directly by the observation of V–T transition curves or evaluated from the behaviour of the V–I characteristics as described in the preceding section.

The width of the hysteresis is determined by the heat transfer properties to the environment of the sample. In the case of whiskers, in principle, there are two mechanisms for heat removal, heat transfer through the surface, and thermal conductivity along the sample. However, the heat transfer through the surface dominates and the heat flow into the contacts can be neglected.

There are two reasons why the heat flow into the contacts is only small. One reason is that the hot spot may not extend over the whole length of the whisker. Since the thermal conductivity in the whisker is much smaller than in pure bulk material (due to the smaller value of the electron mean free path in a whisker)*¹, already a small distance between the hot spot and the contacts (which can be estimated from experiment) is sufficient to reduce the heat flow to the contacts to a negligible value compared to the surface heat transfer. See ref. 480 for a detailed discussion.

Another reason is that for sufficiently low temperatures there is nearly no heat flow due to thermal conductivity via electrons into the strongly superconducting contacts, even if the self-heating normal conducting region

*¹ Note that the main contribution to thermal conductivity at low temperatures comes from the electrons. The thermal conductivity of the phonons is very small.

extends to the contacts. There is a large boundary resistance for the heat transfer through a normal/superconducting boundary. The heat flux through such a boundary has been calculated by Andreev [499, 500]. The problem is discussed in detail in ref. 481.

In the case of Zn and Zn-Ag whiskers usually the whole whisker becomes normal conducting so that the self-heating region extends to the strongly superconducting Wood metal contacts. We evaluated Andreev's result for a Zn whisker, finding that the heat flux through both contacts is ten orders of magnitude smaller than the total energy dissipation [481].

For whiskers of pure In, the In-Pb alloy system and pure Pb (of refs. 480, 483, and section 3.3.1 of ref. 40) the heat flux into the superconducting contacts in nearly all cases would not be sufficiently depressed if the hot spot would extend to the contacts. The reason is that the measuring temperature is not low enough compared to the critical temperature of the contact material [*1]. However, in these samples the hot spot actually does not extend to the contacts. As explained in ref. 480, we estimated the distance between the borders of the hot spot and each contact block and calculated the ratio of the heat flow toward the contacts and the total heat flow. For all samples we find the ratio to be smaller than 0.01. Thus, less than 1% of the total heat flow occurs into the contacts.

For the pure Sn whiskers investigated in ref. 332 (contacted with Wood's metal) it is finally not clear which of the two mechanisms supresses the heat flow into the contacts more effectively. If the hot spot would extend to the contacts for all samples less than 8% of the total heat dissipated in the sample would flow into the contacts. Actually the hot spots do not extend to the contacts. Due to this effect for all samples less than 8% of the total heat flow should occur into the contacts.

Since in our samples the heat transfer through the surface dominates, the heat transfer coefficient and, thus, the Kapitza resistance [501 - 505] can be evaluated easily from measurements of the thermal hysteresis.

Imagine the measurement of a V-T characteristic at fixed current I_F. As already pointed out in the beginning of the preceding section, a temperature hysteresis, $\delta T_H(I_F) = T_c(I_F) - I_R(I_F)$, is observed, because for increasing temperature the sample enters a dissipative state at the transition temperature $T_c(I_F)$, while the specimen for decreasing temperature recovers the superconducting state at a bath temperature $T_R(I_F)$ which is lower than $T_c(I_F)$. If the hysteresis is simply caused by a warming-up of the whisker due to Joule heating, the re-entering of the superconducting state at T_R indicates that at the bath temperature T_R the sample temperature, T_W, is just equal to the transition temperature T_c. Thus, in this moment the difference

[*1] We used squeeze contacts made of In and Wood's metal for pure In whiskers and pure Pb whiskers, respectively. For whiskers of In-Pb alloys usually Pb contacts were used, except for samples In-Pb 26 of ref. 480 and In-Pb 3 of ref. 483 which were contacted with Wood's metal.

between the sample and the bath temperature, $T_w - T_R$, is equal to $T_c - T_R = \delta T_H$, so that the temperature of the whisker is related with the temperature hysteresis by $T_w = T_R + \delta T_H$.

We assume that the warming-up of the sample is caused by the formation of a self-heating hot spot. If the heat transfer through the surface of the sample into the ^{4}He dominates, the hot spot can be approximately regarded as a normal conducting region with sharp edges and a uniform temperature T_w over its whole length, L_{spot}. In this case [505, 506] $\dot{Q} = A_s[C(T_w)T_w^4 - C(T)T^4]$, where $\dot{Q}$ is the heat transmitted through the boundary surface area, A_s, C characterizes the heat transfer properties, and T is the helium bath temperature. This implies the usual simplification that the phonon transmission probability is independent of phonon frequency. If C is a slowly varying function of the temperature, this equation leads to

$$\dot{Q} = C A_s (T_w^4 - T^4) \tag{213}$$

If the temperature difference, δT_w, between sample and bath is very small, then $T_w^4 - T^4 \approx 4T^3 \delta T_w$. This leads to

$$\dot{Q} = \alpha_K A_s (T_w - T) \tag{214}$$

where $\alpha_K = 4CT^3$ is the heat transfer coefficient which is related to the Kapitza resistance, R_K, by $\alpha_K = 1/R_K$. Thus, equation (213) can be rewritten as

$$\dot{Q} = (4 R_K T^3)^{-1} A_s (T_w^4 - T^4) \tag{215}$$

If R_n is the residual resistance of the sample and L its total length, the dissipated energy is given by

$$\dot{Q} = (R_n L_{spot}/L) I_F^2 \tag{216}$$

Assuming that the whisker is a cylindrical filament with radius r_w, it is

$$A_s = 2\pi r_w L_{spot} \tag{217}$$

We thus get

$$(T_w^4 - T^4)\pi r_w L / I_F^2 \, 2 R_n = R_K T^3 \tag{218}$$

as our final result, valid for arbitrary temperature differences between whisker and helium bath.

In situations where the difference between sample and bath temperature is very small it is $(T_w^4 - T^4) \approx 4T^3(T_w - T)$. We furthermore introduce $R_n = \rho_n L/A$ and $\alpha_K = 1/R_K$, where $A = \pi r_w^2$ is the cross-sectional area of the whisker and ρ_n the residual resistivity. Then we get from eq. (218)

$$(2\alpha_K / r_w)(T_w - T) \approx \rho_n (I_F / A)^2 \tag{219}$$

This equation is only valid for small temperature differences between a whisker and the helium bath.

Our most systematic investigations of the heat transfer properties have been performed with Zn whiskers, including a Zn-Ag specimen, in a superfluid ^{4}He bath [481].

For this purpose we determined the left-hand side of eq. (218) at $T=T_R$ from the measurement of the hysteresis width of a specimen (using $T_W=T_R+\delta T_H$) and plotted the values as a function of the temperature. The results for sample Zn 18 are shown in Fig. 66. For temperatures between 0.52 and 0.70 K the quotient has the constant value $14.0\,cm^2K^4W^{-1}$. The left-hand side of eq. (218) is equal to $R_K T^3$. Thus, we have found that for sample Zn 18 in the mentioned temperature interval it is $R_K T^3 = 14.0\,cm^2K^4W^{-1}$, indicating that $R_K \sim T^{-3}$ in this temperature range.

Close to T_{c0} the hysteresis is generated by intrinsic non-thermal mechanisms of the nonequilibrium phase-slip state. In this temperature region the evaluation procedure cannot be applied. Moreover, we do not expect eq. (218) to be valid in the intermediate region in which no constant value of its left-hand side has been observed.

From our experiments the mean value $R_K T^3 = (15.2 \pm 3.1)\,cm^2K^4W^{-1}$ may be calculated, valid for pure zinc whiskers between about 0.5 and 0.7 K.

In ref. 481 we compared our results with theoretical expressions for R_K and measurements on copper at 1 K (in the literature we did not find a value of R_K for zinc). First, some remarks on the theoretical results: The

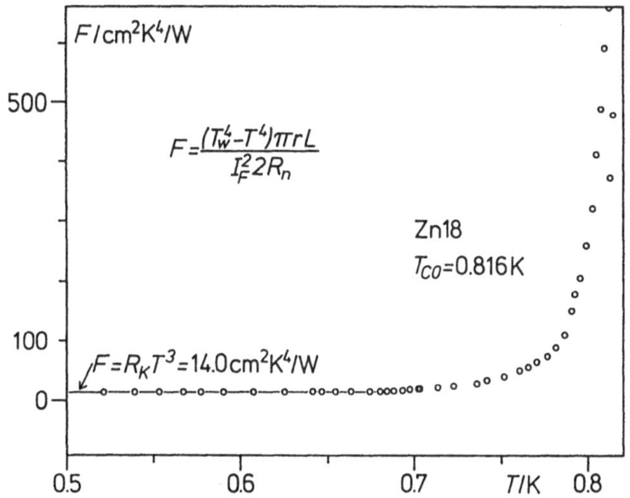

Fig. 66: *Plot of the left-hand side of eq.(218) versus the temperature T for a zinc whisker (sample Zn 18). The sample parameters are given by $R_n = 0.73\,\Omega$ for the residual resistance, $L = 225\,\mu m$ for the length, and $r_W = 0.29\,\mu m$ for the radius. The solid line represents $F = R_K T^3 = 14.0\,cm^2K^4W^{-1}$.*

Kapitza resistance is a thermal impedance that occurs at the boundary between two different materials, leading to a temperature discontinuity across the interface. The microscopic reason for the temperature discontinuity is that phonons on both sides of the interface are only partially transmitted through the boundary [501 - 505].

A theoretical expression for the Kapitza resistance between liquid helium and a solid has been calculated by Khalatnikov within the framework of an acoustic mismatch theory [501, 505]. The theory predicts that $R_K \sim T^{-3}$ as observed for our zinc whiskers. However, the absolute value measured for $R_K T^3$ in our experiments is a factor of 40 smaller than the theoretical prediction [481]. This is no accidental result. Except for very low temperatures ($T \lesssim 0.1\,K$) the experimental values are one or two orders of magnitude smaller than the result as calculated by the acoustic mismatch theory [501, 505]. The explanation is that the acoustic mismatch mechanism is shunted or bypassed by a more efficient heat transfer mechanism. Recent theories on this scope have been proposed which concentrate on excitations associated with the helium at the interface [505, 507, 508]. This additional mechanism becomes more and more important between 0.1 and 1 K, leading to a strong temperature dependence of $R_K T^3$ in this temperature range. Measurements performed with a copper/^{4}He boundary show the predicted behaviour [507, 508]. Our results for zinc between 0.5 and 0.7 K do not show any temperature dependence for $R_K T^3$. The reason may be a shift of the temperature dependent range to lower temperatures.

A lower limit for $R_K T^3$ can be calculated by assuming that there is no reflection of phonons at the interface. This assumption is called 'phonon radiation limit' [509, 510]. The resulting value of $R_K T^3$ for zinc is much (about a factor of 6) smaller than our experimental result [481].

Experimentally it is found that the dependence of R_K on the Debye temperature, Θ, shows the tendency $R_K \sim \Theta$ (see refs. 502 and 511). Since the Debye temperature of copper and zinc are nearly equal [512, 513], $R_K T^3$ of both materials should also be very similar. As no experimental value for R_K of zinc is known from the literature we compared our results with those of copper near 1 K, where $R_K T^3$ for copper is independent of temperature [507, 508, 514]. For copper it is $R_K T^3 \approx 10 - 20\,\mathrm{cm}^2\,K^4\,W^{-1}$, in agreement with the experimental result for zinc.

For whiskers of pure In, In-Pb alloys, pure Pb and pure Sn we do not have such detailed experimental results about the temperature dependence of the Kapitza resistance as for our zinc whiskers. Nevertheless, for a lot of samples a heat transfer coefficient $\alpha_K = 1/R_K$ has been evaluated from measurements of the thermal hysteresis. Results for α_K are given in refs. 480, 483, section 3.3.1 of ref. 40, and in ref. 332.

For these specimens we simply assumed that α_K is a fixed temperature independent quantity for a given sample at all measuring temperatures. This may be allowed, because the factor by which the temperature T_R is varied in these measurements is much smaller than in the case of zinc. For all samples (except the Sn whiskers) the ratio T_R/T_{c0} of the measurements involved in the evaluation of α_K does not fall below 0.98. For the Sn whiskers the minimum value of T_R/T_{c0} is 0.81, 0.93, and 0.91 for samples SnH1, SnH2, and SnH3, respectively.

The usual procedure is then to plot the hysteresis width $\delta T_H(I_F^2)$ and to look for a region where the plot is a straight line which extrapolates back to the origin. Using eq. (219) a fixed value for α_K for this range is then evaluated which is assumed to hold also for all temperatures closer to T_{c0}. The use of eq. (219) should be allowed, because $\delta T_W(T_R) = \delta T_H$ is much smaller than T_R for these specimens.

In this context we should remark that in our early determination of α_K for In-Pb and a pure Pb whisker (ref. 480 and section 3.3.1 of ref. 40) we did not trace the behaviour of $\delta T_H(I_F^2)$ to such high values of I_F^2 for which $\delta T_H(I_F^2)$ is a straight line extrapolating back to the origin. We found, however, that already for lower values of I_F^2 all samples investigated showed a range where $\delta T_H \sim I_F^2 - I_H^2$, where I_H^2 is the extrapolated zero hysteresis width intercept [480]. Then we calculated α_K using eq. (219) but replacing I_F^2 by $I_F^2 - I_H^2$. This implies the idea that in this region there is a certain contribution of the dissipated energy which does not lead to a thermal hysteresis. This procedure was the first empirical attempt to separate Joule heating induced hysteresis effects from intrinsic ones.

We compared our results for α_K with values of $1/R_K$ obtained by extrapolating literature data to 4 K of the Kapitza resistance for In, Pb, Pb with oxidized surface, and Cu as measured in superfluid helium [480]. This comparison shows that our results have a reasonable magnitude.

In Fig. 67 the results for the heat transfer coefficient α_K are summarized for whiskers of In, In-Pb, Pb, and Sn. For specimens of In, In-Pb, and Pb

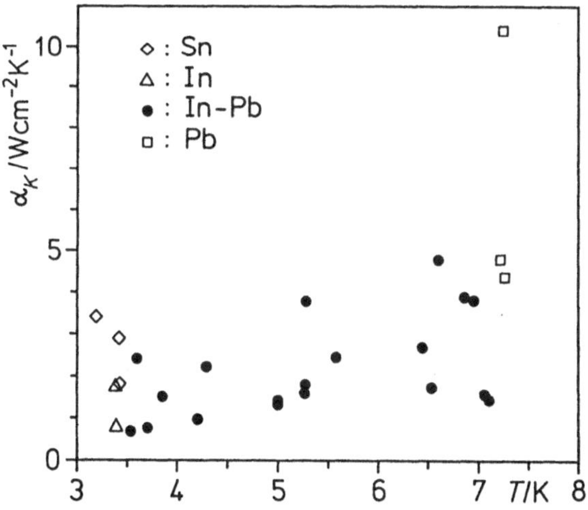

Fig. 67: *Heat transfer coefficients, $\alpha_K = 1/R_K$, as a function of the temperature for whiskers of Sn [332], In [483], In-Pb [40, 480, 483], and Pb [480, 483].*

the heat transfer coefficient has been determined for temperatures very close to T_{c0} (in all cases evaluations have been carried out up to $T_R/T_{c0} \gtrsim 0.99$). Therefore, we plotted α_K as a function of the critical temperature of the sample. In the case of Sn the heat transfer coefficient was calculated for $T_R/T_{c0}=0.809-0.869$, 0.930, 0.914 - 0.926 for samples SnH1, SnH2, and SnH3 of ref. 332, respectively. For these samples we plotted α_K at the temperature border closest to T_{c0}.

The results seem to scatter between 1 and 5 W/cm^{2}K. However, within the values for specimens of In, In-Pb, and Pb there seems to be an increasing tendency from roughly 1.25 W/cm^{2}K at $T \approx 3.4$ K to about 4.5 W/cm^{2}K at $T \approx 7.22$ K. This increase is much weaker than a T^3 law. A possible reason for these observations may be oxide at the surface of the whisker. A different oxidation of the specimens may lead to the scattering of the results. There are measurements reportet in the literature for Pb which indicate a reduction of α_K if the surface of a sample is oxidized [480]. Assuming that the disturbance of heat transfer through the surface of a whisker from the In-Pb system (including the pure materials) becomes stronger for increasing lead content would be a possible explanation for the slow increase of α_K with increasing temperature.

The investigation of the heat transfer properties of a whisker to the surrounding helium bath enables us to estimate the contribution of Joule heating to the hysteresis in the phase-slip regime. Without this information it would not be possible to separate thermal and intrinsic contributions to the hysteresis of a phase-slip center.

11.3. Hysteretic Behaviour of a Tin Whisker

In this section we give an example for the experimental investigations of the hysteretic behaviour of a whisker. For this purpose a pure tin whisker (sample SnH3 of ref. 332) has been chosen.

An overview of the development of $I_c^{2/3}$ and $I_R^{2/3}$ with decreasing temperature has already been given in section 6.1, where we discussed the behaviour of the critical current of our filaments (Fig. 21). All values result from V-I characteristics. In Fig. 68 we plot the vicinity of the critical temperature in more detail. Moreover, in this figure we indicate how the temperature hysteresis, δT_H, at arbitrary fixed current I_F is evaluated. As can be seen from the figure, δT_H depends on the magnitude of I_F and is given by

$$\delta T_H(I_F) = T_c(I_F) - T_R(I_F) \tag{220}$$

In Fig. 69 we plot δT_H as a function of I_F^2. As discussed in the preceding section, a simple heating model with a constant value of the heat transfer coefficient, α_K, would predict $\delta T_H \sim I_F^2$ and, therefore, deviations from simple

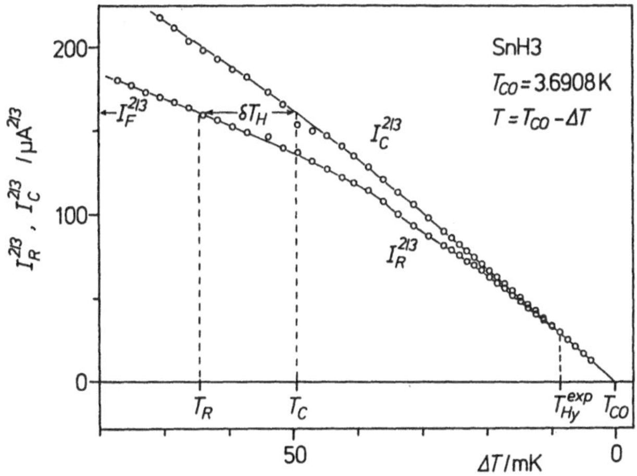

Fig. 68: *Critical current, $I_c^{2/3}$, and jump-back current, $I_R^{2/3}$, as a function of ΔT for a pure tin whisker (sample SnH3 of ref. 332). Large magnification close to T_{c0}*

The full lines indicate the estimated slope through the measured values. The horizontal line between the two branches represents the temperature hysteresis, δT_H, at arbitrary but fixed current I_F. The quantities $T_c(I_F)$ and $T_R(I_F)$ are the transition temperature and the jump-back temperature (into the superconducting state) for fixed current I_F, respectively. The temperature T_{Hy}^{exp} indicates the onset of hysteresis observed experimentally.

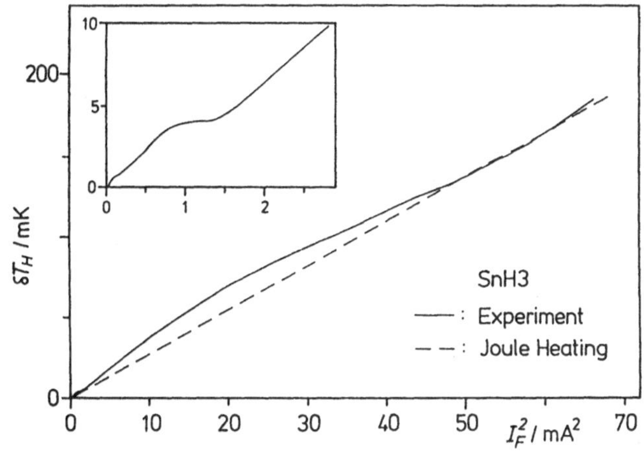

Fig. 69: *Width of the temperature hysteresis, δT_H, as a function of I_F^2 for a tin whisker (sample SnH3 of ref. 332). The inset shows the lower part in larger magnification in order to make conspicuous the footlike structure. The full line is the estimated slope through the measured values. The dashed line is the case of only Joule heating.*

heating should clearly be visible in such a plot. For sample SnH3 there seems to be the expected linear region for $I_F^2 > 50\,mA^2$. For smaller currents δT_H shows a more complicated behaviour. At very small currents the mentioned footlike structure can be seen.

It is this footlike structure which indicates intrinsic hysteresis. In the following section we will present a phenomenological model which is able to explain the hysteretic properties of a phase-slip center in the 'foot-structure region'.

11.4. Phenomenological Hysteresis Model

11.4.1. A Model with Two Fitting Parameters

In this subsection we will discuss the basic ideas which lead to a phenomenological hysteresis model with two fitting parameters. The discussion is based on the ideas given in ref. 332 and the improvements made on this subject in ref. 483. In subsection 11.4.3 we will present a model without fitting parameters. Naturally, the second case is of advantage. The parameter free model is, however, restricted to a smaller current range. The reason is that detailed information about the voltage developed by the phase-slip center is needed in this case. Thus, the 'two fitting parameters version' of the model is still an important tool to get information about the contribution of the different mechanisms which generate the hysteresis of a phase-slip center.

As already pointed out in section 11.1, our model considers three different mechanisms which lead to a hysteretic behaviour: the self heating of the phase-slip center, charge imbalance waves ('KSS mechanism') and an overpopulation of the excitation spectrum with nonequilibrium quasiparticles ('T^* mechanism').

We start with the estimation of the self-heating of a phase-slip center: The phase-slip state is a dissipative state. The energy dissipation in detail is expected to be a rather involved problem. We only consider a very rough model which gives an upper boundary for the effect. Much simpler than the treatment of Stuivinga et al. [515], we assume that a region of the length of the quasiparticle diffusion length Λ at both sides of the phase-slip center is equally warmed up. The entire superconductor remains at the bath temperature T.

Due to eq. (214) it is

$$\dot{Q} = \alpha_K S (T_L - T) \tag{221}$$

where α_K is the heat transfer coefficient, T_L the lattice temperature of the whisker, and S the surface of the warm region through which the dissipated energy is transmitted into the surrounding helium bath. Assuming again a circular-cross sectional area of the whisker with radius r_w, which is the smallest surface possible, it is

$$S = 4\pi r_w \Lambda \tag{222}$$

To calculate the energy dissipation $\dot{Q}$ in the phase-slip center, we remember that in time-average a part I_0 of the total current I is carried as a supercurrent. Therefore $\dot{Q} = V(I)(I - I_0)$, where V(I) is the time averaged voltage generated by the phase-slip center. As discussed in section 7.2 (see eqs. (187) and (188)) it is $V(I) = (2\rho_n\Lambda/A)(I - I_0)$. With $R_n = \rho_n L/A$ it follows

$$\dot{Q} = (2R_n\Lambda/L)(I - I_0)^2 \tag{223}$$

With $\delta T_L := T_L - T$ we, thus, get

$$\delta T_L = (I - I_0)^2 R_n / 2\pi r_w \alpha_K L \tag{224}$$

This expression for the enhancement δT_L of the lattice temperature above the bath temperature may overestimate the warming-up of the lattice, because charge imbalance relaxation does not necessarily include a complete energy relaxation of the excitations. The length Λ, however, is the relaxation length for the quasiparticle charge imbalance. It is remarked that for $I_0 = 0$ eq. (224) yields eq. (219), describing the temperature enhancement in a self-heating hot spot. To calculate δT_L in the phase-slip regime, where I_0 is nonzero, measured values of I_0 are used.

Naturally, the warming-up of the lattice leads to a hysteresis. For fixed current I_F the jump back into the superconducting state takes place at a bath temperature $T = T_R$ for which $T_R + \delta T_L(I_F) = T_c(I_F)$. Thus, $\delta T_L(I_F)$ is equal to the temperature hysteresis $\delta T_H(I_F)$ generated by a self-heating of the phase-slip center.

Considering this mechanism only, there usually is no agreement with the measured $\delta T_H(I_F^2)$. The temperature rise in the phase-slip regime, where the current is typically only some hundreds of μA, is usually much too small, especially in the case of whiskers of pure material.

Next we discuss the <u>charge imbalance wave induced hysteresis ('KSS mechanism')</u>: The KSS model predicts an intrinsic nonthermal hysteresis. The origin of this hysteresis is a reactive effect of the charge imbalance waves excited by the phase-slip center on the phase-slip process. We discussed the KSS model and its hysteretic behaviour in detail in section 5.8.4.

The onset of hysteresis is given by the criterion $\tau_{0R} = \tau_E$, where τ_E is the inelastic electron-phonon collision time and $\tau_{0R} = (\ell / 2 v_F \chi)(1 - T/T_{c0})^{-1}$ the supercurrent response time. (This expression for τ_{OR} is valid for the weak-coupling case, see eq. (112), but will be used for all materials to discuss their hysteretic behaviour.) The condition $\tau_{OR} = \tau_E$ yields the onset temperature T_{Hy}^{KSS} below which the KSS model predicts hysteretic behaviour.

Using the KSS model it is also possible to calculate the temperature hysteresis δT_H as a function of arbitrary fixed current I_F. For this purpose we first calculated the jump-back current, I_R, from the theoretical result for the V-I characteristics at given fixed bath temperature $T = T_R$. For this purpose we use the analytical approximation of the KSS model given in eq.(160) and insert the current from eq.(163), yielding

$$I = I_c e^x/(e^x - 1) + (V/R_{eff}) [1 - (\tau_E/\tau_{OR})^{1/2}] \tag{225}$$

where $x(V) = I_c R_{eff} (\tau_{OR}/\tau_E)^{1/2} V^{-1}$, with $R_{eff} = 2 \Lambda_{Q^* in}/\sigma A$.

This is an equation for the I-V characteristics at fixed temperature yielding reversal V-I characteristics for the hysteretic case where $\tau_{OR} < \tau_E$ (see Fig. 16). From the condition $dI/dV = 0$ we get the voltage V_R where the sample jumps back into the superconducting state. Calculating dI/dV from eq. (225) and setting the result equal to zero, yields the condition

$$x_R^2 e^{x_R}/(e^{x_R} - 1)^2 = 1 - (\tau_{OR}/\tau_E)^{1/2} \tag{226}$$

where $x_R := x(V_R)$. This problem has to be solved by iteration. Reinserting V_R into the I-V characteristic, eq. (225), we get the jump-back current I_R.

Using the results of the GL theory for a quasi-one-dimensional superconductor (see eqs. (15) and (17)) we then get the transition temperature corresponding to I_R as

$$T_c(I_R) = T_{c0} - I_R^{2/3} |d I_c^{2/3}/d T_c|^{-1} \tag{227}$$

The KSS prediction for the temperature hysteresis $\delta T_H(I_F)$ then follows from eq. (220) as

$$\delta T_{KSS}(I_F = I_R) = T_{c0} - I_R^{2/3} |d I_c^{2/3}/d T_c|^{-1} - T_R \tag{228}$$

For an illustration of the different quantities used in this equation see Fig. 68. Usually the KSS mechanism alone does not lead to a quantitative description of the hysteresis, because the KSS model predicts the onset of hysteresis for too low temperatures and the width of the hysteresis is too small. Moreover, the qualitative shape of $\delta T_{KSS}(I_F^2)$ does not show the characteristic footlike structure.

Finally, we explain our attempt to consider a quasiparticle overpopulation induced hysteresis ('T^*mechanism'): During the phase-slip cycle there is a periodic production of nonequilibrium quasiparticles. We assume that these quasiparticles create a charge balanced overpopulation of the excitation spectrum in the core of the phase-slip center so that they are characterized by an effective temperature $T^* > T$ in the sense of Parker's model, widely discussed in section 5.5. In this case the superconductor behaves as if it had the temperature T^*, so that its critical current is reduced to $I_c(T^*) < I_c(T)$. Hysteresis in the V-I characteristics at fixed bath temperature should be observed, because the current has to be lowered to $I_c(T^*)$ before the sample jumps back into the superconducting state.

As we have shown in section 5.5, close to the critical temperature T_{c0} the enhancement of T^* above the bath temperature, T, is approximately given by (see eq. (90))

$$\delta T^* = (2/3)\, T_{c0}\, (N/N_T - 1) \tag{229}$$

where $\delta T^* := T^* - T$, N the actual quasiparticle number per volume, and N_T the quasiparticle number per volume at the temperature T in thermal equilibrium.

The quasiparticle density in thermal equilibrium is given by

$$N_T = N_{max} - n_{se} \tag{230}$$

where n_{se} is the 'superelectron density', that means the number of electrons per volume bound to Cooper pairs, and N_{max} is the maximum quasiparticle density which is simply the total electron density, n, of electrons in the normal state. The ratio n_{se}/n is given in eq. (99). If we furthermore use the expression for the energy gap of a weak-coupling superconductor in the vicinity of the critical temperature (see eq. (69)) it follows from eq. (230)

$$N_T = N_{max}(1 - 2\chi(1 - T/T_{c0})) \tag{231}$$

where χ is given by eq.(8) of the present work.

Now we introduce the two fitting parameters mentioned in the heading of this section by assuming that in the hysteretic case it is

$$N/N_T - 1 = [(N_{max}/N_T)\, \varepsilon_H - 1]\, \beta_H \tag{232}$$

Before we somewhat discuss the meaning of the parameters ε_H and β_H and their determination, we give our final result for δT^* which is

$$\delta T^* = (2/3)\, T_{c0}\, [\varepsilon_H/(1 - 2\chi(1 - T/T_{c0})) - 1]\, \beta_H \tag{233}$$

For fixed current I_F, the jump-back into the superconducting state occurs as soon as the bath temperature is lowered to a value T_R for which $T_R + \delta T^*(T_R) = T_c(I_F)$. Thus, δT^* is equal to the width of the temperature hysteresis $\delta T_H(I_F)$ generated by the T^* mechanism.

The parameter ε_H describes the onset of the T^* hysteresis. It takes into consideration that a quasiparticle overpopulation induced hysteresis may only be observed below a certain temperature T_{Hy}^{op}. (Here 'op' stands for 'overpopulation'.) This temperature has to be calculated from the experimental data. At T_{Hy}^{op} one should have $N = N_T$, requiring $\varepsilon_H N_{max}/N_T = 1$, yielding

$$\varepsilon_H = 1 - 2\chi(1 - T_{Hy}^{op}/T_{c0}) \tag{234}$$

Naturally, this result also follows directly from eq. (233), because $\delta T^* = 0$ at T_{Hy}^{op}.

The existence of a temperature $T_{Hy}^{op} < T_{c0}$ seems to indicate that for temperatures closer to T_{c0} than T_{Hy}^{op} there is no quasiparticle overpopulation. More precisely one should argue, that the overpopulation is so small that no quasiparticle overpopulation induced hysteresis is detected in our experiments.

If ε_H would be equal to unity (that means $T_{Hy}^{op} = T_{c0}$), the parameter β_H would describe which part of the maximum quasiparticle overpopulation which can be generated by the phase-slip process in the core of the phase-slip center (that means which part of $N_{max} - N_T$) contributes to the establishment of the effective temperature T^*. More precisely, β_H gives us the part of $N_{max}\varepsilon_H - N_T$ contributing to T^*. A comparison of our hysteresis model with experimental data leads to values of ε_H which are very close to unity. Thus, the physical meaning of β_H is just that one obtained for the case that ε_H is equal to unity.

The parameter β_H can be evaluated from one point of the measured $\delta T_H(I_F^2)$ curve if ε_H is known.

To get a more detailed understanding of the parameters ε_H and β_H a more detailed description of the generation and healing processes of nonequilibrium quasiparticles during the phase-slip cycle is required. This problem will be dealt with in section 11.4.3.

In the present representation ε_H and β_H are fitting parameters which have to be determined from experimental data. If the hysteresis would only be generated by the T^* mechanism, ε_H could be easily obtained, because then T_{Hy}^{op} would be equal to T_{Hy}^{exp}. Moreover, β_H then could easily be calculated by setting $\delta T^*(T_R)$ equal to a measured value of the $\delta T_H(I_F^2)$ curve.

However, it turns out that usually the T^*mechanism alone is not sufficient to describe the measured slope of $\delta T_H(I_F^2)$.

Our phenomenological model of hysteresis (with two fitting parameters) considers all of the three mechanisms discussed above. The resulting $\delta T_H(I_F^2)$ curves are calculated by a computer using the following procedure:

For fixed bath temperature in the hysteretic range we start our calculations at a total current $I = I_c(T)$. Due to the self-heating of the phase-slip center the lattice temperature, T_L, is enhanced above the bath temperature by an amount of $\delta T_L(I)$. Now we calculate the temperature T^* of the quasiparticle system which is enhanced over T_L by $\delta T^*(T_L)$ caused by the quasiparticle overpopulation. Using the GL theory we calculate $I_c(T^*)$. Now we reduce the current I and repeat this calculation until $I = I_c(T^*)$. Then we check whether the KSS model predicts a hysteresis. If not so, the sample re-enters the superconducting state and the width of the hysteresis $\delta T_H(I_F^2 = I^2)$ is given by the sum of the actual values δT_L and δT^*.

In the other case where the KSS model predicts hysteresis we calculate the jump-back current in the KSS model. - [A problem is, at which temperature the different quantities of the KSS model have to be calculated. Our actual opinion is, due to the discussion given in section 11.4.3, that the temperature T^* is not generated throughout the phase-slip center, but is restricted to its core. Thus, T^* should only enter the expression for the critical current of the core. The other quantities of the model should be taken at the lattice temperature $T_L = T + \delta T_L$. These quantities are τ_{0R} and Λ_{Q^*in}. The reason is that in the transmission line picture of the KSS model these quantities determine the impedance of outside the core.] - Naturally, the jump-back current is smaller than I. Therefore, a further step-by-step reduction of the current is necessary. This reduction changes T_L and T^* and the whole procedure has to be repeated from the beginning until the jump-back current of the KSS model is equal to the actual current I. The total hysteresis $\delta T_H(I_F^2 = I^2)$ is then given by the sum of the actual values δT_L, δT^*, and δT_{KSS}.

For the determination of δT^* we need values for the parameters ε_H and β_H from the experiment. In the general case the fitting procedure has to be performed with respect to the hysteresis caused by the KSS mechanism and the self-heating of the phase-slip center. The procedure is somewhat involved [483]:

In the case of ε_H, first the current I_{F1} where $\delta T^* = 0$ has to be estimated from the shape of the measured $\delta T_H(I_F^2)$ curve. Knowing this current, the related jump-back temperature $T_R(I_{F1})$ can be determined. It can be obtained by a horizontal cut at $I_{F1}^{2/3}$ through an $I_c^{2/3}$, $I_R^{2/3}(\Delta T)$ plot. - [Another possibility is to calculate $T_R(I_{F1})$, using $\delta T_{H1} := \delta T_H(I_{F1}) = T_c(I_{F1}) - T_R(I_{F1})$ and $I_{F1}^{2/3} = |d I_c^{2/3}/dT_c| (T_{c0} - T_c(I_{F1}))$, where $dI_c^{2/3}/dT_c$ is the measured slope of the $I_c^{2/3}(T_c)$ straight line.] - Then, $\delta T_L(I_{F1})$ can be calculated using I_0 as measured for T_R. The onset temperature of the T^* hysteresis then is $T_{Hy}^{op} = T_R(I_{F1}) + \delta T_L(I_{F1})$ and then ε_H is calculated from eq. (234).

In the case of β_H, a point $(I_{F2}^2, \delta T_{H2})$ of the measured $\delta T_H(I_F^2)$ curve is chosen, with I_{F2}^2 somewhat larger than I_{F1}^2. In general, at this point it is $\delta T_{H2} = \delta T_L(I_{F2}) + \delta T^* + \delta T_{KSS}$. The aim is to separate δT^* so that β_H can be calculated from eq. (233) with the known value of ε_H. The first step is to calculate $T_R(I_{F2})$ and $\delta T_L(I_{F2})$ as described in the case of the ε_H fit. The remaining problem is to separate the known width $\delta T_{H2}' = \delta T_{H2} - \delta T_L(I_{F2})$ into the contributions δT^* and δT_{KSS}, where δT_{KSS} depends on δT^*. This problem can be solved by an iteration procedure. As a starting temperature we take the lattice temperature $T_R(I_{F2}) + \delta T_L(I_{F2})$ and calculate the contribution δT_{KSS} for this temperature. Subtracting this result from $\delta T_{H2}'$ gives a first approximation for δT^*. Now the actual temperature is set to $T_R(I_{F2}) + \delta T_L(I_{F2}) + \delta T^*$ and δT_{KSS} is calculated again leading to a better estimate for δT^*. The procedure is repeated until the accuracy of δT^* is better than 0.05 mK. The result for δT^* is inserted into eq. (233) together with the lattice temperature $T_R(I_{F2}) + \delta T_L(I_{F2})$ and the value for ε_H, to obtain the parameter β_H.

In the following section the predictions of our phenomenological hysteresis model with two fitting parameters will be compared with experimental data of different materials. It turns out that the model is able to give a quantitative description of the measured hysteresis including the limits in which one of the mechanisms dominates.

11.4.2. Comparison with Experiment

In this subsection we compare the predictions of the phenomenological hysteresis model with two fitting parameters (which will be briefly called 'fitted version') with measurements of the temperature hysteresis of whiskers from Sn, In, In-Pb alloys, Pb, and Zn. The calculations are performed with the most recent version of the model as described in the preceding subsection. It contains all improvements introduced in ref. 483. Thus, the results given in the present work may be somewhat different from those published priviously [332, 482]. The characteristic properties (including the I_0/I_c profiles, as a function of the temperature, and the results for the parameters ε_H and β_H) of all samples mentioned are summarized in ref. 483.

In Fig. 70 results for the pure tin whisker of section 11.3 are shown *1. Now, circles mark the experimental values, while the full and dashed lines are the results of the phenomenological model and its single contributions, respectively. The footlike structure as well as the subsequent increase are predicted by the model. Over the whole range the quantitative agreement is satisfactory. Close to T_{c0} (low currents) the hysteresis is governed by quasiparticle overpopulation, while Joule heating becomes important for large currents. The hysteresis caused by the KSS mechanism is of minor importance.

In Figs. 71 - 74 we present experiments and calculations for pure In, In-Pb alloys, and pure Pb. For pure In (Fig. 71), $\delta T_H(I_F^2)$ behaves very similar to the case of pure tin. Again the experimental data show a pronounced footlike structure which is well described by the fitted version of the phenomenological model. In the low current region the largest contribution is due to the T^* mechanism, while for larger currents heating effects become more and more important. Again the KSS contribution is only small throughout the investigated current range.

---

*1 The sample is called 'SnH3' in ref. 332 and in the present work, while it is called 'Sn1' in ref. 483.

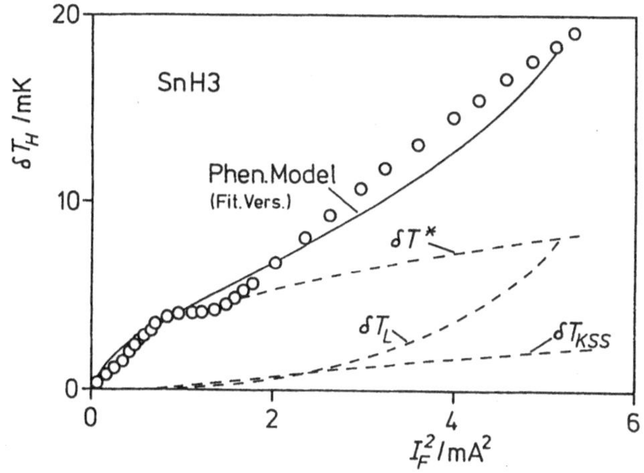

Fig. 70: *Width of the temperature hysteresis, δT_H , as a function of I_F^2 for a tin whisker (sample SnH3). The circles represent measured values while the full line is calculated from the fitted version ('Fit. Vers.') of the phenomenological hysteresis model, using ε_H and β_H as fitting parameters. The dashed lines are the single contributions from the mechanisms of quasiparticle overpopulation (δT^*), self-heating of the phase-slip center (δT_L), and charge imbalance waves from the KSS model (δT_{KSS}). The solid line is given by the sum $\delta T_L + \delta T^* + \delta T_{KSS}$. For this sample it is $\varepsilon_H = 0.9956$ and $\beta_H = 0.1199$.*

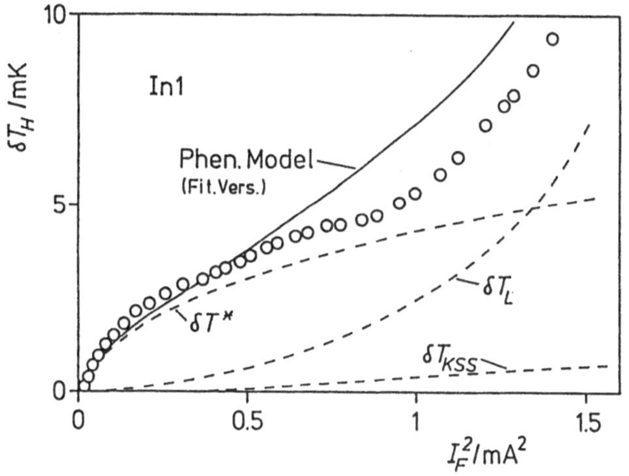

Fig. 71: *Measured width of the temperature hysteresis, δT_H, as a function of I_F^2 for an indium whisker (sample In 1, $T_{c0} = 3.3938\,K$) in comparison with the fitted version of the phenomenological hysteresis model. Same notation as used in Fig. 70. For this sample it is $\varepsilon_H = 0.9970$ and $\beta_H = 0.1645$.*

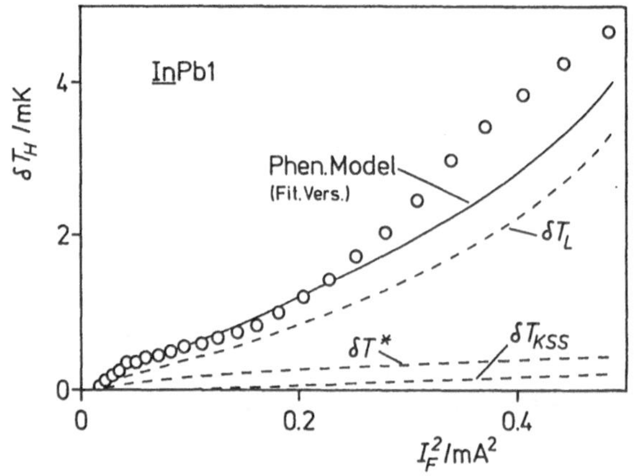

Fig. 72: *Measured width of the temperature hysteresis, δT_H, as a function of I_F^2 for an indium-lead alloy whisker (sample In Pb 1 with $c_{Pb} = 2.8$ at%, $T_{c0} = 3.5311$ K) in comparison with the fitted version of the phenomenological model. Same notation as used in Fig. 70. For this sample it is $\varepsilon_H = 0.9997$ and $\beta_H = 0.2218$.*

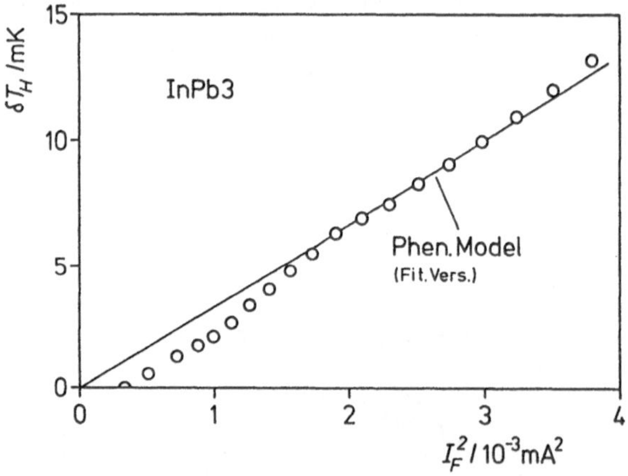

Fig. 73: *Measured width (circles) of the temperature hysteresis, δT_H, as a function of I_F^2 for an indium-lead alloy whisker (sample In Pb 3 with $c_{Pb} = 78.5$ at%, $T_{c0} = 6.8693$ K) in comparison with the fitted version of the phenomenological model (full line). In this case Joule heating is the only contribution to the hysteresis.*

By alloying Pb to In, the footlike structure becomes smaller with increasing Pb content (that means decreasing electron mean free path ℓ) and the influence of the self-heating grows. Sample In Pb 1 only shows a small δT^* and heating is the largest contribution over the whole measured range (Fig. 72). The limiting case is represented by sample In Pb 3, where only Joule heating occurs (Fig. 73). In this sample no voltage steps have been

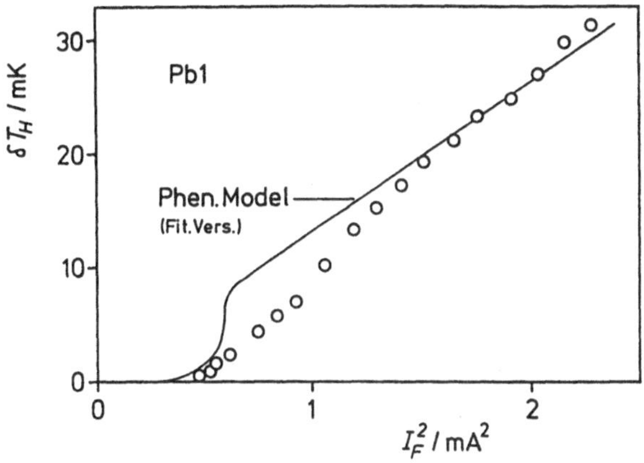

Fig. 74: *Measured width (circles) of the temperature hysteresis, δT_H, as a function of I_F^2 for a lead whisker (sample Pb1) in comparison with the fitted version of the phenomenological model (full line). In this case Joule heating is the only contribution to the hysteresis.*

observed and the $\delta T_H(I_F^2)$ plot does not show any footlike structure. The prediction of the phenomenological model is a straight line through the origin because we set $I_0=0$.

In Fig. 74 we show results obtained for a pure Pb whisker (sample Pb1). Since the V-I characteristics of sample Pb1 show linear portions close to T_{c0} without any voltage jumps and without any hysteresis, $T_H(I_F^2)$ is equal to zero in the low current range. For lower temperatures a voltage step and hysteresis occur. However, no intrinsic hysteresis (absent footlike structure) is seen and the whole width δT_H is well described by heating.

Finally, results for pure zinc whiskers are shown in Figs. 75 and 76. The examples given are representative for the hysteretic behaviour of long (sample Zn 25, Fig. 75) and short (sample Zn 18, Fig. 76) samples. In both cases a footlike structure indicates an intrinsic hysteresis.

For the long sample, Zn 25, the KSS mechanism explains the hysteresis for small currents completely. For higher currents the quasiparticle overpopulation induced hysteresis becomes important. The warming up of the lattice is very small for all currents.

For the short sample, Zn18, the hysteresis width is dominated by the T^*mechanism. The absolute value of $\delta T_H(I_F^2)$ is more than an order of magnitude larger compared to sample Zn 25. The reason for such large δT^* is probably the Andreev reflection of the quasiparticle excitations at the strongly superconducting Wood metal contacts (compare the discussion given in section 8.3). Thus, the quasiparticle excitations are trapped within the whisker.

The various materials investigated enable us to test the validity of our hysteresis model in the limits where only one mechanism governs the

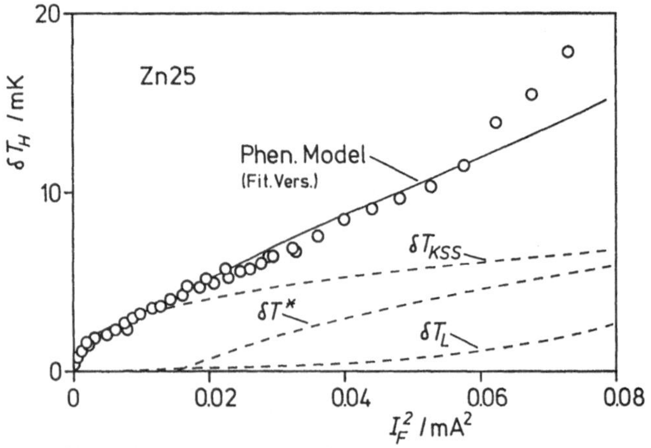

Fig. 75: *Measured width of the temperature hysteresis, δT_H, as a function of I_F^2 for a (long) zinc whisker (sample Zn 25, $L = 1500\,\mu m$, $T_{c0} = 0.8102\,K$) in comparison with the fitted version of the phenomenological model. Same notation as used in Fig. 70. For this sample it is $\varepsilon_H = 0.9764$ and $\beta_H = 0.3512$.*

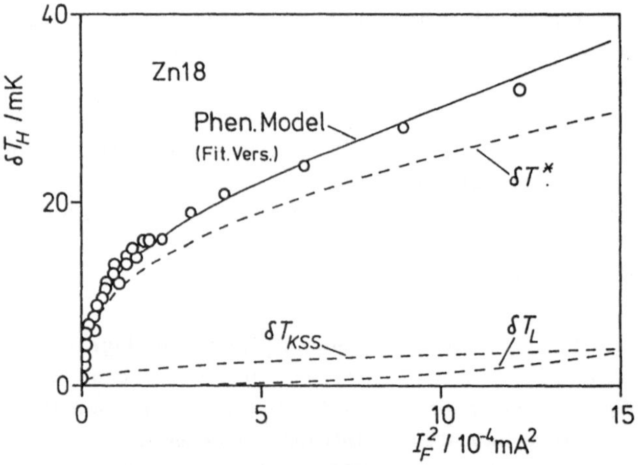

Fig. 76: *Measured width of the temperature hysteresis, δT_H, as a function of I_F^2 for a (short) zinc whisker (sample Zn 18, $L = 225\,\mu m$, $T_{c0} = 0.8161\,K$) in comparison with the fitted version of the phenomenological model. Same notation as used in Fig. 70. For this sample it is $\varepsilon_H = 1$ and $\beta_H = 0.8038$.*

hysteresis: The limit of quasiparticle overpopulation (T^*mechanism) induced hysteresis occurs for whiskers from pure Sn or In in the low current range. Charge imbalance wave (KSS mechanism) induced hysteresis generates the hysteresis in the low current range of long whiskers from pure Zn. The only heating case is observed in dirty In-Pb whiskers and specimens of pure Pb. In all cases the model calculation leads to a satisfying description of the experimental results.

Joule heating effects occur in all samples. While these effects are negligibly small in clean samples (for not too high currents), they become the largest contribution to the hysteresis in dirty whiskers. From our phenomenological model it can be seen that the thermal hysteresis does not only mask the intrinsic hysteresis but also suppresses its magnitude. The reason is that the hysteresis caused by intrinsic mechanisms increases with decreasing temperature. Joule heating, however, enhances the lattice temperature. For Pb whiskers we measured $I_0/I_c=1$ close to T_{c0}, leading to $\delta T_L=0$ at $I=I_c$, so that no heating induced hysteresis can occur, in agreement with the experiment.

Since the measured hysteresis is well described by our phenomenological hysteresis model, we believe that the model contains the essential mechanisms which generate the hysteretic behaviour of a phase-slip center in a quasi-one-dimensional superconductor. Naturally, one would desire to have a model without fitting parameters. An attempt to come to a fitting-parameter-free description is given in the following subsection.

11.4.3. A Model without Fitting Parameters

For the development of a hysteresis model without fitting parameters a more detailed description of the generation and healing processes of quasiparticles during the phase-slip cycle is needed. The description must be explicit enough to derive an expression for the T^*contribution to the hysteresis which does not contain any fitting parameter. The following attempt (elaborated in ref. 483) will yield such a result. A comparison with the former expression for δT^* leads to analytical results for ε_H and β_H.

A phase-slip center has a 'core region' of some GL coherence lengths, $\xi(T)$, where the relaxation oscillations of the order parameter lead to strong variations of the Cooper-pair density. In this region the nonequilibrium quasiparticles are produced during the phase-slip cycle. Then these quasiparticles diffuse away from the core region into the surrounding superconductor, while they relax to reach an equilibrium occupation of the excitation spectrum. Depending on the polarity of the transport current (that means of the applied battery) more hole-like quasiparticles and more electron-like quasiparticles diffuse into opposite directions. The consequence is that a charge imbalance is created at both sides of the center ('core') of the core region. In the core itself, the paths of both kinds of quasiparticles cross, compensating their charge so that there is no charge imbalance but a charge balanced quasiparticle overpopulation which will be described by an effective temperature T^* in the sense of Parker's model (discussed in detail in section 5.5).

It is assumed that the temperature T^* in the core determines whether the superconductor recovers the superconducting state or remains in the dissipative phase-slip state. The jump back into the superconducting state cannot occur as long as $I>I_c(T^*)$, where I is the total current.

The overpopulation of the excitation spectrum in the core (and therefore in principle also T^*) is time dependent, because it depends on the generation, the migration and the relaxation of nonequilibrium quasiparticles. For a rough estimate we take the spatial quasiparticle distribution in the moment of the phase-slip event and trace its time evolution without considering the further development of the order parameter. Thus, we assume that the quasiparticles are generated instantaneously at the beginning of the phase-slip cycle. Furthermore we assume that the charge is balanced in this moment at each point.

In the moment of the phase-slip event the absolute value $|\psi|$ of the superconducting order parameter is zero in the core (located at $x=0$) and increases exponentially with the characteristic length $\xi(T)$ on both sides of the core to its equilibrium value $|\psi_T|$ far away from the phase-slip center. Thus,

$$|\psi(x)| = |\psi_T|(1-\exp(-|x|/\xi(T)) \tag{235}$$

where $\xi(T)$ is given by eq. (6). This shape of $|\psi(x)|$ is supported by results of the TDGL theory (see Fig. 5 of ref. 305).

Identifying the square of the absolute value of the order parameter with the density of Cooper pairs leads to an expression for the local value of the quasiparticle overpopulation in the moment of the phase-slip event ($t=0$), namely

$$N_{op}(x,t=0) = (N_{max}-N_T)(1-(|\psi(x)|/|\psi_T|)^2) \tag{236}$$

yielding with $|\psi(x)|$ from eq. (235)

$$N_{op}(x,t=0) = (N_{max}-N_T)(2e^{-|x|/\xi} - e^{-2|x|/\xi}) \tag{237}$$

For $t=0$ both branches of the excitation spectrum are equally populated so that there is neither a branch imbalance nor a charge imbalance. For $t>0$ everywhere in the core region the quasiparticles are divided into two halves, diffusing into opposite directions depending on the effective charge of the quasiparticles. At the core both halves cross and we get

$$N_{op}(x=0,t) = (N_{max}-N_T)(2e^{-\Lambda(t)/\xi} - e^{-2\Lambda(t)/\xi}) \tag{238}$$

where $\Lambda(t)=((1/3)\ell v_F t)^{1/2}$ is the distance over which a quasiparticle excitation travels along the x-axis in the time, t, by random walk.

Until now we have neglected that N_{op} also relaxes by a recombination of the quasiparticles. However, actually on their way to the core the number of excess quasiparticles relaxes by recombination. In the sense of a relaxation time approximation we assume that there is an exponential decay of the number of travelling quasiparticles so that the quasiparticle overpopulation in the core is given by

$$N_{op}(x=0,t) = (N_{max}-N_T)(2e^{-\Lambda(t)/\xi} - e^{-2\Lambda(t)/\xi})e^{-t/\tau_{eff}} \qquad (239)$$

The effective quasiparticle recombination time, τ_{eff}, is the product of the intrinsic recombination time, τ_r', and the phonon trapping factor which takes into account that phonons created by recombination processes can break a pair again and, therefore, delay the relaxation of N_{op} (see for instance refs. 103, 104, 110, 131, and 132). Kaplan et al. [103] have calculated the time τ_r between two recombination events in the dirty limit. It is $\tau_r' = \tau_r/2$, because τ_r' characterizes the decay of a quasiparticle population and two quasiparticles of this population vanish in each recombination event. The time τ_r depends on the temperature and on the quasiparticle energy. For T approaching T_{c0} the limiting value for τ_r of a quasiparticle at the gap edge is given by $\tau_r = \tau_0/4.2$. The scaling time τ_0 is related to τ_E by $\tau_0 = 8.4\,\tau_E$ [80] leading to $\tau_r = 2\,\tau_E$ or $\tau_r' = \tau_E$.

The use of this limiting value should be a good approximation in our experimental situation some millikelvin below T_{c0}. Close to T_{c0} the temperature dependence of τ_r is very weak. Moreover, it is reasonable to assume that the quasiparticles created by the phase-slip process have energies not very far away from the gap edge, much smaller than the zero energy gap $\Delta(0)$. Even if a quasiparticle would have an energy $\Delta(0)$, τ_r would only be about a factor of two smaller [103].

The problem of phonon trapping has been investigated in detail for thin films [104, 110, 131, 132, 516, 517]. In our model phonon-trapping effects are neglected so that $\tau_{eff} = \tau_r'$ leading to $\tau_{eff} = \tau_E$.

To relate the time dependent quasiparticle overpopulation in the core with an effective temperature T^*, we average $N_{op}(x=0,t)$ over one period, τ_{PSC}, of the phase-slip cycle. The time τ_{PSC} is related to the time-averaged voltage V at the contacts of the sample by the Josephson relation so that $\tau_{PSC} = h/2eV$.

Furthermore, using $\delta T^* = (2/3)T_{c0}(N/N_T - 1)$ from eq.(229), where N is the actual total quasiparticle density, $N = N_{op} + N_T$, leads to

$$\delta T^* = \frac{2}{3} T_{c0} \frac{N_{max}-N_T}{N_T} \frac{1}{\tau_{PSC}} \int_0^{\tau_{PSC}} (2e^{-\Lambda(t)/\xi(T)} - e^{-2\Lambda(t)/\xi(T)})e^{-t/\tau_E}\,dt \qquad (240)$$

Due to eq. (231) it is $(N_{max}-N_T)/N_T = 1/[1-2\chi(1-T/T_{c0})]-1$. Thus, comparing the result for δT^* given in eq.(240) with that one of eq. (233) yields

$$\varepsilon_H = 1 \qquad (241)$$

and

$$\beta_H = \frac{1}{\tau_{PSC}} \int_0^{\tau_{PSC}} (2e^{-\Lambda(t)/\xi(T)} - e^{-2\Lambda(t)/\xi(T)})e^{-t/\tau_E}\,dt \qquad (242)$$

Thus, ε_H is unity and β_H becomes a function of the voltage and the temperature.

In the presence of Joule heating, T has to be replaced by the lattice temperature, T_L, which is enhanced over the bath temperature, T, by an amount δT_L. Since $\varepsilon_H=1$, the onset of the T^* hysteresis (that means the exceeding of a certain threshold so that the effect can be detected in our experiments) and the further development of its magnitude are determined by β_H which relates the effective temperature T^* with the migration and recombination behaviour of the quasiparticles in a phase-slip center.

For a comparison with the experiment, the same computing method as before is used (see section 11.4.1), but ε_H is set to unity and β_H from eq. (242) is inserted into eq. (233). To calculate the contribution of Joule heating we again insert I_0 from the experiment. Moreover, the voltage $V(I)=(dV/dI)_1(I-I_0)$ is also taken from the measurement. Here $(dV/dI)_1$ is the differential resistance of the V-I characteristics after the first voltage jump. With this voltage the time τ_{PSC} is known, so that the integral in β_H can be evaluated by numerical methods.

Although the experimental shape of V(I) is used, the decision for the jump back into the superconducting state is given by the model calculation. In the calculations made for our samples the jump-back occurs between I_c and I_0. Therefore, there is no problem with the point $I=I_0$, where δT_L becomes zero due to the influence of an experimental parameter and δT^* also vanishes, because $V(I)=0$, leading to $\tau_{PSC}=\infty$.

In the following subsection we will compare the predictions of our phenomenological model in the 'analytical version' (i. e. with $\varepsilon_H=1$ and the 'analytical' expression for β_H from eq. (242)) with experimental data. A satisfactory agreement will be obtained which is remarkable, because in the analytical version the model does not contain any fitting parameter.

11.4.4. Comparison with Experiment

Using the hysteresis model without fitting parameters, we have performed calculations for all materials investigated which show a hysteresis not only caused by Joule heating [483].

For all samples (except In 2, the characteristic properties of this sample are given in ref. 483), 'fitted' model calculations, using the fitting parameters ε_H and β_H are also shown in the present work (see section 11.2).

The 'analytical' computations, using $\varepsilon_H=1$ and the 'analytical' expression for β_H from eq. (242) are restricted to a smaller range of I_F^2 than the 'fitted' computations, because reliable results for the differential resistance are only available in a restricted range of temperatures.

Throughout there is a satisfactory agreement between the experiment and the analytical computations (Figs. 77 - 80). This is remarkable, because no fitting parameters are used.

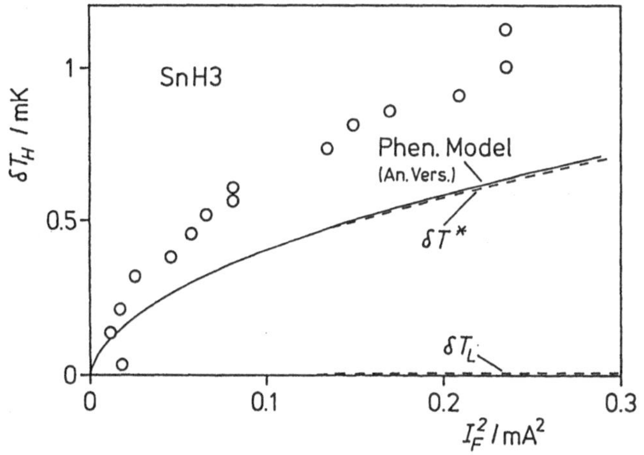

Fig. 77: *Measured width of the temperature hysteresis, δT_H, as a function of I_F^2 for a tin whisker (sample SnH3) in comparison with the 'analytical' version ('An. Vers.') of the phenomenological hysteresis model, using $\varepsilon_H = 1$ and the analytical expression for β_H. The notation used is the same as introduced in Fig. 70.*

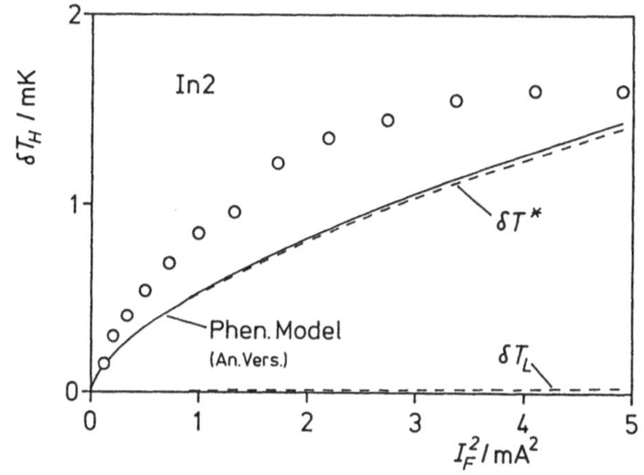

Fig. 78: *Measured width of the temperature hysteresis, δT_H, as a function of I_F^2 for an indium whisker (sample In 2, $T_{c0} = 3.3938\,K$) in comparison with the 'analytical' version ('An. Vers.') of the phenomenological model. The notation used is the same as introduced in Fig. 70.*

For the pure whiskers SnH 3, In 2, and Zn 18 the calculated hysteresis is somewhat smaller than the measured one, while there is an exact agreement in the case of the alloy whisker In Pb 1.

The 'analytical' computation for the long Zn whisker (sample Zn 25) is not shown. There is an agreement between the calculated and the measured

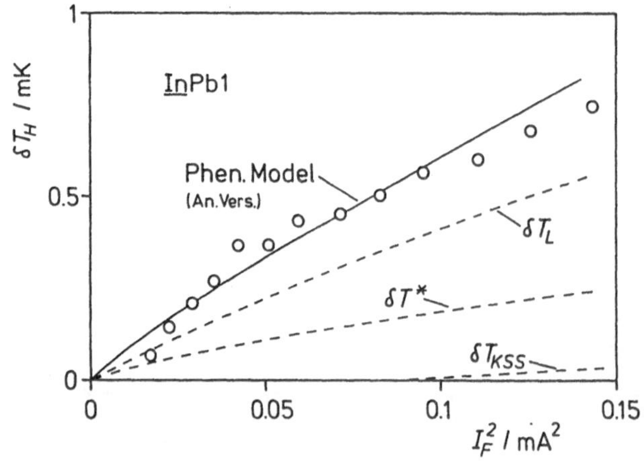

Fig. 79: *Measured width of the temperature hysteresis, δT_H, as a function of I_F^2 for an indium-lead alloy whisker (sample InPb1 with c_{Pb}=2.8 at%) in comparison with the 'analytical' version ('An. Vers.') of the phenomenological model. The notation used is the same as introduced in Fig. 70.*

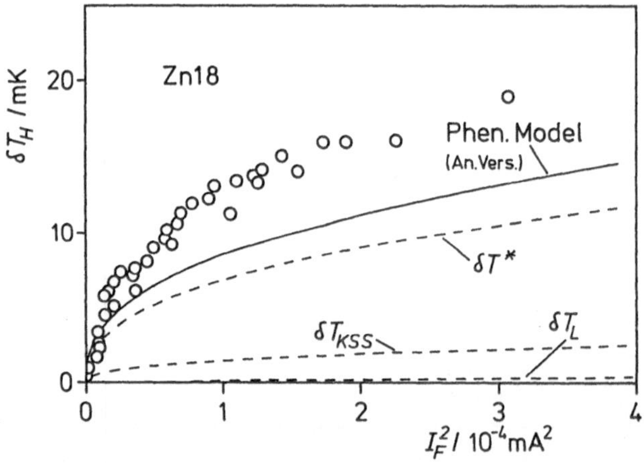

Fig. 80: *Measured width of the temperature hysteresis, δT_H, as a function of I_F^2 for a (short) zinc whisker (sample Zn 18) in comparison with the 'analytical' version ('An. Vers.') of the phenomenological model. The notation used is the same as introduced in Fig. 70.*

hysteresis width. In the calculated range of small currents I_F the hysteresis is given by the KSS mechanism as in the 'fitted' model calculation. There is a vanishing small contribution of δT^*. The reason is that I becomes close to I_0, leading to a very small V(I), so that δT^* becomes small due to the large value of τ_{PSC}. This result shows that the relation between the voltage and δT^* is reasonable. However, in this case the limit of the experimental V-I characteristics strongly influences the result.

For sample Zn 18 we can explain why the calculated hysteresis width is smaller than the measured one. The reason is that this sample is very short so that nonequilibrium quasiparticles are Andreev-reflected at the strongly superconducting Wood metal contacts and return to the core (compare the discussion given in section 11.4.2). As the reflection occurs at both contacts, reflected quasiparticles of both charges cross at the core and create an additional contribution to δT^* which has not been included in the model but is present in the experiment.

11.5. The Hysteresis in the TDGL Theory

The TDGL theory predicts an intrinsic hysteresis for the phase-slip center, because for fixed temperature oscillatory phase-slip solutions exist between a minimal and a maximal current density, called j_{min} and j_{max} (see Fig. 19 in section 5.9) *1. The dissipative state is entered at the critical current density j_c of the time independent GL theory.

However, for an isolated phase-slip center in a homogeneous filament it is $j_{max} < j_c$ as long as $\gamma < \gamma_c \approx 5.5$, while for $\gamma > \gamma_c$ it is $j_{max} = j_c$ (where γ is the pair-breaking parameter of the TDGL theory). For the first voltage step of our V-I characteristics, which may be regarded as being generated by an isolated phase-slip center, we, therefore, should observe no hysteresis below γ_c. Above γ_c the hysteresis should be given by $j_c - j_{min}$.

Kramer and Rangel (KR) in Tab. 1 of ref. 316 give values of j_{min} for a periodic array of phase-slip centers. As KR point out that j_{min} is rather insensitive to the spatial period of the phase-slip centers, we may use these values for a comparison with our experiments.

If we make a log-log plot of j_{min} against γ, we obtain for the region $10 < \gamma < 80$ an approximately straight line, leading to (in physical units *2)

$$j_{min} = (j_c / 0.385) \cdot 0.48\, \gamma^{-0.265} \tag{243}$$

Since $j_c^{2/3} \sim \Delta T$ and $\gamma \sim \Delta T^{1/2}$, we thus find out, that $j_{min}^{2/3}$ is predicted to be proportional to $\Delta T^{0.91}$ that means, it is a nearly straight line with a softly decreasing absolute value of its slope with decreasing bath temperature T (i.e. increasing $\Delta T = T_{c0} - T$).

---

*1 In section 5.9 we usually used normalized units, while physical units are used throughout the present section. In the TDGL theory the current densities are normalized with j_0 as given in eq. (171). It is $|j_0| = (1/0.385)\, j_c$, where j_c is the GL critical current density.

*2 In Tab. 1 of ref. 316 values for j_{min} are given in normalized units and, therefore, they were multiplied with $j_c / 0.385$ in order to get current densities in physical units.

To compare the predictions of the TDGL theory with our experiments, we determined the values $\gamma_{Hy}^{exp} = \gamma(\Delta T_{Hy}^{exp} = T_{c0} - T_{Hy}^{exp})$, at which the onset of hysteresis has been observed experimentally. Already in section 11.1. we mentioned that for whiskers of In-Pb alloys [40] γ_{Hy}^{exp} is found to be larger than γ_c, indicating that the measured onset of hysteresis occurs further below T_{c0} than predicted by KR. The same observation is made for the In-Pb whiskers of ref. 483 as well as for whiskers of Sn [332], In [483], and Zn [483]. Also in these samples the onset of hysteresis is predicted for temperatures which are too close to T_{c0}. No correlation between this discrepancy and the ratio ℓ/ξ_0 (that means the 'dirtyness' of a sample) has been observed.

For the Pb whiskers, Pb1 and Pb2, and the dirty sample InPb3 of ref. 483 the TDGL theory predicts an onset of the intrinsic hysteresis at very low temperatures, out of the measured range. The measured hysteresis for these samples is indeed due to Joule heating only [483] (see Figs. 73 and 74 in section 11.4.2).

For a quantitative comparison of the hysteresis width a prediction for δT_H has to be evaluated from the TDGL theory. For this purpose we plot the $I_c^{2/3}(\Delta T)$ straight line observed in our experiment together with the KR prediction $I_{R,TDGL}^{2/3} = (A\, j_{min})^{2/3}$, using eq. (243). Here, A is the cross-sectional area of the sample. From this plot the width δT_H at arbitrary fixed current I_F can be determined so that a prediction of the TDGL theory for $\delta T_H(I_F^2)$ is obtained.

In Fig. 81 results for a pure tin whisker (sample SnH3) are shown. In the whole range plotted we used eq. (243), although for γ below 10 and above 80 (corresponding to $\Delta T = 0.3$ and 20 mK, respectively) we do not know whether this approximation is still valid. The width of the temperature hysteresis, $\delta T_H(I_F^2)$, is shown in the inset. Although there is no quantitative agreement between the KR prediction and our measurements, the shape of the $I_R^{2/3}$ curve and the foot structure in the $\delta T_H(I_F^2)$ plot are similar to the experimental results. One reason for the disagreement may be that sample SnH3 is a pure superconductor while the KR theory is only valid in the dirty limit.

The observation that the hysteresis predicted by the TDGL theory is much too large was not only observed for pure tin whiskers [332] but also for samples of pure In [483] and In-Pb alloys [483] which are more dirty so that $\ell < \xi_0$. With decreasing ℓ/ξ_0 the agreement becomes better. In Fig. 82 results for sample InPb2 are shown, for which $\ell/\xi_0 = 0.28$. This result is the best agreement we found for the samples investigated in ref. [483]. Again we used eq. (243) in the whole range plotted. In this case the condition that γ should be between 10 and 80 is fulfilled for all values plotted.

Very recently KR published a short note dealing with the disagreement between the hysteresis width as predicted by the TDGL theory and as observed in the experiments [518]. KR argue that the disagreement is not really alarming because intrinsic fluctuations and extrinsic fluctuations (that

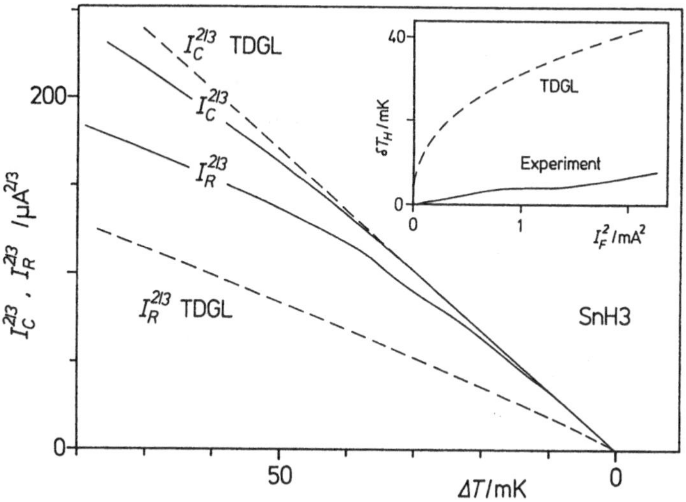

Fig. 81: *Critical current $I_c^{2/3}$ and jump-back current $I_R^{2/3}$ for a pure tin whisker (sample SnH3) as a function of ΔT [332]. The full lines indicate the estimated slope through the measured values. The dashed lines mark results of the TDGL theory as calculated from the results of KR (for details see the text).*

In the inset the width of the temperature hysteresis, δT_H, as a function of I_F^2 is compared with the prediction of the TDGL theory.

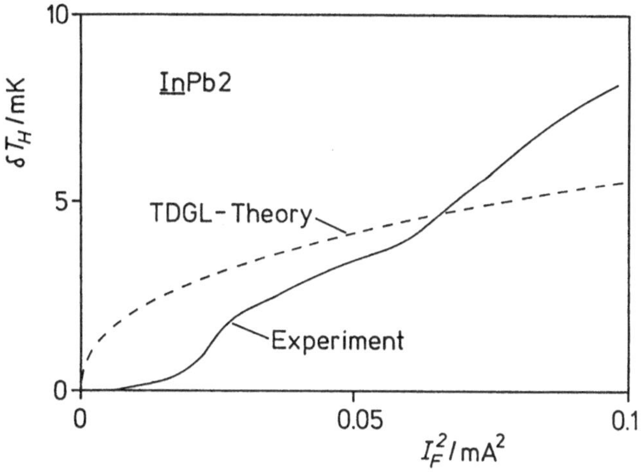

Fig. 82: *Measured width of the temperature hysteresis, δT_H, as a function of I_F^2 for an indium-lead alloy whisker (sample InPb2 with c_{Pb} = 3.5 at% and T_{c0} = 3.5881 K) in comparison with the prediction of the TDGL theory [483].*

means external noise) may reduce the hysteresis. The idea is that the ideal lower limit j_{min} is not reached, because these disturbances convert the stable phase-slip solution below j_c into a metastable state with finite lifetime. KR pointed out that an inclusion of fluctuations in the theory would be important. Moreover, KR propose an idea for an experiment to test their claim. This experiment needs, however, a fast registration method for the voltage at the filament and it seems to be impossible to realize this in the μV range.

Naturally, fluctuations may influence the hysteresis width. However, it seems to us as if the disagreement between the predicted and measured hysteresis width had a more fundamental origin:

A lower border for the jump-back current I_R of the first voltage step of a V-I characteristic is given by I_0 (see Fig. 25). This statement is in agreement with the TDGL theory (see Fig. 19). Thus, I_R/I_c should be larger than I_0/I_c.

If we calculate $I_{R,TDGL}/I_c = j_{min}/j_c$ from eq. (243) for a given γ, we find that $I_{R,TDGL}/I_c$ is larger than the prediction of the TDGL theory for I_0/I_c (as expected), but is smaller than the measured ratio I_0/I_c. To be more explicit, it is $I_{R,TDGL}/I_c = 0.68$, 0.56, 0.51, 0.47, 0.44, and 0.42 for $\gamma = 10$, 20, 30, 40, 50, and 60, respectively. These values are smaller than the experimental results for I_0/I_c, lying close to the lower border of the measurements plotted in Fig. 63.

Since the experimental value of I_0 is larger than $I_{R,TDGL}$, the measured hysteresis width can never have the magnitude predicted by the TDGL theory. The experimental width would be smaller than the theoretical width even if the hysteresis in the experiment would extend down to I_0. Therefore, even if fluctuations would force the sample to re-enter the superconducting state at a larger current than in the undisturbed situation, this effect would not be sufficient to explain the disagreement between the TDGL theory and the experiments. Since not only the prediction of the TDGL theory for $I_{R,TDGL}/I_c$ but also that one for I_0/I_c is lower than observed experimentally, it seems to us as if there was a more rudimentary difference between the behaviour of an array of phase-slip centers in the TDGL theory of KR (only for an array of phase-slip centers results for $I_{R,TDGL}$ and I_0 are available) and a single phase-slip center in our samples.

12. Tunable Weak Links

The regular voltage step structure which characterizes the current-induced breakdown of superconductivity in a quasi-one-dimensional superconductor is generated by spatially localized phase-slip centers. The first phase-slip center appears at the point of weakest superconductivity in the sample. The next phase-slip center occurs at a point, where the local critical current has its smallest value in the presence of the first phase-slip center and so on.

In an ideal homogeneous filament with strongly superconducting contacts in the absence of fluctuations, the first phase-slip center is expected to develop in the middle of the filament and the appearance of further phase-slip centers is only governed by stabilizing and destabilizing effects of the phase-slip centers already present.

In a real specimen there may be material inhomogeneities leading to places of (more or less) weakened superconductivity. The first phase-slip center then appears at the weak link with the lowest critical current. The development of subsequent phase-slip centers is then governed by stabilizing and destabilizing effects and the strength of superconductivity in the other weak links.

In (long) microbridges there are inhomogeneities which influence the appearance of phase-slip centers in these samples. In whiskers, inhomogeneities are of minor importance. These samples are very homogeneous. If we, nevertheless, assume that also in whiskers the phase-slip centers develop at localized weak links, one should keep in mind that the superconducting properties of these weak links differ only very slightly from those of the whole whisker. A weak link in whiskers only seems to be a kind of 'pinning center' for the phase-slip center (see sections 7.2, 7.4, 7.5, and chap. 10 for a detailed discussion).

The weak links pinning the phase-slip centers in a quasi-one-dimensional superconductor usually have a fixed and in most cases fortuitous strength. Therefore, the question how the behaviour of the phase-slip center depends on the properties of the weak link usually cannot be answered.

However, already in section 7.2 and chap. 10 we discussed that the grade of homogeneity of a specimen may influence the behaviour of a phase-slip center and that such an influence is also expected by the TDGL theory widely discussed in section 5.9.

Therefore, we performed measurements of the V-I characteristics of a phase-slip center developing in a whisker at a weak link with a continuously adjustable strength of superconductivity ('tunable weak link', abbreviation 'TWL'). Naturally, such an experimental arrangement is a superconducting device. It may be not only a tool for basic research but also for application purposes in superconducting electronic circuits.

In the present chapter we only report about the results of a tunable weak link in a tin whisker [23]. We refer to chap. 13 for remarks on ongoing experimental investigations dealing with tunable weak links in whiskers of tin, indium, and zinc.

12.1. Experiments

The sample, results of which are given in the present chapter, was made using two tin whiskers. Each whisker is held by two isolated Wood metal blocks which are superconducting during measurements (see Fig. 1d or the inset of Fig. 83). At the cross-over the whiskers were connected by a kind of spot welding procedure: The filaments are brought into touch with each other and then a discharge current is allowed to flow across the touching region, leading to a metallic contact between the two whiskers.

Now one whisker is used as 'measuring whisker' while the other one is the 'controlling whisker'. In the measurements shown in the present chapter, whisker 1 is the measuring whisker (voltage and current denoted by V and I, respectively) and whisker 2 is the controlling whisker (voltage and current denoted by V_t and I_t, respectively). Through the controlling whisker a 'control current' (or 'tuning current'), I_t, flows, influencing the superconducting properties of the measuring whisker near the contact region. As the control current can be continuously adjusted, a weak-link structure with tunable strength ('TWL') can be induced in the measuring whisker.

Without control current ($I_t = 0$), the intrinsic properties can be studied. In order to investigate the behaviour of the TWL one then may for instance measure V-I characteristics of whisker 1 with T fixed and I_t as a parameter.

An example for such a measurement is shown in Fig. 83: For $I_t = 0$, the V-I characteristics show the typical intrinsic voltage-step structure. For $I_t \neq 0$, the basic observation is now, that for a certain magnitude of the control current, I_t, an additional voltage step appears before the onset of the intrinsic characteristic. This additional voltage step is caused by the TWL. It vanishes when the control current is switched off. With increasing value of the control current the critical current of the TWL decreases and the slope of its V-I charateristics changes. For very high values of I_t the characteristics start with zero critical current from the origin. The onset of the intrinsic characteristics, however, remains nearly unaltered.

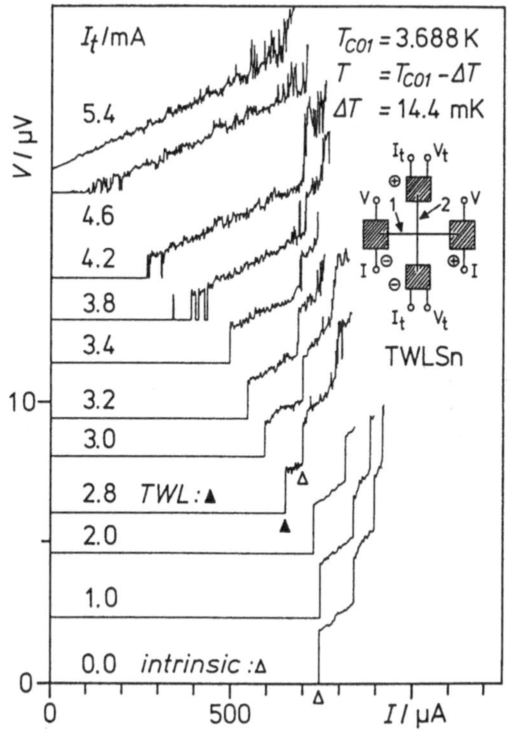

Fig. 83: *V-I characteristics of a tunable weak link (sample TWL Sn) at fixed temperature for different values of the control current, I_t [23].*

The inset shows the sample geometry: Whisker 1 is the 'measuring whisker' while whisker 2 acts as 'controlling whisker'. The symbols '⊖' and '⊕' indicarte the polarity of the current. The left part of whisker 1 (from the left Wood metal contact to the crossing point of both whiskers) is 250 µm long, while the length of the right part is 563 µm. The length of the lower part of whisker 2 is 400 µm and that one of the upper part is 225 µm. Here, T_{c01} is the critical temperature of whisker 1. The critical temperature of whisker 2 is 3.5 mK higher. Further characteristic properties are given in ref. 23.

For a certain amount of the current I_t an additional voltage step with tunable critical current ('TWL') appears before the first natural ('intrinsic') voltage step. The zero voltage point varies for the different characteristics. It is indicated by the fine horizontal line which a characteristic starts from. The current I is increasing.

These measurements were carried out for several fixed temperatures. The critical current of the TWL and the critical current belonging to the onset of the intrinsic characteristic are plotted in Fig. 84. While the critical current of the intrinsic characteristic is nearly not changed by the control current (except of a slight decrease for large values of I_t), the critical

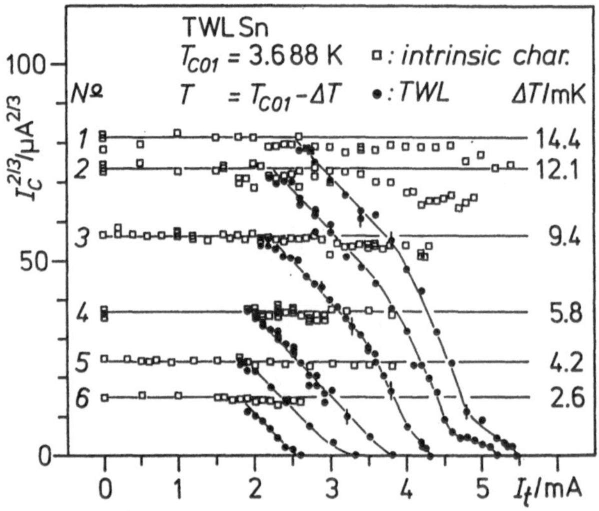

Fig. 84: *$I_c^{2/3}$ as a function of the control current, I_t, for the intrinsic characteristics of whisker 1 and for the tunable weak link (TWL) of sample TWL Sn for several different fixed temperatures [23].*

The full lines indicate the estimated slope through the measured critical currents. The horizontal line holds for cases in which the phase-slip center at the TWL does not appear before the onset of the intrinsic characteristics. The downward bending full lines indicate the critical current of the TWL.

current of the TWL depends strongly on I_t. For all temperatures investigated $I_c^{2/3}$ initially depends linearly on I_t. For temperatures not too close to T_{c01} the linear dependence changes into a more complicated behaviour before I_c goes to zero. With decreasing temperature both the control current at which the voltage step caused by the TWL appears in the V-I characteristics and the value belonging to zero critical current of the TWL are shifted to higher values.

Not only the critical current, but also the other characteristic properties of the voltage step generated by the phase-slip center at the TWL depend on the control current, I_t, that means on the strength of superconductivity in the tunable weak link. As can be seen from Figs. 83 and 84, this strength decreases with increasing control current.

In Fig. 85 we plotted the differential resistance $(dV/dI)_1$ of the first voltage step of the V-I characteristics as well as results for the ratio I_0/I_c. For lower values of I_t the V-I characteristics start with the intrinsic voltage step, whereas for higher I_t, the phase-slip center at the TWL generates the first voltage step. Both quantities are clearly smaller for the phase-slip center at the TWL, with a downward tendency for increasing control current. It is remarkable that for the intrinsic voltage step as first step of the characteristics these properties are not changed by the control current which flows through the TWL.

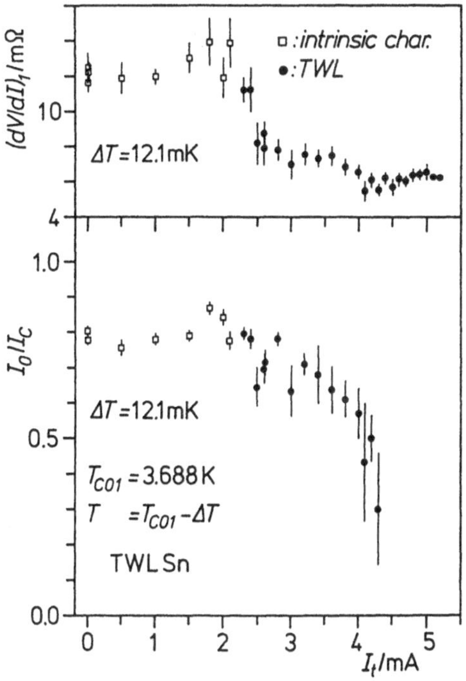

Fig. 85: *Differential resistance $(dV/dI)_1$ and ratio I_0/I_c for the first voltage step of the V-I characteristics of whisker 1 of sample TWL Sn [23].*

For lower control currents, I_t, the V-I characteristic starts with an intrinsic voltage step, while for larger values of I_t the first voltage step of the characteristic is generated by the phase-slip center at the TWL.

We also investigated the hysteretic properties of the phase-slip center at the TWL. Our experimental observation is that the hysteresis width, $I_c - I_R$, of the first voltage step of the V-I characteristics seems to be independent of I_t as long as the step is generated by an intrinsic phase-slip center. However, $I_c - I_R$ decreases rapidly with increasing I_t for the phase-slip center at the TWL [23].

An interesting question is whether the temperature dependence of $(dV/dI)_1$ and I_0/I_c of the TWL differs from that one of the intrinsic voltage step. To answer this question, the intrinsic V-I characteristic for $I_t = 0$ has to be measured at several fixed temperatures as well as the V-I characteristic for a fixed control current $I_t \neq 0$ which is sufficiently large so that the phase-slip center at the TWL generates the first voltage step of the characteristic.

Naturally, the desired information can be obtained by performing measurements of the kind plotted in Fig. 83 for several fixed temperatures. Another possibility (which may be more comfortable) is to measure V-I characteristics with I_t fixed and T as a parameter (Fig. 86).

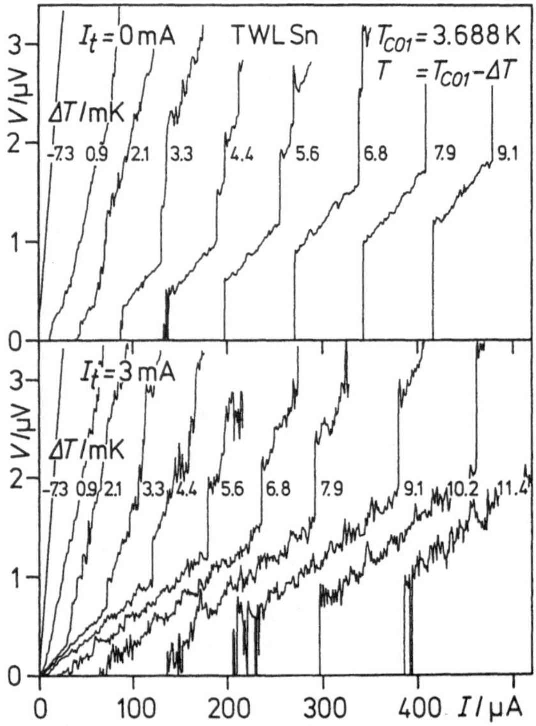

Fig. 86: *V-I characteristics of whisker 1 of TWL Sn for several fixed temperatures, without a control current (upper part of the figure) and with a control current $I_t = 3.0\,mA$ flowing (lower part of the figure). The current I is increasing [23].*

The V-I characteristics with an applied control current look similar to those without control current, but they show more noise and changed characteristic properties:

Without a control current the critical current shows the typical dependence $I_c^{2/3}(T_c) \sim \Delta T$. With a control current, the first intrinsic voltage step shows the same temperature dependence with an absolute value which is only very slightly reduced. The behaviour of the TWL, however, is changed by I_t. Its critical current does not follow an $I_c^{2/3}(T_c)$ straight line and shows a reduced value depending on the magnitude of the control current [23].

The differential resistance $(dV/dI)_1$ and the ratio I_0/I_c of the TWL both are smaller than for the intrinsic phase-slip center at $I_t = 0$. As well for the intrinsic phase-slip center at $I_t = 0$ as also for the TWL, both quantities do not depend on temperature [23].

This statement is valid for temperatures which are not too close to T_{c0}. In the case of $(dV/dI)_1$ we considered only characteristics of the TWL with a nonzero critical current. Closer to T_{c0}, the differential resistance somewhat increases with increasing temperature (see the lower part of Fig. 86 for

$\Delta T < 4.4$ mK). In the case of I_0/I_c we only considered characteristics of the TWL for which a voltage jump is visible at the critical current. Closer to T_{c0}, where the characteristics more or less continuously start at I_c, the ratio I_0/I_c is close to unity or not defined for vanishing critical current (see the lower part of Fig. 86 for $\Delta T < 7.9$ mK). We should, however, remark, that there are very recent experiments performed with tin whiskers showing a temperature dependent differential resistance. This temperature dependence is present in the intrinsic case and for the TWL. The strength of the temperature dependence of the differential resistance of the TWL is altered by the control current. We refer to chapter 13 for a detailed discussion.

The height of the first voltage step, $V_1 = V(I_c)$, depends linearly on the critical current I_c, following different straight lines for the intrinsic voltage step at $I_t = 0$ and for the TWL [23]. It is remarkable that all values for the TWL join the same straight line, independent of the magnitude of I_t, that means of the weakness of the TWL.

Since $V(I_c)$ can be written as $V(I_c) = (dV/dI)_1(1 - I_0/I_c)I_c$, this observation indicates that the product of $(dV/dI)_1$ and $(1 - I_0/I_c)$ does not, or only weakly, depend on the strength of the TWL, although both quantities change with increasing weakness of the TWL. This constancy of the product seems to be characteristic for a phase-slip center under conditions where $(dV/dI)_1$ and I_0/I_c change. It has also been observed for indium whiskers, where both quantities grow if the temperature approaches the critical temperature (see section 7.2).

12.2. Interpretation

The experimental results obtained for tunable weak links are not easy to interpret. In the following we will try to give an explanation for the weakening mechanism and somewhat speculate about an explanation for the behaviour of the characteristic properties of a phase-slip center occurring at a TWL. However, most of our results must remain empirical statements at the moment.

One of the basic questions concerns the weakening mechanism which is responsable for the weakening of superconductivity in the TWL. There are several effects of the control current which may lead to a weakening of superconductivity and, therefore, have to be discussed, namely current-induced depairing, pair breaking by the magnetic field, simple heating, and so called 'nonequilibrium heating' which is nothing else but the excitation of the energy mode (discussed in section 5.5) by an overpopulation of quasiparticle excitations.

In the literature several methods of continuously weakening the properties of a superconductor are reported: illumination [104, 231 - 236, 519], quasiparticle injection through a tunnel contact [198, 200 - 206, 520 - 524], phonon injection from a normal conducting heater strip crossing the sample [224 - 230], a locally applied magnetic field [525], and the weakening of superconductivity by a transport current [108, 526]. In several cases controllable weak links were generated [201, 202, 204, 206, 224 - 230, 233, 235, 519 - 526]. Also their application to a dc squid is reported [522, 524].

The cited experimental works show results which we also observe in our experiments: With increasing weakness of the superconductor its critical current is reduced and the current-hysteresis of the transition becomes smaller. However, there is no measurement of the reduction of the critical current with increasing weakness parameter which shows the same slope as we have obtained in Fig. 84. Therefore, an explanation for the weakening mechanism cannot be found by a comparison of our results with an experiment described in the literature. The reason may be that in our experiments we generate a tunable weak link in a quasi-one-dimensional superconductor. The samples in the cited literature are not quasi-one-dimensional (except perhaps those used in ref. 526).

Therefore, we tried to find an explanation for the weakening mechanism in our experiment by a discussion of all relevant weakening processes, however, restricting our analysis to the region of control currents where $I_c^{2/3}$ depends linearly on I_t. It turns out [23], that our observations cannot be explained by simple heating, where the temperature of the sample is locally enhanced over the bath temperature and all phonons are characterized by this higher temperature. Moreover, the magnetic field of the control current can be excluded as well as current-induced depairing due to a part of the control current flowing across the measuring whisker.

In our opinion 'nonequilibrium heating' caused by quasiparticle injection is a probable explanation for the weakening mechanism in our experiment. The appearance of the phase-slip center at the TWL before the first intrinsic phase-slip center in the V-I characteristics occurs at control currents for which the controlling whisker has nearly reached the normal conducting state. Therefore, we assume that nonequilibrium quasiparticles or normal electrons which are present in the controlling whisker are injected into the measuring whisker, leading to an overpopulation of the excitation spectrum. The injection can be caused by a diffusion through the metallic contact between both whiskers or by the effect that a part of a normal current tries to take its path through the superconducting measuring whisker.

The weakening of superconductivity by a quasiparticle overpopulation has been widely dealt with in section 5.5. In that section two quasi-thermal models are discussed, the so called μ^* model of Owen and Scalapino and the T^* model of Parker. Both models predict a decrease of the gap parameter with increasing excess quasiparticle density. While in the μ^* model the gap shows a step-like transition to zero at high quasiparticle densities, the T^* model predicts a continuously decreasing gap. Since in our experiment the critical current continuously goes to zero with increasing control current, I_t, we prefer to apply the T^* model to our problem. In this model the properties of the nonequilibrium superconductor are obtained as the thermal equilibrium properties of that superconductor at the temperature T^*.

What we, therefore, need is a relation between the control current, I_t, and the temperature T^*. Then the critical current of the TWL can be calculated for a given I_t as GL critical current of the measuring whisker at the temperature T^*.

As far as the critical current of the TWL is nonzero, T^* can only have values between the bath temperature, T, and the critical temperature, T_{c0}. Since T is only some millikelvins below T_{c0}, $\delta T^* = T^* - T$ cannot exceed a few millikelvin. As T is several kelvins, this implies $\delta T^* \ll T$. As we pointed out in section 5.5, Parker's model then yields

$$\delta T^* \approx (2/3)\, T\, N_{op}/N_T \tag{244}$$

where $N_{op} = N - N_T$ is the quasiparticle overpopulation of the excitation spectrum.

This equation gives us the relation between the actual quasiparticle number per volume, N, and the quasiparticle temperature, T^*. Here, N_T is the thermal equilibrium quasiparticle number per volume at the bath temperature, T. If we furthermore would use $T \approx T_{c0}$, we would get eq. (90) of section 5.5 which we applied for the description of the T^* contribution to the hysteresis.

To relate the quasiparticle overpopulation N_{op} of the excitation spectrum in the TWL with the control current I_t , we assume that

$$N_{op} = \beta_{TWL}(I_t - I_{tA}) \tag{245}$$

where β_{TWL} characterizes the quasiparticle production due to the control current.

This equation expresses the experimental observation, that a certain control current I_{tA} is needed until the phase-slip center at the TWL appears before the onset of the intrinsic V-I characteristic. We, thus, assume that only the amount of the control current, I_t, flowing above the 'appearance current', I_{tA}, creates a quasiparticle overpopulation which is large enough to lead to a remarkable enhancement of the effective temperature, T^*, above the bath temperature, T.

According to the GL theory the critical current of a quasi-one-dimensional superconductor is given by (see section 5.1)

$$I_c^{2/3}(T_c) = -(dI_c^{2/3}/dT_c)(T_{c0} - T_c) \tag{246}$$

where T_c is the transition temperature at current I_c.

Usually T_c is equal to the bath temperature T at which the V-I characteristics is measured. In the TWL, however, it is $T_c = T + \delta T^*$ so that $T_{c0} - T_c = \Delta T - \delta T^*$, where $\Delta T = T_{c0} - T$. With eqs. (244) and (245) it follows for the critical current of the tunable weak link

$$I^{2/3}_{c,TWL}(\Delta T, I_t) = -\left(\frac{d I_c^{2/3}}{d T_c}\right)_0 \left(\Delta T - \frac{2}{3}\frac{T}{N_T}\beta_{TWL}(I_t - I_{tA})\right) \tag{247}$$

Here, the index '0' expresses, that the value has to be taken for the intrinsic first voltage step at $I_t = 0$.

For fixed bath temperature, $I^{2/3}_{c,TWL}$ depends linearly on I_t as observed in our experiment for the initial change of the critical current of the TWL. From eq. (247) we get the slope

$$\frac{d I^{2/3}_{c,TWL}(\Delta T, I_t)}{d I_t} = \left(\frac{d I_c^{2/3}}{d T_c}\right)_0 \frac{2}{3}\frac{T}{N_T}\beta_{TWL} \tag{248}$$

Since $T \approx T_{c0}$ for all measurements, we should get the same initial slope for all measurements independent of the special ΔT. This was indeed observed experimentally (Fig. 84).

The quantity β_{TWL} is a fitting parameter which has to be determined from the experiment. For this purpose one may use N_T from eq. (231) and apply eq. (247) or (248).

It may be interesting to know something about the magnitude of δT^*. This information can be directly obtained from Fig. 84: Let us, for instance, discuss the critical current of the TWL plotted in measurement no. 2 performed at a bath temperature $T_2 = T_{c01} - 12.1$ mK. For a certain control current the critical current of the TWL becomes equal to the critical current of the first intrinsic voltage step at $I_t = 0$ of measurement no. 3, occurring at the bath temperature $T_3 = T_{c01} - 9.4$ mK. This fact indicates that the effective temperature T^* in the TWL is enhanced to the bath temperature of measurement no. 3. Thus, $T_2 + \delta T^* = T_3$ or $\delta T^* = T_3 - T_2 = 2.7$ mK.

It is also possible to calculate δT^* for arbitrary value of I_t in the linear region. Using eqs. (244), (245), and (248) we get

$$\delta T^* = \frac{d I^{2/3}_{c,TWL}(\Delta T, I_t)}{d I_t} \left(\frac{d I_c^{2/3}}{d T_c}\right)_0^{-1} (I_t - I_{tA}) \tag{249}$$

Naturally, this formula also yields a value for δT^* in the situation described above. From Fig. 84 we get $d I^{2/3}_{c,TWL}/d I_t = -20.5\,\mu A^{2/3}/mA$. Furthermore, [23], $(d I_c^{2/3}/d T_c)_0 = 6121\,\mu A^{2/3}/K$. At the point where the measurement no. 2 yields a critical current of the TWL equal to the critical current of the first intrinsic voltage step of measurement no. 3, it is $I_t - I_{tA} = 0.8$ mA. As expected we get $\delta T^* = 2.7$ mK from eq. (249).

We think to have found a probable explanation for the weakening mechanism, whereas the other properties of the phase-slip center at the TWL remain unexplained:

The reduced differential resistance of the phase-slip center at the TWL seems to indicate that the inelastic scattering time τ_2 for the quasiparticles has a smaller value than in the intrinsic case (see section 7.2 for the relation of τ_2 and $(dV/dI)_1$). This may be caused by an enhanced quasiparticle recombination rate due to the quasiparticle overpopulation [206], but can also be generated by the magnetic field of the control current (see the discussion of the mechanisms contributing to τ_{Q*} given in section 5.4).

The last explanations must remain speculative, because there do not exist any theoretical calculations concerning the properties of a phase-slip center with a weak link of tunable strength as its origin which can be directly compared with the present experiment.

The only theoretical approach to such a problem known to the author has been recently done by Kramer and Rangel (KR) in the framework of the TDGL theory (see ref. 316 and section 5.9) for localized weak links with reduced mean free path for the electrons ('MFP-type weak region') or reduced critical temperature ('T_{c0}-type weak region').

The calculations lead to a smaller critical current and a smaller normal-like length (and, thus, to a smaller $(dV/dI)_1$) for the weaker link. These results are obtained for both kinds of weak regions (except for large reductions of the electron mean free path in MFP-type weak regions, where the normal-like length is predicted to increase). They are in qualitative agreement with our measurements.

The ratio I_0/I_c is decreased by MFP-type weak regions and appears to be increased by T_{c0}-type weak regions. Thus, in the case of the ratio I_0/I_c a qualitative agreement with the predictions of KR for MFP-type weak regions may be remarked.

For T_{c0}-type weak regions KR found that "the situation is shifted toward the static limit". In addition to the mentioned reduction of the absolute value of the normal-like length this prediction implies a weaker temperature dependence of this quantity. Also for MFP-type weak regions KR predict a reduction of the temperature dependence of the normal-like length. In the experiments discussed in the present chapter we found that $(dV/dI)_1$ (and, thus, the normal-like length) and I_0/I_c do not depend on temperature, but we refer to the following section for further experiments carried out with whiskers of Sn and other materials.

Thus, there is some qualitative agreement between the calculations of KR and our measurements. One should, however, keep in mind that the theory does not consider the weakening mechanism which we expect to govern the properties of the TWL in our experiment. Furthermore, there remains the problem that the calculations are carried out in the dirty limit, while our samples are superconductors being in the clean limit.

Finally, we should remark that the tunable weak links are the subject of current investigations and we refer to chapter 13 for additional results on this scope.

13. Remarks on Ongoing Work

In this chapter we will briefly sketch the problems we are dealing with at the moment. These are the investigation of the low temperature behaviour of a phase-slip center, experiments on tunable weak links in whiskers of Sn, In, and Zn, studies of microcontacts between two whiskers and measurements of the interaction of phase-slip centers in Zn whiskers. In most cases we can only present our experimental results, because momentary there is no interpretation available.

The research work reported in the following has been done by X. Yang (and the author) in our laboratory and has not yet been published elsewhere.

13.1. Low-Temperature Behaviour of a Phase-Slip Center

Usually, detailed investigations of the behaviour of a phase-slip center in our whiskers are restricted to a temperature range in the direct vicinity of the critical temperature. In the case of pure tin whiskers the temperature range does not exceed about 15 mK below T_{c0}.

An interesting problem should be to trace the behaviour of a phase-slip center down to much lower temperatures. Then, the sample leaves the quasi-one-dimensional case more and more.

In section 6.1 we discussed measurements of the critical current of a tin whisker (sample SnH3) which were carried out down to more than 300 mK below T_{c0} (Fig. 21). In section 11.3 we showed the hysteretic behaviour of that tin whisker for the whole temperature range (Fig. 69). In the present section we will discuss the behaviour of the characteristic properties of the first voltage step in the V-I characteristics generated by a single phase-slip center in that sample. We extended our measurements down to about 40 mK below T_{c0}.

The characteristic properties of the first voltage step are (besides the critical current, I_c, and the jump-back current, I_R) the differential resistance $(dV/dI)_1$, the extrapolated zero voltage intercept, I_0 (and, thus, the ratio I_0/I_c), the height of the voltage jump at the critical current, $V_1 = V(I_c)$, and the jump-back voltage, $V_R = V(I_R)$. A sketch of the first voltage step is given in section 7.2 (see Fig. 25).

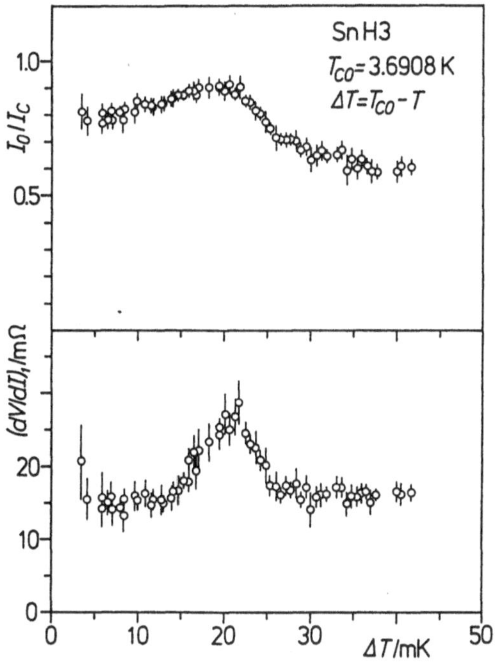

Fig. 87: *Differential resistance $(dV/dI)_1$ and ratio I_0/I_c for the first voltage step in the V-I charateristic of a tin whisker (sample Sn H3), traced from temperatures very close to the critical temperature T_{c0} down to 'low temperatures'.*

In Fig. 87 we plotted $(dV/dI)_1$ and I_0/I_c as a function of temperature. The differential resistance is constant until ΔT exceeds about 14 mK. Then, $(dV/dI)_1$ increases, reaches a maximum value at $\Delta T \approx 21$ mK, and then decreases toward a constant value. The magnitude of the differential resistance in the temperature independent regime at low temperatures (large ΔT) is slightly higher than close to T_{c0} (small ΔT).

The behaviour of I_0/I_c is very similar. While the maximum is observed at the same temperature, the increase is observed somewhat closer to T_{c0} and the decrease extends to somewhat lower temperatures. Moreover, the constant behaviour is not so clearly expressed as in the case of the differential resistance.

To get additional information we made a log-log plot of $(dV/dI)_1$ against ΔT. This plot shows that the increase of $(dV/dI)_1$ is proportional to $\Delta T^{5/4}$, while the decrease occurs proportional to ΔT^{-2}.

Furthermore, we evaluated the height of the first voltage jump, V_1, and plotted it as a function of the critical current, I_c. We observed that there is a linear increase of $V_1(I_c)$ with increasing critical current up to a magnitude of I_c which corresponds to that one at the temperature of the maximum of the differential resistance. At this critical current there is a sudden change in the slope of the straight line which now increases with several times the inclination as before.

Finally, close to the temperature of the maximum of the differential resistance also the behaviour of the jump-back voltage, V_R, changes. As long as the jump-back current, I_R, is smaller than the value corresponding to that temperature, $V_R(I_R)$ shows the typical saturating behaviour (there may be a slight increase with increasing jump-back current). As soon as I_R somewhat exceeds the mentioned value, $V_R(I_R)$ increases linearly.

The behaviour reported is not a fortuitous event of a single specimen, but seems to be typical for the low temperature behaviour of a phase-slip center in a tin whisker. We performed the same measurements with a different tin whisker and obtained the same results. The only difference is that the maximum of the differential resistance is shifted some millikelvin toward lower temperatures and that a slight increase of $(dV/dI)_1$ is observed for very small values of ΔT (that means very close to T_{c0}).

At the moment we do not know which mechanism generates the observed behaviour.

13.2. Tunable Weak Links

Also tunable weak links (introduced in chap. 12) are the subject of current research. On the one hand we continued to investigate the properties of phase-slip centers occurring at a tunable weak link (TWL) in pure Sn whiskers, on the other hand we extended our measurements to other materials, such as In and Zn.

First let us discuss recent experiments performed with tin whiskers: To get additional information about the weakening mechanism in the TWL we built a sample where the measuring whisker and the controlling whisker (both pure tin whiskers) touch each other but where there is no metallic contact between both whiskers, because no discharge current is allowed to flow across the touching region. In this case the application of a current through the controlling whisker does not generate a tunable weak link in the measuring whisker. The only effect is that in the case of very large control currents the characteristics of the measuring whisker become somewhat noisy. The reason may be that the heat dissipation of the control whisker leads to disturbances of the temperature constancy in the environment of the measuring whisker.

This experiment establishes our interpretation that nonequilibrium quasiparticles generated by injection processes through the metallic microcontact between the measuring and the controlling whisker are responsible for the weakening of superconductivity in the tunable weak link.

We, therefore, continued to investigate samples with a metallic microcontact between both tin whiskers. A lot of experimental data were obtained for these specimens. They behave similar to sample TWL Sn of chapter 12. There are, however, some remarkable new results.

To give an example, we shall discuss the measurements performed with sample TWL Sn 10: The sample consists of two pure Sn whiskers and was prepared as described in chapter 12. The geometry (that means the size of the different lengths) is comparable with sample TWL Sn. The critical temperatures are $T_{c0} = 3.6995\,K$ and $T_{c0t} = 3.6998\,K$ for the measuring and the controlling whisker, respectively. Moreover, it may be interesting to know some properties of the measuring whisker. One part is 263 μm long, while the length of the other part is 488 μm. The cross-sectional area is $A = 3.94\,\mu m^2$, the orientation is [001], the residual resistance is $R_n = 0.0455\,\Omega$, and the electron mean free path is $\ell = 4.2\,\mu m$. The slope of the $I_c^{2/3}(T_c)$ straight line of the measuring whisker without a control current is $dI_c^{2/3}/dT_c = -6800\,\mu A^{2/3}/K$.

An example for the V-I characteristics of sample TWL Sn 10 at fixed temperature but for different control currents is given in Fig. 88. In this specimen (and in some of the others) we succeeded to trace the behaviour of the voltage step generated by the phase-slip center at the TWL while this step still appears within the V-I characteristics above the critical current of the first few intrinsic voltage steps. Again the controlling whisker is in the dissipative state (close to being fully normal conducting) as soon as the reduction of the critical current of the TWL with increasing control current, I_t, can be traced in Fig. 88. We should remark that there are also measurements of sample TWL Sn 10 (at lower temperature) where the movement of the phase-slip center at the TWL within the intrinsic characteristic can be already observed while the controlling whisker is still in the superconducting state. In all measurements, however, the controlling whisker is in the dissipative state when the critical current of the TWL is smaller than the intrinsic critical current so that the phase-slip center at the TWL appears prior to the first intrinsic phase-slip center.

The results for $I_c^{2/3}(I_t)$ are similar to those given in Fig. 84 for TWL Sn. The difference is, naturally, that in the case of TWL Sn 10 the critical current of the TWL can already be traced while it is smaller than that one of the onset of the intrinsic characteristic. The influence of the control current on the critical current of the first intrinsic voltage step is somewhat stronger than in sample TWL Sn. The reason may be that the first intrinsic phase-slip center is localized at a place which is closer to the TWL than in sample TWL Sn. As soon as a control current is applied the intrinsic critical current is slightly reduced. This reduction becomes stronger with increasing magnitude of I_t. There is a stabilizing interaction between the phase-slip center at the TWL and the first intrinsic phase-slip center. As soon as the phase-slip center at the TWL appears prior to the first intrinsic voltage step a jump-like increase of the critical current of the first intrinsic phase-slip center and a reduction of its control current induced downward tendency is observed.

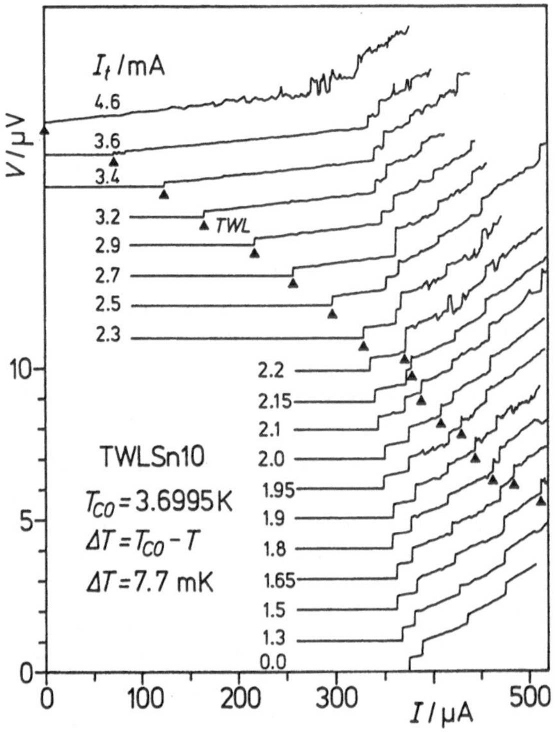

Fig. 88: *V-I characteristics of a tunable weak link (sample TWLSn10) at fixed temperature for different values of the control current, I_t. For a sketch of the sample geometry see Fig. 1d or the inset of Fig. 83. The measuring whisker and the controlling whisker both are pure tin whiskers.*

In this specimen a voltage step with tunable critical current ('TWL') can already be traced while it is still within the intrinsic characteristic.

The zero voltage point varies for the different characteristics. It is indicated by the fine horizontal line from which a characteristic starts. The current I is increasing. Here, T_{c0} is the critical temperature of the measuring whisker.

As in the case of sample TWL Sn of chapter 12, also for TWL Sn 10, the differential resistance $(dV/dI)_1$ and the ratio I_0/I_c of the voltage step generated by the phase-slip center at the TWL (for constant temperature) are smaller than for the first intrinsic voltage step and a downward tendency is observed for increasing value of I_t.

There is, however, a very remarkable difference in the temperature dependence of the differential resistance (Fig. 89). Already the differential resistance of the first intrinsic voltage step (for $I_t = 0$) is only temperature independent for temperatures which are not too close to T_{c0}. In the direct vicinity of T_{c0} a strong increase of $(dV/dI)_1$ is observed. Also the differential resistance of the voltage step generated by the phase-slip center at the TWL is only temperature independent somewhat further away from T_{c0}, but

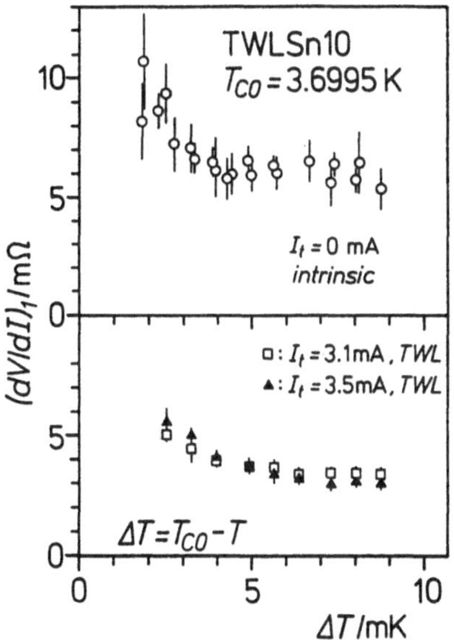

Fig. 89: *Differential resistance $(dV/dI)_1$ of the first voltage step of the V-I characteristics of the measuring whisker of sample TWL Sn 10. For $I_t = 0$, the first voltage step is generated by an intrinsic phase-slip center. This case is shown in the upper part of the figure. In the lower part the first voltage step is generated by the TWL. Results for two different control currents, $I_t = 3.1$ and 3.5 mA are shown. Here, T_{c0} is the critical temperature of the measuring whisker.*

increases when the critical temperature is approached. Compared to the intrinsic voltage step, the onset of the increase of the differential resistance of the TWL occurs at lower temperatures (larger ΔT) and is shifted to even lower temperatures for larger I_t. Moreover, the increase of its differential resistance with increasing temperature (decreasing ΔT) becomes stronger for larger control current.

The ratio I_0/I_c is temperature independent for the intrinsic first voltage step ($I_t = 0$). In cases where a constant control current, I_t, is flowing and the first voltage step is generated by the phase-slip center at the TWL, its ratio I_0/I_c remains temperature independent somewhat further away from T_{c0}, but decreases if the temperature becomes very close to the critical temperature.

Our measurements show that the kind and strength of the temperature dependence of the differential resistance (and, thus, of the normal-like length, L_{An1}) depend on the strength of superconductivity in the weak link at which a phase-slip center develops. The role of inhomogeneities on the behaviour of phase-slip centers has been discussed in several sections of the present work (see sections 5.9, 7.2, 7.4, 7.5, and chapters 10 and 12). With some caution, we now draw the following conclusion: The differential resistance of a phase-slip center in an ideal homogeneous or nearly homogeneous filament of pure tin is temperature independent. If the phase-slip centers develop at weak links in the sample with superconducting properties which do not only differ very slightly from those of the whole whisker, they generate voltage steps with a differential resistance which is more or less temperature dependent, depending on the weakness of superconductivity in the weak link.

This conclusion implies, that in sample TWL Sn10 also the first intrinsic phase-slip center develops at a weak link with superconducting properties which deviate somewhat stronger from the properties of the sample than this is the case in other specimens, in which $(dV/dI)_1$ is not temperature dependent.

This indeed may be the reason: The tin whiskers used for the experiments reported in the present section are 'new' whiskers while in other experiments 'old' whiskers are used. New whiskers are freshly grown specimens while old whiskers were grown several years before they are picked up from the screw (see chap. 2) and used for a measurement. In the new whiskers the phase-slip centers seem to be pinned at weak links with properties which deviate somewhat more from those of the whole whisker as in the case of old whiskers. May be that defects in a whisker are healing while the filaments are kept at room temperature. This interpretation is supported by the observation that new whiskers used for the investigation of the low temperature behaviour of a phase-slip center (section 13.1) show an increase of the differential resistance $(dV/dI)_1$ very close to T_{c0}.

As already mentioned in the beginning of this section we not only investigated tunable weak links in tin whiskers but also in whiskers of indium and zinc. In both cases measuring and controlling whisker are of the same material, that means in the case of indium the specimen is made from two pure In whiskers, while in the case of zinc two pure Zn whiskers are used. The fabrication procedure of the TWL is the same as before, except that in the case of indium the contact blocks which hold the whiskers are made of indium.

Let us first discuss the indium specimen, TWL In4: The behaviour of the critical current $I_c^{2/3}(I_t)$ at fixed temperature looks similar to the case of TWL Sn. The difference is that the control currents needed to shift the critical current of the TWL to zero are much smaller. For $\Delta T = 9.4$ mK the critical current of the TWL becomes smaller than that one of the first intrinsic voltage step already for $I_t \approx 0.2$ mA and a control current of only $I_t = 1$ mA is needed to suppress the critical current to zero. The controlling whisker is in the dissipative state in situations where the properties of the measuring whisker are influenced by the control current.

For constant temperature the differential resistance of the voltage step generated by the phase-slip center at the TWL is somewhat smaller than that one of the intrinsic first voltage step for $I_t = 0$. The same observation is made for the ratio I_0/I_c.

As usually observed for indium whiskers, the differential resistance of the first intrinsic voltage step is temperature dependent and strongly increases if the temperature approaches T_{c0}. Contrary to the observation for tin specimens reported above, the differential resistance of the voltage step generated by the phase-slip center at a TWL in an indium whisker for constant I_t shows a much weaker temperature dependence than that one of the first intrinsic voltage step.

The behaviour of the ratio I_0/I_c for constant I_t is similar to that one of the differential resistance: In the intrinsic case ($I_t = 0$) it increases if T approaches T_{c0}. The increase is weaker for the phase-slip center at the TWL than for the first intrinsic phase-slip center.

Finally, we shall discuss the properties of a TWL in a pure zinc whisker: We may first give some information about the specimen called TWL Zn 3. The lengths of the two parts of the measuring whisker are 500 µm and 337.5 µm, respectively. Its cross-sectional area is $A = 5.93\,\mu m^2$, the orientation is [11$\overline{2}$0], the resistance at room temperature is $R_{298K} = 8.26\,\Omega$, the residual resistance is $R_n = 0.03\,\Omega$, and its electron mean free path as calculated from the residual resistance is $\ell = 10.35\,\mu m$. Furthermore, for the measuring whisker it is $dI_c^{2/3}/dT_c = -8100\,\mu A^{2/3}/K$. The critical temperature of the measuring whisker is $T_{c0} = 0.8182$ K. The critical temperature of the controlling whisker, T_{c0t}, is 0.8182 K, too. Also in this case the controlling whisker is in the dissipative state in situations where the control current influences the properties of the measuring whisker.

Experimental results for the V-I characteristics of sample TWL Zn 3 at fixed temperature but different control currents, I_t, are shown in Fig. 90. The behaviour is very different from that one observed for specimens of tin and indium. No real TWL seems to establish in a zinc whisker. Instead of this the whole characteristic is shifted toward smaller critical currents. In Fig. 90 we traced the behaviour until the critical current of the first voltage step is completely suppressed.

In our opinion the long life-time of nonequilibrium quasiparticles in zinc is responsible for the observed effect. The whole measuring whisker seems to be flooded by nonequilibrium quasiparticles entering the whisker through the microcontact to the controlling whisker, because their relaxation is so slow.

Nevertheless, we evaluated the properties of the first voltage step with applied control current and compared the results with the intrinsic case where $I_t = 0$. We found that very small control currents do not change the critical current at which the first onset of voltage occurs. For somewhat larger control currents I_c is then more and more reduced (Fig. 91). The behaviour of $I_c^{2/3}(I_t)$ then is similar to that one obtained for the critical current of the TWL in sample TWL Sn. It may be interesting to know that the controlling whisker for $\Delta T = 3.8$, 3.2, 3.5, and 2.0 mK, respectively, enters the dissipative state at $I_{t,c} = 110$, 80, 55, and 40 µA and the completely normal conducting state at $I_{t,R} = 265$, 252, 235, 222 µA.

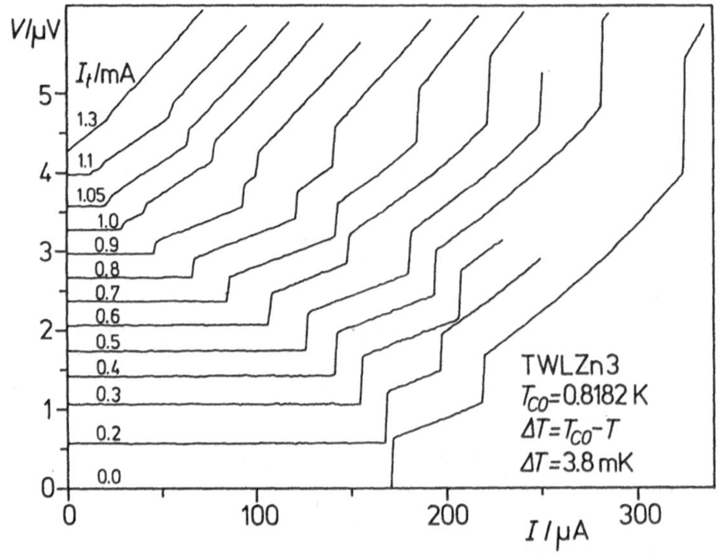

Fig. 90: *V-I characteristics of sample TWLZn3, a tunable weak link arrangement with two pure zinc whiskers used as measuring whisker and controlling whisker, respectively. The measurements were carried out at fixed temperature and increasing current, for several control currents,* I_t.

Here, T_{c0} *is the critical temperature of the measuring whisker. The zero voltage point varies for the different characteristics. It is indicated by the horizontal line which a characteristic starts from.*

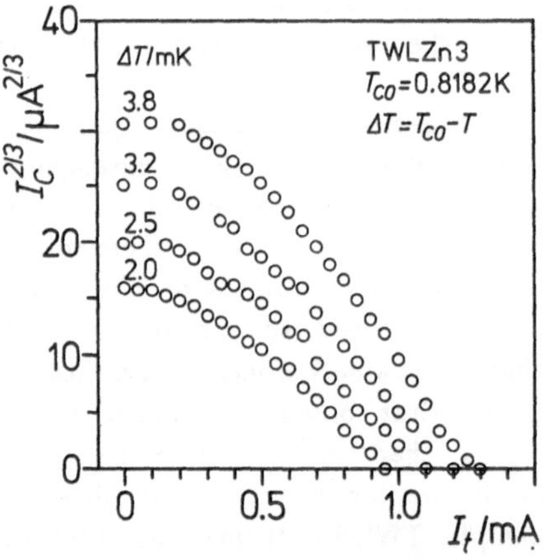

Fig. 91: $I_c^{2/3}$ *as a function of the control current,* I_t*, for the onset of the characteristics of the measuring whisker of sample TWL Zn 3 for several fixed temperatures. Here,* T_{c0} *is the critical temperature of the measuring whisker.*

For constant temperature, the differential resistance $(dV/dI)_1$ and the ratio I_0/I_c are nearly independent of the control current. There may be a slight increase of both quantities with increasing I_t.

Furthermore, we investigated the temperature dependence of $(dV/dI)_1$ and I_0/I_c. In the intrinsic case ($I_t = 0$) these quantities do not depend on temperature. With applied control current, both quantities become temperature dependent. Their temperature dependence becomes stronger for larger control current. An example for the intrinsic ($I_t = 0$) and the quasiparticle overflooded case ($I_t \neq 0$) is plotted in Fig. 92.

The experiments performed with zinc establish our assumption that the weakening mechanism in our experiments is due to quasiparticle overpopulation. Somewhat fortuitously an experiment has been performed which shows that the metallic connection between controlling and measuring whisker is needed to control the properties of the measuring whisker. In this experiment the measuring whisker was a pure zinc whisker and two controlling whiskers were applied at two different places. As controlling whiskers we used Zn-Ag whiskers which are normal conducting during the measurements. By a fortuitous event the metallic connection between one of the controlling whiskers and the measuring whisker was destroyed during the measurements. Before the destruction of the microcontact, the properties of the measuring whisker could be controlled very effectively by the mentioned controlling whisker. Only small control currents already lead to substantial shifts of the critical current of the measuring whisker. After the disconnection of the microcontact no shift of I_c by the control current was observed.

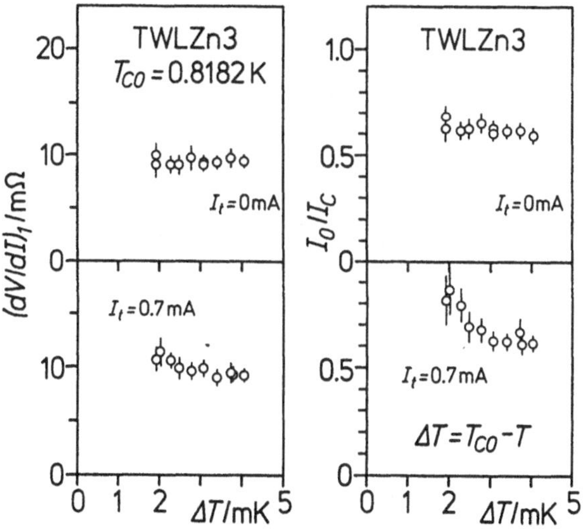

Fig. 92: *Differential resistance $(dV/dI)_1$ and ratio I_0/I_c for the first voltage step of sample TWL Zn 3 in the intrinsic case ($I_t = 0$) and with applied control current $I_t = 0.7\,mA$. Here, T_{c0} is the critical temperature of the measuring whisker.*

This experiment also shows that the shift of the whole characteristic observed for zinc is not caused by Joule heating effects due to a heat transfer from the controlling whisker through the superfluid helium to the measuring whisker. In this experiment the controlling whisker is normal conducting. Therefore, Joule heating already occurs for small control currents. The heat generation is still present after the destruction of the metallic connection between the whiskers. The dissipated heat does, however, not influence the measuring whisker.

13.3. Microcontacts

In specimens used for tunable weak link measurements, there is a metallic microcontact between the measuring and the controlling whisker (see chaps. 2.2, 12, and 13.2). This microcontact is located at the crossing point of the two filaments. It is created by a discharge current which is allowed to flow across the touching region of the whiskers. This discharge current destroys the oxide layer on the surface of the whiskers and generates a spot welding of the filaments. Already in our early studies of tunable weak links [23] we mentioned that the V-I characteristics of such a microcontact seem to indicate its metallic character. Recently we started a more detailed investigation of the properties of these microcontacts.

For this purpose we measured the V-I characteristics at several fixed temperatures for microcontacts between two whiskers of Sn, In, and Zn, respectively. In these experiments the measuring current I enters the sample (for instance) through the measuring whisker, flows through the microcontact into the controlling whisker, and then leaves the sample through this whisker. The voltage V is then detected between the measuring and the controlling whisker at the two free squeeze contacts which no current flows through.

As an example, the V-I characteristics of a microcontact between two Sn whiskers (sample TWLSn8) are shown in Fig. 93. The geometry of the sample is sketched in the inset. The lengths of the left and right side of whisker 1 and those of the lower and upper part of whisker 2 are 400, 413, 288, and 425 μm, respectively. Furthermore, it is for whisker 1 (whisker 2): $R_{298K} = 8.9\,\Omega$ ($34.5\,\Omega$), $R_n = 0.031\,\Omega$ ($0.077\,\Omega$), $A = 9.05\,\mu m^2$ ($3.15\,\mu m^2$), $\ell = 2.90\,\mu m$ ($2.94\,\mu m$), $dI_c^{2/3}/dT_c = -12020\,\mu A^{2/3}/K$ ($-7180\,\mu A^{2/3}/K$), and the orientations are [100] ([001]).

For all characteristics shown, both whiskers are in the superconducting state. This is clear for the non-current-carrying parts, because the measuring temperatures are lower than the critical temperatures of the whiskers. Also the current-carrying parts remain fully superconducting, because the transport currents are not large enough to cause their transition into the dissipative state.

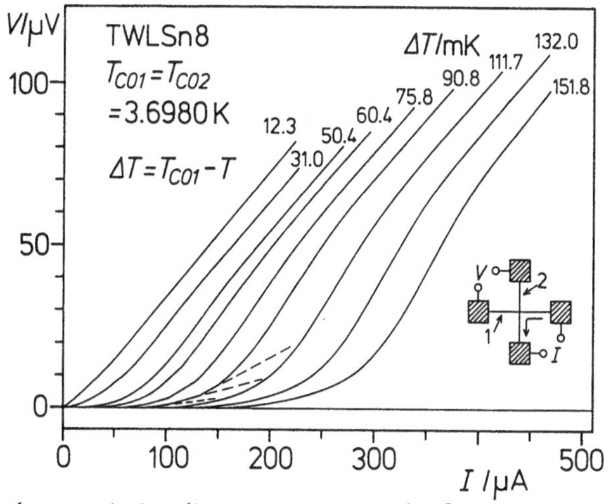

Fig. 93: *V-I characteristics (increasing current) of a microcontact between two tin whiskers (sample TWLSn 8) for several fixed temperatures. The inset shows the geometry of the sample.*

Here T_{c01} and T_{c02} are the critical temperatures of whisker 1 and 2, respectively. In this sample the critical temperatures are equal for both whiskers. Both whiskers are superconducting throughout the measurement. The dashed lines are guides to the eyes to make the portions of constant differential resistance visible.

Up to a certain critical current the microcontact is superconducting. Then there is a continuous transition into the voltage carrying state. There is no step-like change of the voltage at the critical current. A remarkable observation is that some of the V-I characteristics of the microcontact in the low voltage region exhibit straight parts of constant differential resistance (see, for instance, Fig. 93, $\Delta T = 75.8$ and 90.8 mK).

The effect is qualitatively similar to the linear portion phenomenon observed in Pb whiskers (chap. 9). Also in the case of the microcontacts, the differential resistance of a linear portion at higher current seems to be an integer multiple of those at lower currents. For the characteristic at $\Delta T = 90.8$ mK in Fig. 93 three linear portions are visible with a differential resistance $dV/dI = 43$, 89, and 178 mΩ, respectively. The differential resistance of the second linear portion is roughly two times that one of the first linear portion (which can only be approximately determined), and the differential resistance of the third linear portion is exactly two times that one of the second linear portion.

The experiments on microcontacts between two In or two Zn whiskers yield qualitatively similar results (including the linear portion phenomenon). In Fig. 94 some V-I characteristics of a microcontact between two In whiskers are shown (sample TWLIn 4). The geometry of the sample is

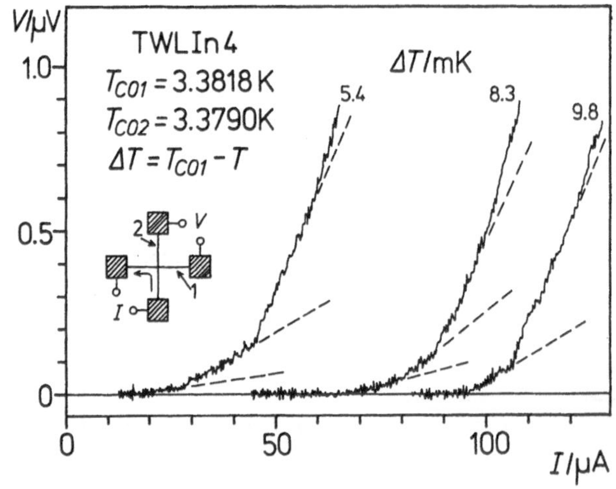

Fig. 94: *V-I characteristics (increasing current) of a microcontact between two indium whiskers (sample TWLIn4) for several fixed temperatures. The inset shows the geometry of the sample.*

Here, T_{c01} and T_{c02} are the critical temperatures of whisker 1 and 2, respectively. The dashed lines are guides to the eyes to make the portions of constant differential resistance visible.

sketched in the inset. The lengths of the left and right side of whisker 1 and those of the lower and upper part of whisker 2 are 438, 288, 412, and 338 μm, respectively. Furthermore, it is for whisker 1 (whisker 2): R_{298K} = 34.1 Ω (61.4 Ω), R_n = 0.448 Ω (0.980 Ω), A = 1.95 μm² (1.13 μm²), ℓ = 1.23 μm (1.01 μm), and $dI_c^{2/3}/dT_c = -4580\,\mu A^{2/3}/K$ ($-3360\,\mu A^{2/3}/K$). It is remarked that whisker 1 is used as measuring whisker in the TWL experiments mentioned in the preceding section.

For all characteristics shown the non-current-carrying parts of the whiskers are in the superconducting state, so that the voltage across the microcontact is measured with superconducting probes. Also the current-carrying part of whisker 1 remains fully superconducting during the measurements, while the current-carrying part of whisker 2 probably changes into the dissipative state.

The characteristics in Fig. 94 were recorded with high resolution and, therefore, show the linear portion phenomenon most clearly. The dashed lines indicate our estimated slope through the sections of constant differential resistance. For the characteristics at ΔT = 5.4 mK our estimate for the differential resistance of the indicated linear portions is (1.87 ± 0.30), (7.43 ± 0.7), and (30.0 ± 1.3) mΩ. For the characteristics at ΔT = 8.3 mK the values are (3.5 ± 0.7), (10.6 ± 1.6), and (28.2 ± 1.5), while at ΔT = 9.8 mK we get (7.8 ± 0.7) and (30.9 ± 1.0) mΩ. Within our limits of accuracy, we, thus, see that the obtained differential resistances are multiples of the smallest differential resistance observed in a characteristic: For ΔT = 5.4 mK the larger differential resistances are 4 and 16 times the lowest differential resistance. For $\Delta T = 8.3$ mK we observe 3 and 8 times the lowest differential resistance, whereas for $\Delta T = 9.8$ mK the larger differential resistance is equal to 4 times the smaller one.

These results are not understood at present.

13.4. The Interaction of Phase-Slip Centers in Zn Whiskers

The influence of an active phase-slip center in a whisker on the superconducting properties of the filament is one of the unresolved problems in the studies of nonequilibrium superconductivity in quasi-one-dimensional superconductors. Before we discuss our very recent experimental results obtained for Zn whiskers, we should briefly summarize our knowledge about this phenomenon.

Experiments on tin whiskers performed with multi-potential-probe contacts clearly show the existence of such an influence: There is a mutual activation of phase-slip centers leading to voltage steps appearing synchronously at the same current (section 7.4). There is a stabilization of superconductivity by active phase-slip centers leading to an enhancement of the critical current of other phase-slip centers. Sometimes also depression effects are observed. It seems as if present phase-slip centers might change the differential resistance of other phase-slip centers (section 7.5).

For other materials we did not perform such detailed investigation of the influence of active phase-slip centers on the superconducting properties of the filament. Nevertheless, also for whiskers of Zn there is evidence for a stimulation of superconductivity by an active phase-slip center. To show this, we simply evaluated the distance in current between the first and second voltage jump in the V-I characteristic of a Zn whisker for several fixed temperatures. We found that this distance is very small for temperatures very close to T_{c0}. It first increases and then decreases again if the temperature is lowered (section 8.2).

A similar behaviour is found for the width of the first linear portion of a Pb whisker. In a different Pb whisker the width also increases with decreasing temperature but then shows a saturating behaviour. In this sample the decrease is not observed in the measured temperature range (section 9.2.3).

Also In whiskers show an increase of the distance between the first two voltage jumps of their V-I characteristics, if the temperature is lowered (Fig. 3 of ref. 16). Similar effects are observed for <u>Sn</u>-In and <u>In</u>-Pb whiskers. For sample <u>Sn</u>-In 7 the distance grows with decreasing temperature (Fig. 4 of ref. 15) while an increase with a subsequent decrease is observed for sample <u>In</u>-Pb 7 (Fig. 1 of ref. 40).

Finally, there may be evidence for an interaction of phase-slip centers from the behaviour of Sn and In whiskers in a HF radiation field. These specimens show inverse-ac-Josephson-like effects also in cases where there is more than one phase-slip center present. One idea to explain this observation might be based on the assumption of a periodic stimulation (or weakening) of the superconducting properties of the filament by present phase-slip centers (section 7.3).

Up to now the mechanism responsible for the influence of an active phase-slip center on the properties of the filament is not understood. The main problem is that our experiments performed with Sn whiskers show that this influence acts over so large distances that it cannot be explained by effects due to quasiparticle diffusion currents.

In this situation we recently decided to start detailed investigations of the interaction of phase-slip centers in Zn whiskers, using multi-potential-probe contacts. In this material the quasiparticle relaxation times are very long. Therefore quasiparticle diffusion induced effects are expected to be large in this material. In the following we will briefly sketch the main results obtained up to now.

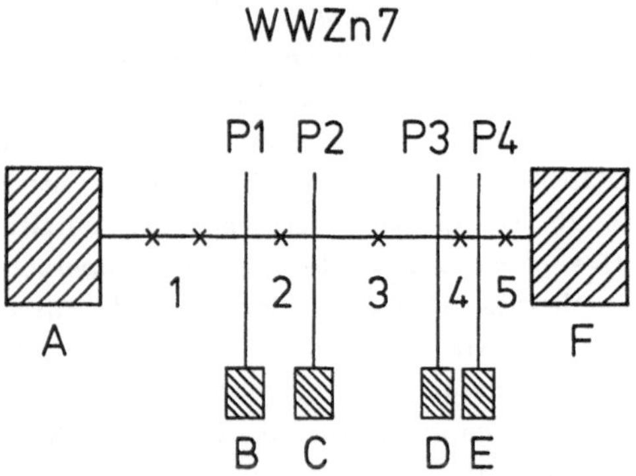

Fig. 95: *Sketch of sample WWZn 7, a zinc whisker with four additional potential probes made of tin whiskers.*

The zinc whisker is held by the Wood metal contacts A and F, while the potential probes P1 - P4 are held by the Wood metal contacts B - E. The numbers 1 - 5 denote the whisker parts.

The crosses indicate all phase-slip centers which may appear, two in part 1 and one in each of the other parts. The phase-slip centers will be referred to as PSC 1 a, PSC 1 b, PSC 2, PSC 3, PSC 4, and PSC 5. Here, the number indicates the whisker part which the phase-slip center is located in. In part 1 there may be two phase-slip centers. As it is not known which one is the left and which one is the right one, they are characterized by their appearance current which is smaller for PSC 1 a than for PSC 1 b.

The lengths of the whisker parts as measured with the optical microscope are: $L_1 = (413^{+38}_{-13})\,\mu m$, $L_2 = (194 \pm 6)\,\mu m$, $L_3 = (344 \pm 6)\,\mu m$, $L_4 = (125 \pm 6)\,\mu m$, *and* $L_5 = (150^{+38}_{-13})\,\mu m$.

For these experiments we used multi-potential-probe contacts, where the Zn whisker is held by two Wood metal squeeze contacts and where there are four additional potential probes made from tin whiskers. At the measuring temperatures both the Wood metal contacts and the tin whisker potential probes are strongly superconducting. The potential probes are somewhat pressed against the Zn whisker and then spot welded with the whisker by a discharge current across the touching point.

Several specimens of this kind were investigated. All results given below are from sample WWZn 7. The geometry of the sample is sketched in Fig. 95. The lengths of the different whisker parts are given in the figure caption. The resistance of the whole zinc whisker at room temperature is $R_{298K} = 32.3\,\Omega$, its residual resistance is $R_n = 0.18\,\Omega$. Its cross-sectional area is $A = 2.23\,\mu m^2$, its mean free path is $\ell = 6.74\,\mu m$, and its crystallographic orientation is $[11\bar{2}0]$. Furthermore, it is $dI_c^{2/3}/dT_c = -4500\,\mu A^{2/3}/K$. The critical temperature as extrapolated from the $I_c^{2/3}(T_c)$ straight line is $T_{c0} = 0.8187\,K$. The transition width of a V-T characteristic at small currents is about 4.8 mK for $1\,\mu A$.

First we studied the transition behaviour of the whole Zn whisker by measuring some of its V-T transition curves and several of its V-I characteristics (Fig. 96). During these measurements a current I_{AF} is flowing

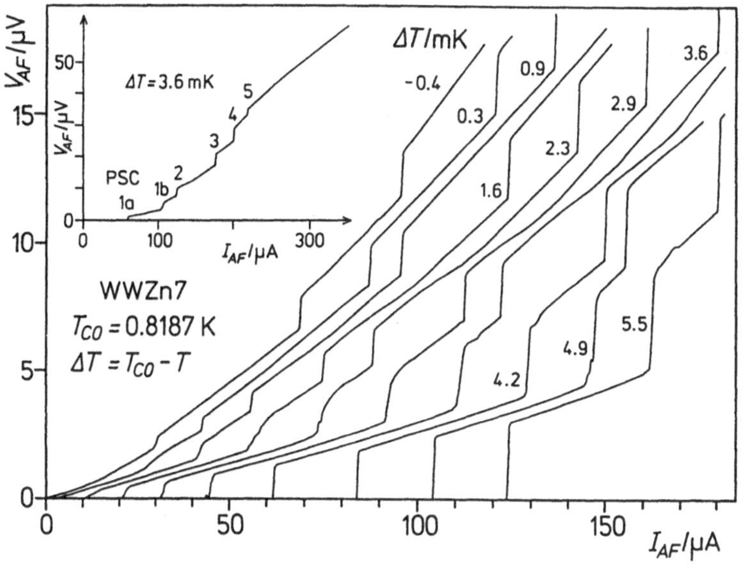

Fig. 96: *V_{AF}-I_{AF} characteristics of sample WWZn 7 (a Zn whisker with 4 additional Sn whiskers as superconducting potential probes) at several fixed temperatures for increasing current.*

The main figure only shows the lower part of the characteristics. The whole characteristic for one of the temperatures is plotted in the inset. The numbers at this characteristic indicate which one of the phase-slip centers sketched in Fig. 95 generates a certain voltage jump.

through the Zn whisker entering the specimen through contact A and leaving it through contact F, and a voltage V_{AF} is detected between contacts A and F (conventional current and voltage as introduced in section 7.6).

The V_{AF}-I_{AF} characteristics of the whole Zn whisker exhibit six voltage steps until the completely normal conducting state is reached. To find out in which part of the specimen the phase-slip centers are situated which generate these voltage steps, the V-I_{AF} characteristics of all parts of the whisker (that means the V_{AB}, V_{BC}, V_{CD}, V_{DE}, V_{EF}-I_{AF} characteristics) have been measured. The result is that for increasing current the first two voltage steps are generated by the two phase-slip centers in part 1, called PSC 1 a and PSC 1 b, while the third, fourth, fifth, and sixth voltage step is generated by PSC 2, PSC 3, PSC 4, and PSC 5, respectively. There are also cases in which PSC 5 appears before PSC 4, so that the fifth voltage step is generated by PSC 5 and the sixth voltage step by PSC 4. Thus, for increasing current I_{AF}, the phase-slip centers in WW Zn 7 (except perhaps those in part 1) are usually switched on from left to right.

With this knowledge about the transition behaviour of our sample, we performed measurements concerning the interaction of neighbouring phase-slip centers. At first we investigated how the current at which a phase-slip center appears (the 'appearance current' or 'critical current of the phase-slip center') is influenced by an active phase-slip center in its neighbourhood.

To see how the first phase-slip center, PSC 1 a, influences the appearance current of the second phase-slip center, PSC 1 b, we evaluated the distance in current, ΔI, between the voltage jumps generated by these phase-slip centers in the V_{AF}-I_{AF} characteristic (Fig. 97). For decreasing measuring temperature (increasing ΔT) the distance ΔI first increases, then

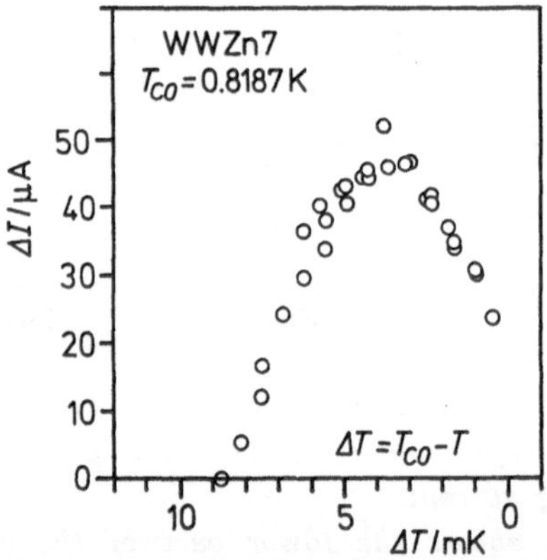

Fig. 97: *Difference ΔI between the currents at which the first and second voltage step appear in the V-I characteristics of sample WWZn 7, plotted as a function of the temperature.*

decreases, falling below its (nonzero) value observed extremely close to T_{c0}, and finally becomes zero. Thus, for decreasing temperature the critical current of the second phase-slip center is first enhanced, then the enhancement disappears until finally a depression of this current is observed. The nonzero value of ΔI extremely close to T_{c0} indicates that a certain distance in current between the first two voltage steps has its origin in a slightly incomplete homogeneity of the whisker.

To see, how the two phase-slip centers in part 1, PSC 1a and PSC 1b, influence the appearance current of the phase-slip center in part 2, PSC 2, a different method was used: For fixed temperature, we first applied a current I_{AF} and measured the V_{AF}-I_{AF} characteristic. In this case the current flows through the whole sample and PSC 2 appears under the influence of the two phase-slip centers in part 1. Then we applied a current I_{BF} and measured the V_{AF}-I_{BF} characteristic. In this case the current enters the Zn whisker through the first potential probe so that PSC 1a and PSC 1b are not activated and PSC 2 appears without any influence of other phase-slip centers.

The same method was used to study how PSC 2 influences the appearance current of PSC 3: In a V_{AF}-I_{BF} characteristic the current enters the Zn whisker through the first potential probe and PSC 3 appears under the influence of PSC 2. In a V_{AF}-I_{CF} characteristic the current enters the Zn whisker through the second potential probe, so that PSC 2 is not activated and PSC 3 appears without any influence of other phase-slip centers.

With this method also the influence of PSC 3 on PSC 4 was studied (by comparing V_{AF}-I_{CF} and V_{AF}-I_{DF} characteristics) and that one of PSC 4 on PSC 5 (by comparing V_{AF}-I_{DF} and V_{AF}-I_{EF} characteristics).

In the following we will give some quantitative results about the influence of PSC 1a and PSC 1b on the appearance or critical current $I_{c,PSC2}$ of PSC 2 and about the influence of PSC 2 on the critical current $I_{c,PSC3}$ of PSC 3. In these cases the effects are clearly expressed.

Before proceeding, we would like to remark that we plotted not only $I_c^{2/3}(T_c) = I_{AF,c}^{2/3}(T_c)$ to extrapolate the critical temperature T_{c0} of the whisker (which we also may call $T_{c0,\,part1}$, because the first phase-slip centers appear in part 1 of the whisker), but also $I_{BF,c}^{2/3}(T_c)$ and $I_{CF,c}^{2/3}(T_c)$. These critical currents characterize the uninfluenced onset of voltage in parts 2 and 3, respectively. Also in these cases a straight line behaviour is obtained and one may extrapolate also these straight lines to vanishing critical current, to get $T_{c0,\,part2}$ and $T_{c0,\,part3}$, respectively. The result is that $T_{c0} = T_{c0,\,part1} < T_{c0,\,part2} < T_{c0,\,part3}$ with $T_{c0,\,part2} - T_{c0,\,part1} = 1.5$ mK and $T_{c0,\,part3} - T_{c0,\,part1} = 2.4$ mK.

The measurements described above show that the presence of PSC 1a and PSC 1b leads to a change of $I_{c,PSC2}$ and that the presence of PSC 2 leads to a change of $I_{c,PSC3}$. Close to T_{c0} the critical currents are enhanced while somewhat further away from T_{c0} they are depressed.

In Fig. 98 some characteristics are shown which demonstrate the enhancement of $I_{c,PSC2}$ and $I_{c,PSC3}$, while in Fig. 99 we plotted some

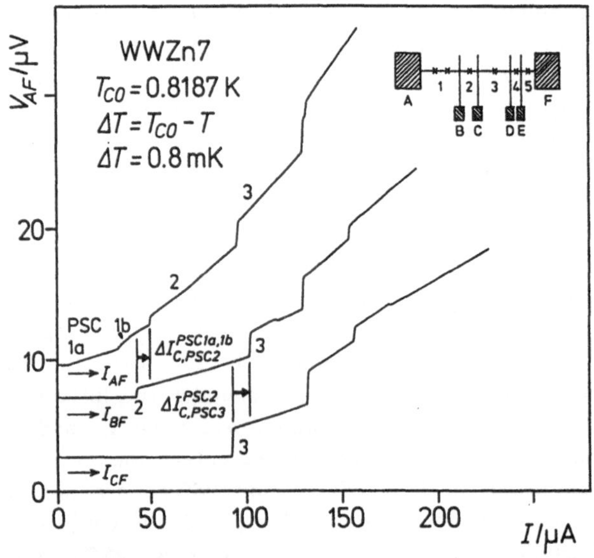

Fig. 98: *V_{AF}-I characteristics (with I = I_{AF}, I_{BF}, and I_{CF}) of sample WWZn7 (increasing current).*

The zero voltage point is different for the characteristics. It is given by the horizontal line which a characteristic starts from.

This plot demonstrates that, very close to the critical temperature, T_{c0}, of the Zn whisker, the phase-slip centers PSC 1a and 1b enhance the critical current of PSC 2, and that PSC 2 enhances the critical current of PSC 3.

The changes in the critical currents are called $\Delta I^{PSC\,1a,1b}_{c,PSC2}$ and $\Delta I^{PSC2}_{c,PSC3}$, respectively. Here, the subscript indicates which critical current is changed while the exponent indicates which phase-slip center (or centers) generates this change.

characteristics demonstrating the depression of both critical currents. In Fig. 100 we summarize the results obtained for different temperatures.

The idea of the notation used to characterize the change in the critical currents in the figures is that the subscript gives information about the phase-slip center, the critical current of which is shifted, while the exponent indicates the phase-slip center (or centers) which generates the change in the critical current. More explicitly, it is $\Delta I^{PSC\,1a,1b}_{c,PSC2} = I^{PSC\,1a,1b}_{c,PSC2} - I_{c,PSC2}$ and $\Delta I^{PSC2}_{c,PSC3} = I^{PSC2}_{c,PSC3} - I_{c,PSC3}$. The critical currents with an exponent are those of the phase-slip center in the subscript measured in the presence of the phase-slip centers in the exponent. There is no exponent if the phase-slip center mentioned in the subscript appears as the first one in a characteristic in the absence of any other phase-slip center in the whisker.

In the experiments reported above, we deal with phase-slip centers in their 'natural sequence' . The only thing we do, is to feed the current through different contacts, so that some of the phase-slip centers do not appear. Below, we will discuss experiments which also concern critical

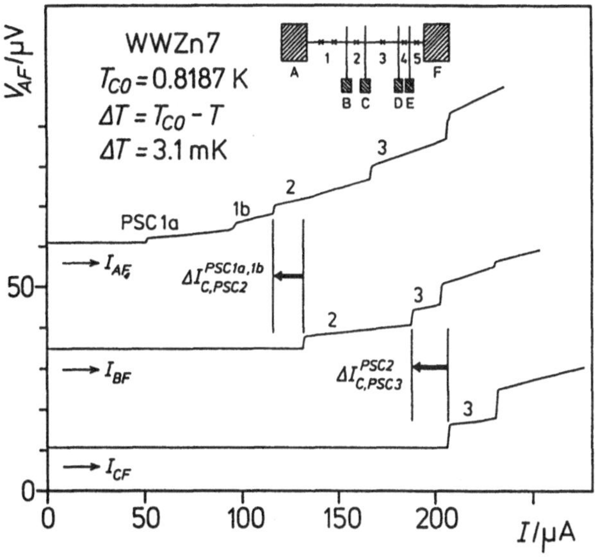

Fig. 99: *V_{AF}-I characteristics (with $I = I_{AF}$, I_{BF}, and I_{CF}) of sample WWZn 7 (increasing current).*

The zero voltage point is different for the characteristics. It is given by the horizontal line which a characteristic starts from.

This plot demonstrates that, somewhat further from T_{c0}, the phase-slip centers PSC 1a and 1b reduce the critical current of PSC 2 and that PSC 2 reduces the critical current of PSC 3.

The notation used to characterize the critical current changes has been explained in Fig. 98 and in more detail in the text.

current shifts of phase-slip centers by neighbouring phase-slip centers, but in which the current in the influencing phase-slip center is held fixed and in which we study the critical current of the influenced phase-slip center as a function of the magnitude and the polarity of the fixed current.

These experiments were performed with the phase-slip centers PSC 2 and PSC 3, situated in part 2 and 3 of sample WWZn7, respectively. First we studied the influence of PSC 2 on the critical current of PSC 3. In Fig. 101 we show V_{AF}-I_{CD} characteristics with I_{BC} as a parameter. For $I_{BC} = 0$ the voltage V_{AF} is zero until PSC 3 appears at a certain critical current $I_{CDc,PSC3}$. In this case PSC 2 is inactive. For $I_{BC} = 120\,\mu A$, PSC 2 is active already before the appearance of PSC 3. Therefore, a constant voltage $V_{AF} > 0$ is observed until a voltage jump indicates the appearance of PSC 3 at a critical current $I_{CDc,PSC3}$ shifted to a somewhat larger value. For $I_{BC} = -120\,\mu A$, again PSC 2 is active, but now a constant voltage $V_{AF} < 0$ is observed until a voltage jump indicates the appearance of PSC 3 at a critical current $I_{CDc,PSC3}$ shifted to a smaller value.

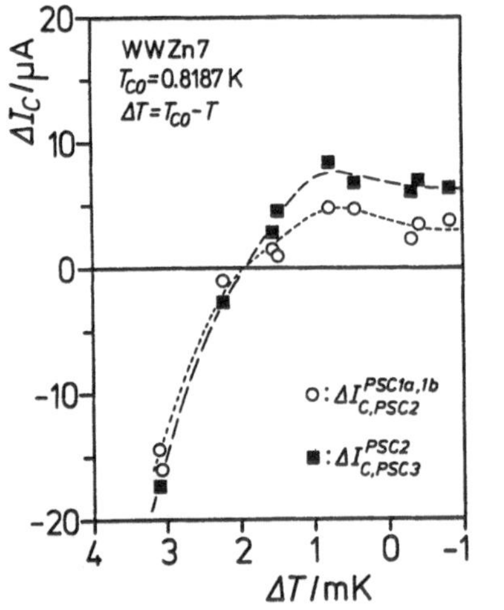

Fig. 100: *Change of the critical current of phase-slip centers PSC 2 and PSC 3 (of sample WWZn 7) by PSC 1 a, 1 b and PSC 2, respectively, as a function of the temperature.*

The notation used has been introduced in Fig. 98 and in more detail in the text. The dashed and dotted lines are estimated slopes through the measured values.

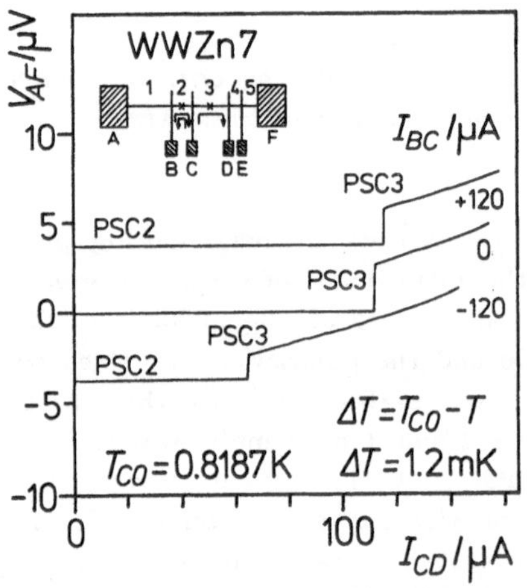

Fig. 101: *V_{AF}-I_{CD} characteristics of sample WW Zn 7 (increasing current I_{CD}) with I_{BC} as a parameter, to demonstrate how the activation of a phase-slip center in part 2 of the whisker (PSC 2) influences the critical current at which a phase-slip center in part 3 (PSC 3) appears.*

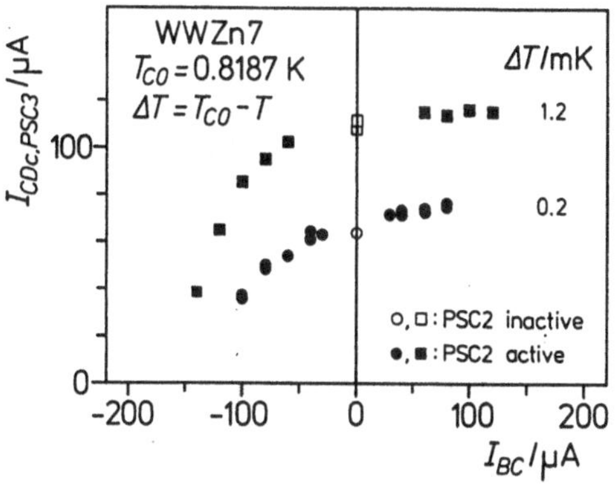

Fig. 102: *Critical current $I_{CDc,PSC3}$ of phase-slip center PSC 3 in part 3 of sample WWZn 7 as a function of the current I_{BC} through part 2 of the sample, for two different fixed temperatures.*

We performed these experiments at two different fixed temperatures for several values of the fixed current I_{BC}. The results are summarized in Fig. 102. This figure clearly shows that the critical current of PSC 3 is enhanced by PSC 2 if $I_{BC} > 0$ (that means in the 'series case', where I_{BC} and I_{CD} are flowing in the same direction), while the critical current of PSC 3 is depressed if $I_{BC} < 0$ (that means in the 'opposed case', where I_{BC} and I_{CD} are flowing in opposite directions).

In a similar experiment we studied the influence of PSC 3 on the critical current of PSC 2. Now V_{AF}-I_{BC} characteristics are measured with I_{CD} as a parameter (Fig. 103). The critical current of PSC 2 is enhanced by PSC 3 in the

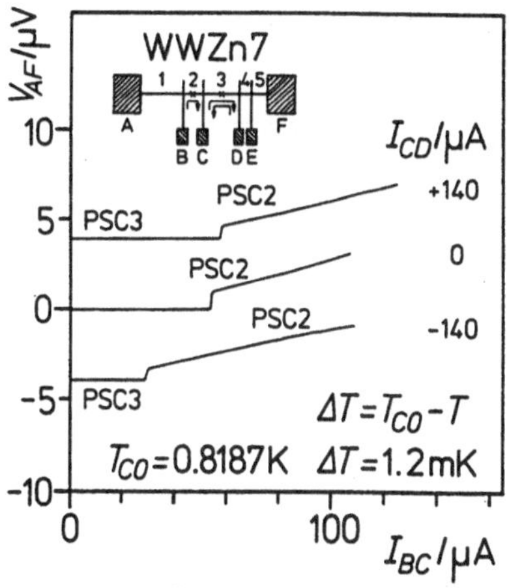

Fig. 103: *V_{AF}-I_{BC} characteristics of sample WWZn 7 (increasing current I_{BC}) with I_{CD} as a parameter to demonstrate how the activation of a phase-slip center in part 3 of the whisker (PSC 3) influences the critical current at which a phase-slip center in part 2 (PSC 2) appears.*

series case and depressed in the opposed case. Results for three different fixed temperatures are summarized in Fig. 104. In this figure a measuring point is shown, for which I_{CD} is nonzero, but not large enough to activate PSC 3 (see the light circle at $\Delta T = 0.32$ mK). In this case no shift of the critical current of PSC 2 is observed, indicating that the current I_{CD} itself does not influence the critical current of PSC 2 as long as it is carried as a supercurrent and PSC 3 is not activated.

Up to now we have not done a detailed quantitative interpretation of the critical current measurements discussed above. Therefore, at the moment we can only make some statements concluded from the qualitative behaviour. The discussion given in section 7.5 in mind, we expect that for the interaction effects observed for neighbouring phase-slip centers in our Zn whisker diffusing nonequilibrium quasiparticles are of importance. This assumption would explain the different shift of the critical current in the series and opposed case observed in Figs. 101 - 104. Besides the asymmetric quasiparticle diffusion effect, there seems to be a symmetric effect (perhaps a T^*effect) present which leads to a depression of the critical current, independent of the relative polarity of the currents flowing through the phase-slip centers. The presence of this effect would explain why the enhancement effects of the critical current are smaller than the depression effects.

The same explanation may hold for the experiments shown in Figs. 97 - 100, if one assumes that the symmetric effect becomes stronger at lower temperatures and finally overcompensates the asymmetric effect. However, it may also be that the explanation of these experiments with its enhancement of the critical current very close to T_{c0} and the depression somewhat further away from the critical temperature is much more complicated due to the excitation of charge imbalance waves. Then there are not simply spatially decaying quasiparticle currents around the phase-slip

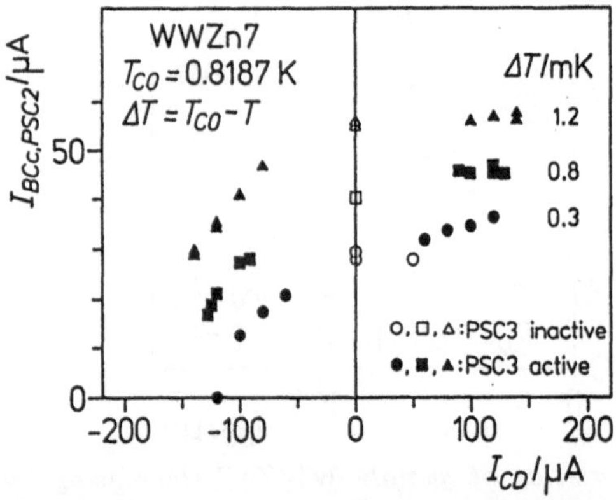

Fig. 104: *Critical current $I_{BCc,PSC2}$ of phase-slip center PSC 2 in part 2 of sample WWZn 7 as a function of the current I_{CD} through part 3 of the sample, for three different fixed temperatures.*

center flowing into the same direction at any time. In this case one may speculate about an alternative explanation of the effects shown in Fig. 100 by assuming that the influencing phase-slip center causes quasiparticle currents at the locus of the influenced phase-slip center which are flowing parallel to the supercurrent very close to T_{c0} but antiparallel at somewhat lower temperature.

The calculations of the time averaged electrochemical quasiparticle potential $\langle\mu\rangle$ in the presence of collective excitations for a tin whisker given in Fig. 39 of section 7.6 seem to indicate that something like that should be possible. This figure shows that the gradient of $\langle\mu\rangle$ changes its sign somewhat away from the phase-slip center, where $\langle\mu\rangle$ has a minimum. Following SBT (section 5.8.3) the quasiparticle current is $\langle \underline{j_n} \rangle = (1/e\rho_n)(d\langle\mu\rangle/dx)\hat{\underline{x}}$ and, therefore, should change its direction if the gradient of $\langle\mu\rangle$ changes its sign. As can be seen from Fig. 2 of ref. 474, the site of the minimum of $\langle\mu\rangle$ depends on the measuring temperature and the voltage generated by the phase-slip center. The minimum moves somewhat closer to the core of the influencing phase-slip center for lower measuring temperature and higher voltage. Let us assume that the (fixed) position of the influenced phase-slip center very close to T_{c0} is somewhere between the core of the influencing phase-slip center and the minimum of $\langle\mu\rangle$, not too far away from the minimum. Then a quasiparticle current flows at the site of the influenced phase-slip center which is parallel to the supercurrent. If now the temperature is lowered, the minimum of $\langle\mu\rangle$ moves across the site of the influenced phase-slip center so that now the minimum of $\langle\mu\rangle$ is between the core of the influencing phase-slip center and the site of the influenced phase-slip center. Then a quasiparticle current flows at the site of the influenced phase-slip center which is opposite to the supercurrent. This mechanism would lead to an enhancement of the critical current of the influenced phase-slip center very close to T_{c0} but to a depression of its critical current at lower temperatures. At the temperature where the minimum of $\langle\mu\rangle$ is located at the site of the influenced phase-slip center, there would be no change of its critical current.

Now we will discuss the influence of an active phase-slip center on the voltage and the differential resistance of a neighbouring phase-slip center. For these investigations the V-I_{AF} characteristics of two neighbouring parts were measured at the same run.

Measuring V_{AB}-I_{AF} and V_{BC}-I_{AF} characteristics, we found that the onset of PSC 2 reduces the voltage and the differential resistance of PSC 1a and PSC 1b. From measurements of V_{BC}-I_{AF} and V_{CD}-I_{AF} characteristics we see that PSC 3 has a similar effect on the characteristics of PSC 2. In Fig. 105 we show an example of such a measurement. If we measure V_{CD}-I_{AF} and V_{DE}-I_{AF} characteristics it is seen that PSC 4 generates the same effect in the characteristics of PSC 3. Finally, measurements of V_{DE}-I_{AF} and V_{EF}-I_{AF} characteristics show that also PSC 5 influences the characteristics of PSC 4 in the same way. Sometimes PSC 5 becomes active before PSC 4 appears and

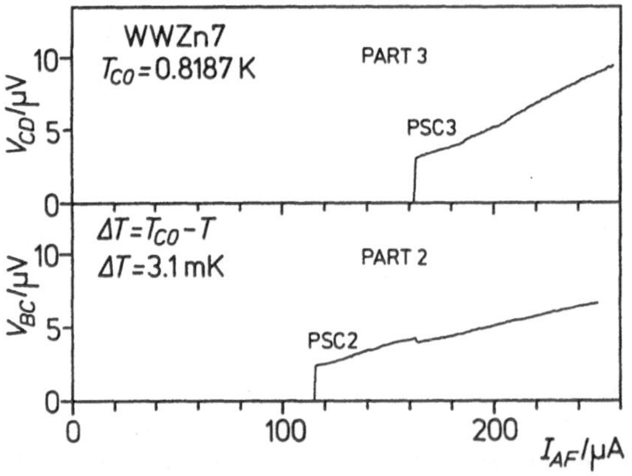

Fig. 105: *V_{BC}–I_{AF} and V_{CD}–I_{AF} characteristics of sample WWZn 7 measured at the same run for increasing current. The appearance of a phase-slip center in part 3 (PSC 3) leads to a change in the voltage and differential resistance of the characteristic of the phase-slip center which is already active in part 2 (PSC 2).*

then the onset of PSC 4 generates the described effect in the characteristics of PSC 5. An example for the last group of measurements is shown in Fig. 106.

Finally, we performed some experiments, where the interacting phase-slip centers are appearing in parts of the whisker which are not in direct neighbourhood. There are, for instance, measurements of V_{AB}–I_{AF} and V_{CD}–I_{AF} characteristics which show that the appearance of PSC 3 in part 3 leads to an increase of the voltage and the differential resistance of the characteristics of PSC 1a and PSC 1b which are already active in part 1 of the sample.

Furthermore, there are some interesting results concerning the interaction of PSC 2 and PSC 5 which are located in part 2 and 5 of sample WWZn7, respectively, and, thus, have a distance of at least $L_3 + L_4 = 469\,\mu m$. Probably, the real distance is even larger. If we assume that PSC 2 and PSC 5 appear in the center of parts 2 and 5, respectively, their distance would be $641\,\mu m$.

While PSC 2 does not seem to influence the critical current and the voltage and differential resistance of PSC 5, the presence of PSC 5 leads to a depression of the critical current of PSC 2 as well in the series as also in the opposed case and, moreover, it changes the voltage and the differential resistance of PSC 2.

To investigate the influence of PSC 5 on the voltage and the differential resistance of PSC 2, a fixed current I_{BC} is applied which only flows through part 2 of sample WWZn7. The magnitude of I_{BC} is chosen sufficiently large, so that PSC 2 is switched on. The voltage developed by PSC 2 is measured by detecting the voltage V_{AD}. Then a variable current I_{EF} is applied which only

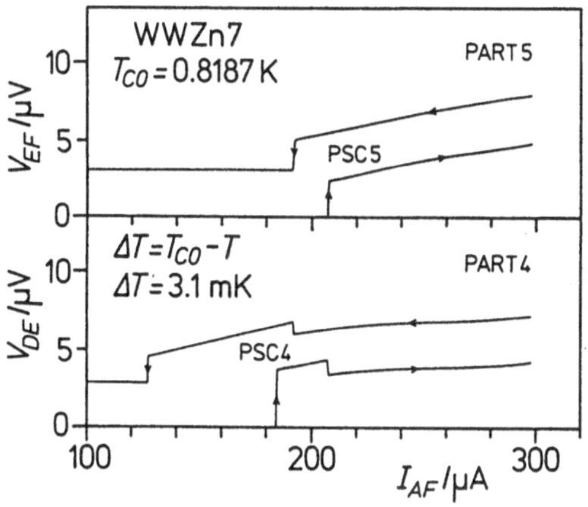

Fig. 106: *V_{DE}-I_{AF} and V_{EF}-I_{AF} characteristics of sample WWZn 7 measured at the same run, for increasing and decreasing current. For decreasing current the zero voltage point is shifted for clearness. It is indicated by the thin line to which the voltage jumps down for decreasing current.*

The appearance and disappearance of a phase-slip center in part 5 (PSC 5) leads to changes in the voltage and differential resistance of the characteristic of the phase-slip center in part 4 (PSC 4).

flows through part 5 and which switches on PSC 5 if it has become sufficiently large. The voltage generated by PSC 5 is measured by detecting V_{DF}.

Then, we measured the V_{DF}-I_{EF} characteristics of PSC 5 and at the same run also recorded the voltage V_{AD} of PSC 2 as a function of I_{EF} with different positive and negative fixed currents I_{BC} as a parameter. As can be seen from Fig. 107, the voltage developed by PSC 2 due to the constant current I_{BC} is only constant until PSC 5 appears. Then the voltage (and the differential resistance) changes. For both polarities of I_{BC} the absolute value of the voltage (and of the differential resistance) generated by PSC 2 is enhanced. The enhancement becomes larger with increasing current I_{EF} up to a maximum value and then rapidly decreases again if I_{EF} is further increased.

The interpretation of this effect is not clear at the moment. Several of the measurements described were performed for different temperatures and currents I_{BC}. In all cases the same qualitative behaviour is found, but the magnitude of the effect depends on the parameters. As no quantitative evaluation of the curves has been done up to now, no statement about the parameter which governs the effect can be made.

By changing the current through PSC 5, also the voltage generated by this (influencing) phase-slip center is changed and, therefore, also the repetition frequency of the phase-slip process (via the Josephson relation). The voltage developed by the influenced phase-slip center (PSC 2) depends on I_{BC} and is further changed by the influence of PSC 5. Thus, not only the

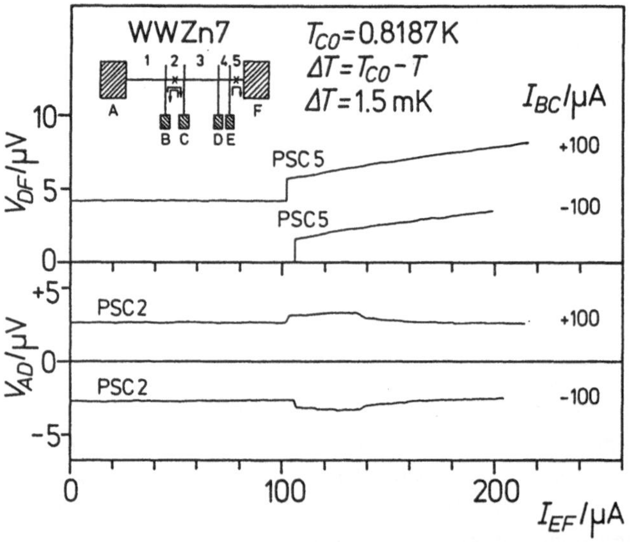

Fig. 107: *V_{DF}-I_{EF} and V_{AD}-I_{EF} characteristics of sample WWZn 7 for increasing current I_{EF}, with a fixed current I_{BC} as a parameter. Characteristics with the same I_{BC} were measured at the same run.*

In the V_{DF}-I_{EF} characteristics the voltage is zero until a phase-slip center (PSC 5) appears in part 5 of the specimen. The zero voltage point of the upper characteristic has been shifted for clearness and is indicated by the horizontal line which the characteristic starts from.

In the V_{AD}-I_{EF} characteristics there is a constant nonzero voltage (generated by PSC 2 in part 2 which is activated by the current I_{BC}) until PSC 5 appears which changes the voltage generated by PSC 2.

temperature and the currents have to be considered. It may also be possible that the observed effect is generated by a frequency dependent long range interaction between the two phase-slip centers.

We have not performed a measurement to answer the question whether also neighbouring phase-slip centers exhibit the reported effect. This is not possible, because we do not have enough measuring lines from the sample to the measuring electronics (there are voltage and current leads at contacts A and F, but contacts B - E are only supplied with one line each).

These experiments about the interaction of phase-slip centers in Zn whiskers are our most recent results and we are ending our remarks on ongoing work at this point.

14. Conclusions

In the present work we report about the investigation of current-induced nonequilibrium phenomena in quasi-one-dimensional superconductors, from the early experiments to the actual problems in this research field.

To perform the experiments, tiny metallic filaments have to be prepared and handled. Whiskers or evaporated long microbridges are used. Some of the specimens have several additional potential probes. The typical diameter of a filament is only one micrometer and the total length of the specimen is about one millimeter. In the early experiments the microbridges were cut from thin films with a diamond knife. Today a microntechnology is available and the microbridges are , therefore, prepared by photolithography techniques. Whisker samples have to be prepared by hand under the light microscope. This has been the technique in the early experiments and this method is also used today. Therefore, in the case of whiskers the complexity of the samples which can be made somewhat depends on human properties.

For the measurements on whiskers, we developed a special low temperature technique, so that now experiments can be performed in the whole temperature range between 0.45 K and 7.5 K. In all cases the specimen is directly immersed in a helium bath and highly stabilized temperatures can be adjusted. Thus, whiskers from very different materials can be investigated.

The basic phenomenon observed is that a current carrying quasi-one-dimensional superconductor changes from the superconducting to the normal state in a series of regular voltage steps. Today we know that this phenomenon is generated, because at the critical current the filament enters a dissipative state governed by interacting phase-slip centers and, thus, represents a complicated problem of nonequilibrium superconductivity appearing above the critical current.

To see this, first investigations of the equilibrium properties of the filament such as the critical current and the critical temperature have been performed. Then detailed studies followed, concerning the single phase-slip center and the interaction of phase-slip centers in filaments of different materials.

To understand these results a lot of theoretical work had to be done. The description needed is beyond the time independent Ginsburg-Landau theory and the BCS theory. An understanding of charge imbalance phenomena, the energy mode and collective excitations and the combination of these effects is needed to develop models for a phase-slip center in a quasi-one-dimensional superconductor. The famous SBT model considers charge imbalance phenomena and the KSS model, moreover, includes collective excitations. On the other hand, there is the systematic generalization of the time-dependent Ginsburg-Landau (TDGL) theory and its application to the quasi-one-dimensional case. The TDGL theory is the only approach which is directly based on the microscopic theory of superconductivity. Furthermore, some model calculations are presented in the present work to describe some special phenomena such as the slope of the electrochemical quasiparticle potential of a phase-slip center, the behaviour of a phase-slip center in a HF-radiation field, and the hysteresis of a phase-slip center.

The basic mechanism of a single phase-slip center (generating the first voltage step in a characteristic) seems to be well understood: There is a periodic collapse and subsequent re-establishment of the order parameter in the core of a phase-slip center. This relaxation oscillation repeats at Josephson frequency. In each collapse the phase difference across the filament is reduced by 2π. This leads to a stationary state, because this phase-slip mechanism leads to a phase loss which is equal to the increase of phase difference during the phase-slip cycle. During each phase-slip cycle nonequilibrium quasiparticles are generated creating a charge imbalance on both sides of the phase-slip center. The diffusion and relaxation of these nonequilibrium quasiparticles governs the quasiparticle currents flowing through the phase-slip center and, thus, the voltage developed.

In detail, properties are more involved: Charge imbalance waves can be excited by the phase-slip center which act back to the phase-slip process. Moreover, a phase-slip center does not only excite the charge imbalance mode but also the energy mode, leading to a symmetric overpopulation of the excitation spectrum of the superconductor. Finally, Joule heating effects may occur.

Charge imbalance waves seem to be important for the slope of the electrochemical quasiparticle potential near a phase-slip center. Also the slope of the characteristics of strong-coupling Pb whiskers (in which the voltage steps are degenerated to linear portions not separated by voltage jumps) may be understood by the consideration of charge imbalance waves. Moreover, experiments on the hysteretic behaviour of a phase-slip center can only be understood by considering quasiparticle overpopulation effects in its core, Joule heating effects, and charge imbalance waves.

Some observations can only be described within the TDGL theory: The normalized normal-like length, L_{An1}/ξ, calculated from the differential resistance of a voltage step, and the ratio I_0/I_c of the extrapolated zero voltage intercept and the critical current show a systematic development with the pair-breaking parameter, γ, of the theory. In a certain γ range the experimental results of L_{An1}/ξ are in qualitative and reasonable quantitative agreement with the theory. Moreover, a decreasing tendency of I_0/I_c as predicted by the TDGL theory is observed in a restricted γ range.

For a fixed sample the TDGL theory predicts that the normal-like length, L_{An1}, and, thus, the differential resistance should run through different temperature laws with increasing strength if the temperature approaches the critical temperature. For filaments of certain materials this behaviour is indeed observed experimentally and all temperature laws given by the theory are appearing in our measurements in their predicted sequence.

Naturally, there are also experimental results which are not understood at the moment: There are filaments of several materials which show a temperature independent differential resistance of their first voltage step yielding a temperature independent normal-like length throughout the investigated temperature range. Moreover, also filaments of those materials which exhibit the mentioned temperature dependent normal-like lengths close to the critical temperature, change to a temperature independent behaviour at lower temperatures, somewhat further away from T_{c0}. Only some first attempts are made to come to a theoretical description of this phenomenon. It seems to have something to do with the fact that the specimen leaves the range of the local equilibrium approximation used for the derivation of the generalized TDGL equations.

Another important feature not yet understood is the long range interaction of phase-slip centers in whiskers. There is no model which predicts the influence of an active phase-slip center on other parts of the sample which are so far away as observed in our experiments.

Next, there are experiments given in our 'remarks on ongoing work' which remain unexplained. It is not clear which mechanism is responsible for the low temperature behaviour of a phase-slip center. Also the change of the characteristic properties of a phase-slip center due to the change of the strength of superconductivity in our tunable weak-link experiments is not understood.

Finally, not much is known about the core region of a phase-slip center. The inverse-ac-Josephson-like effects observed in a HF-radiation field seem to be generated in the core region. Moreover, the properties in the core region seem to be important for the voltage-current characteristic and for the hysteretic behaviour of a phase-slip center. Charge imbalance waves act back to the core region, influence the V-I characteristics, and generate hysteresis. Furthermore, the magnitude of the quasiparticle overpopulation at the core is important for the magnitude of the hysteresis of a phase-slip center.

The detailed structure of the core has not yet been investigated. Here the phase-slip processes occur. The TDGL theory yields a completely continuous phase-slip process with the order parameter becoming periodically zero at one point. This theory is the only approach which can describe the core region. Phase-slip models use certain assumptions about the behaviour in this region.

The SBT model predicts that the time average of the electrochemical pair potential changes step-like at the core. The experiment of Dolan and Jackel establishes this assumption. Nevertheless, it is not clear, how sharp this step is. To see this, a much higher spatial resolution which is much better than the Ginsburg-Landau coherence length would be required.

We expect that our ongoing research activities will somewhat contribute to clarify the still unresolved problems in the field of current-induced nonequilibrium phenomena in quasi-one-dimensional superconductors reviewed in the present work.

A. Appendix

A.1. Different Definitions of the Charge Imbalance

In eq. (33) we defined the charge imbalance, Q^*, by

$$Q^* = (-2e/\Omega)\sum_{\underline{K}}(u_{\underline{K}}^2 - v_{\underline{K}}^2)\tilde{f}_{\underline{K}} \qquad \text{(A1)}$$

which is equal to

$$Q^* = -2e\int_{-\mu_{c,p}}^{\infty}(u_{\underline{K}}^2 - v_{\underline{K}}^2)\tilde{f}_{\underline{K}}(\varepsilon_{\underline{K}})N_n(\varepsilon_{\underline{K}})\,d\varepsilon_{\underline{K}} \qquad \text{(A2)}$$

In equilibrium, where $\tilde{f}_{\underline{K}}(E_{\underline{K}}(\varepsilon_{\underline{K}}))$ is equal to the Fermi function, $f_{\underline{K}}$, it follows $Q^*=0$. Thus, only the sum (or integral) over the deviations $\delta\tilde{f}_{\underline{K}} = \tilde{f}_{\underline{K}} - f_{\underline{K}}$ determines Q^* and $\tilde{f}_{\underline{K}}$ may be replaced by $\delta\tilde{f}_{\underline{K}}$ in eqs. (A1) and (A2). Since the energy of a quasiparticle excitation (and, thus, the corresponding value for $\varepsilon_{\underline{K}}$) usually is close to μ_F, also $\delta\tilde{f}_{\underline{K}}$ will only be nonzero close to μ_F so that $N_n(\varepsilon_{\underline{K}})$ in eq. (A2) can be replaced by N_0 and the lower border $-\mu_{c,p}$ by $-\infty$. Thus, eq. (A2) yields

$$Q^* = -2eN_0\int_{-\infty}^{\infty}(u_{\underline{K}}^2 - v_{\underline{K}}^2)\,\delta\tilde{f}_{\underline{K}}(\varepsilon_{\underline{K}})\,d\varepsilon_{\underline{K}} \qquad \text{(A3)}$$

The definition of the expression for the charge imbalance in the literature differs. In the present work Q^* is a charge density, whereas a particle density $Q^*_N = Q^*/(-e)$ is introduced in refs. 56 and 65 - 67.

In ref. 64 an energy density Q^*_ε is introduced by equating due to charge neutrality

$$0 = 2N_0(\mu_{c,p} - \mu_F) + 2\int_{-\infty}^{\infty}(u_{\underline{K}}^2 - v_{\underline{K}}^2)\,g(\varepsilon_{\underline{K}})N_n(\varepsilon_{\underline{K}})\,d\varepsilon_{\underline{K}} \qquad \text{(A4)}$$

where $g(\varepsilon_{\underline{K}}) = \delta\tilde{f}_{\underline{K}}(E_{\underline{K}}(\varepsilon_{\underline{K}}))$. Replacing $N_n(\varepsilon_{\underline{K}})$ by N_0 due to the reasons discussed above it follows

$$Q^*_\varepsilon = -(\mu_{c,p} - \mu_F) \qquad \text{(A5)}$$

where

$$Q^*_\varepsilon = \int_{-\infty}^{\infty}(u_{\underline{K}}^2 - v_{\underline{K}}^2)\,g(\varepsilon_{\underline{K}})\,d\varepsilon_{\underline{K}} \qquad \text{(A6)}$$

Thus, $Q^*_\varepsilon = Q^*/(-2eN_0)$. Using, [2], $N_n(\varepsilon_{\underline{K}})d\varepsilon_{\underline{K}} = N_q(E_{\underline{K}})dE_{\underline{K}}$ in eq. (A4), where $N_q(E_{\underline{K}}) = N_0E_{\underline{K}}(E_{\underline{K}}^2 - \Delta^2)^{-1/2}$ is the quasiparticle density of states, leads to

$$Q^*_\varepsilon = \int_{E_{\underline{K}}}(u_{\underline{K}}^2 - v_{\underline{K}}^2)\,g(E_{\underline{K}})\,E_{\underline{K}}(E_{\underline{K}}^2 - \Delta^2)^{-1/2}\,dE_{\underline{K}} \qquad \text{(A7)}$$

Here, $E_{\underline{K}}$ runs over the two branches of the excitation spectrum from ∞ to Δ and then from Δ to ∞. In the first case, however, $dE_{\underline{K}} < 0$, so that

$$Q^*_\varepsilon = 2\int_\Delta^\infty (u_{\underline{K}}^2 - v_{\underline{K}}^2) g(E_{\underline{K}}) E_{\underline{K}} (E_{\underline{K}}^2 - \Delta^2)^{-1/2} |dE_{\underline{K}}| \qquad (A8)$$

The representation of eq. (7) is used in eq. (30) of ref. 24, where, however, the factor $E_{\underline{K}}(E_{\underline{K}}^2 - \Delta^2)^{-1/2}$ has been lost.

A. 2. Remarks on the Calculation of the Quasiparticle Chemical Potential

The sum in eq. (41) can be evaluated in the following way:

$$S_u := (-2/\Omega e) \sum_{\underline{K}} Q_{\underline{K}}^2 (-\partial f_{\underline{K}}/\partial E_{\underline{K}}) = (-2/e) \int_{-\mu_{c,q}}^{\infty} Q_{\underline{K}}^2 (-\partial f_{\underline{K}}/\partial E_{\underline{K}}) N_n(\varepsilon_{\underline{K}}) d\varepsilon_{\underline{K}} \qquad (A9)$$

Since $\partial f_{\underline{K}}/\partial E_{\underline{K}}$ is a bell-shaped function centered at $\varepsilon_{\underline{K}} = 0$ which is already very small at $-\mu_{c,q}$, the integral can be extended to $-\infty$. Then the integral is split into two parts running from $-\infty$ to 0 and from 0 to ∞. Next, the quasiparticle density of states [2], N_q, is introduced by $N_n(\varepsilon_{\underline{K}}) d\varepsilon_{\underline{K}} = N_q(E_{\underline{K}}) dE_{\underline{K}}$, considering that $d\varepsilon_{\underline{K}}$ and $dE_{\underline{K}}$ have opposite sign in the first integral and same sign in the second integral, so that both integrals are equal. Thus,

$$S_u = (-4/e) \int_\Delta^\infty Q_{\underline{K}}^2 (-\partial f_{\underline{K}}/\partial E_{\underline{K}}) N_q(E_{\underline{K}}) dE_{\underline{K}} \qquad (A10)$$

With $Q_{\underline{K}}^2 = (\varepsilon_{\underline{K}}^2/E_{\underline{K}}^2)(-e)^2$, $\varepsilon_{\underline{K}}^2 = E_{\underline{K}}^2 - \Delta^2$, $N_q(E_{\underline{K}}) = N_0 E_{\underline{K}} (E_{\underline{K}}^2 - \Delta^2)^{-1/2}$ it is

$$S_u = -2 N_0 e Z(T) \qquad (A11)$$

where

$$Z(T) = 2\int_\Delta^\infty [(E_{\underline{K}}^2 - \Delta^2)^{1/2}/E_{\underline{K}}](-\partial f_{\underline{K}}/\partial E_{\underline{K}}) dE_{\underline{K}} \qquad (A12)$$

Asymptotic values for $Z(T)$ for $\Delta \ll kT$ and $\Delta \gg kT$ are given in the appendix of ref. 80. In the first case it is

$$Z(T) = 1 - \pi\Delta/4kT + (7\zeta(3)/4\pi^2)(\Delta/kT)^2 \ldots \qquad (A13)$$

where $\zeta(3) = 1.202\ldots$

A. 3. Inelastic Electron-Phonon Scattering Time

In Tab. A1 measured values for the inelastic electron-phonon collision time are summarized for several materials. For a critical discussion see chap. 5.4. It turns out that for dirty Al films the measured quantity actually has not to be interpreted as τ_E but as τ_{ee}. The table also contains values calculated from eqs. (72), (74), and (87) of the present work. The material parameters needed for this calculation are given in Tabs. A2 and A3. To characterize and comment the tabulated values several abbreviations are used.

The first group of abbreviations characterizes the measuring method:

TI - n, s	Charge imbalance created by tunnel injection using an NC1/I/SC/I/NC2 contact (n) or an SC1/I/SC2/I/SC3 contact (s)
TEH	Enhancement of the energy gap due to tunnel injection or extraction of quasiparticles
LVR	Low voltage resistance of an SC/I/NC tunnel junction
OPRL - i, t, a, c	Order parameter relaxation time measurements by: real time response to laser pulse illumination (i), time delay of the voltage response to a supercritical current pulse (t), critical current measurements with superimposed ac current (a), conductivity measurements applying a large chopped dc and a small continuous ac current (c)
QPRK - m, c	Quasiparticle recombination time as measured (m) or calculated from the mean free path for phonon reabsorption (c)
PSC - VI	Differential resistance of the V-I characteristics of a phase-slip center
PSC - μ	Spatially resolved measurements of the electrochemical quasiparticle potential near a phase-slip center
PSC - te	Time evolution of the voltage at a phase-slip center
SC/NC	Boundary resistance of SC/NC interfaces
FF	Flux flow behaviour
Foot	Foot structures in the characteristics of short weak links
HF	Low frequency border of microwave radiation-stimulated superconductivity
CALC	Values calculated from a theory

The second group of abbreviations indicates remarks to the given values and evaluation procedures:

$*_1$ Since T_{c0} of the mentioned sample is not given in the reference, the value for pure bulk material is given in parentheses.

$*_2$ The relaxation time is given for $\Delta/kT=4$. With $2\Delta(0)=4.3\,kT_{c0}$ for Pb it follows (similar to remark '$*_{10}$') $T/T_{c0}\approx 0.5$, so that from Fig. 7 of ref. 103 it follows $\tau_r(\mathrm{Pb})\approx 0.45\,\tau_0(\mathrm{Pb})$, where $\tau_0=8.4\,\tau_E$.

$*_3$ For the calculation eq. (74) of the present work has been used, that means $\tau_E=\tau_0/8.4$.

$*_4$ Using Tinkham's estimate as given in eq. (72) of the present work

$*_5$ Calculated from eq. (87) of the present work

$*_6$ Values for τ_E in parentheses were calculated considering that the gap is reduced by the transport current.

$*_7$ For the calculation of τ_E, eqs. (61) and (70) of the present work were used.

$*_8$ The value for τ_E has been calculated by using the theoretical expression for the order parameter relaxation time in the presence of a gap (see for instance eq. (22) of ref. 9).

*9 It is $\tau_r = 2.0 \times 10^{-9}$s for $\Delta/kT = 4$.

*10 The time τ_r is given for $\Delta/kT = 4$, that means for a weak-coupling superconductor with $2\Delta(0) = 3.5\,kT_{c0}$ it is $(T/T_{c0})(\Delta/\Delta(0))^{-1} = 3.5/8$. From the BCS result for the dependence of the gap on the temperature (see for instance Fig. 36 of ref. 1) it follows $T/T_{c0} \approx 0.43$ and then from Fig. 2 of ref. 103 (for a quasiparticle energy of $\Delta(T)$) it is $\tau_r/\tau_0 \approx 2$, where $\tau_0 = 8.4\,\tau_E$.

*11 For the calculation $\tau_{Q^*in}(0) = 0.42\,\tau_E$ as introduced in eq. (71) of the present work has been used.

*12 For the values of τ_E see Tab. 2 of ref. 72 together with eq. (4) of that work.

*13 Besides the values for ℓ the film thickness, d, is given in parentheses as long as $\ell > d$.

*14 Electron mean free path calculated from the residual resistance [15] with $\rho_n \cdot \ell$ for pure indium [40]

*15 See Fig. 4-7 of ref. 33 and use eqs. (61) and (70) of the present work.

*16 Work contains results for several samples with different values of T_{c0} and ℓ. Here, the ranges for T_{c0}, ℓ, and τ_E, respectively, are given.

*17 Averaged value for the cleanest films ($T_{c0} < 1.3$ K, $\ell > 49$ nm) at an averaged critical temperature of 1.24 K.

*18 Since ℓ is not given in that work, the thickness of the film is given in parentheses.

*19 It is $\tau_r = 2 \cdot 2.86 \cdot (1.4 \pm 0.1)\,\mu s$ for $\Delta/kT = 6$. For the factor 2 see chap. 5.4 of the present work, for the factor 2.86 see remark 63 of ref. 103. With $\Delta \approx \Delta(0)$ from $\Delta/kT = 6$ we get $T/T_{c0} \approx 3.5/12$. Then it follows with eq. (14) of ref. 103, $\tau_0/\tau_r = 5.44 \cdot 10^{-2}$, where $\tau_E = \tau_0/8.4$.

*20 According to ref. 148 the value for τ_E is less reliable.

*21 In that work ℓ has been replaced by the diameter of the sample for clean samples where ℓ is much larger than the diameter.

*22 There is no measured value for c_L and c_T of indium available. As In is close to being cubic (see p. 404 of ref. 347) we calculated both quantities from the elastic constants of indium using relations which are valid for a cubic crystal [530].

Finally, we remark, that for several samples from alloys it could not be identified in the literature whether at% or wt% are given. For <u>Sn</u>-In and <u>Pb</u>-Bi alloys both quantities are, however, nearly equal.

Tab. A1: *Measured and calculated values for the life-time τ_E of electrons in the normal state at the Fermi level due to electron-phonon collisions ('inelastic collision time').* ▶

For dirty Al the measured τ_E has probably to be interpreted as the electron-electron collision time, τ_{ee}. Furthermore, the critical temperature, T_{c0}, the electron mean free path, ℓ, and the measuring method are given.

Remarks indicated by '' are mentioned in the text.*

Material	T_{c0} / K	ℓ / nm	τ_ε / 10^{-10} s	Method	Ref. / Remarks
Pb	(7.23)	-	0.38	QPRK-c	132 Tab. 3/*[1,2]
Pb	7.19	5090	0.20	SC/NC	93
$Pb_{0.99}Bi_{0.01}$	7.16	49	0.25	SC/NC	93
$Pb_{0.98}Bi_{0.02}$	7.21	26.5	0.35 ± 0.07	SC/NC	138
Pb	7.19	dirty	0.23	CALC	103 Tab. 1/*[3]
Pb	7.23	pure bulk	0.0555	CALC	*[4]
Pb	7.23	∞	21.6	CALC	$\tau_{\varepsilon\infty}$, *[5]
Pb	7.23	10	1.73	CALC	Fig. (A1), pres. work
Ta	4.38	1000	1	SC/NC	139
Ta	4.12	1.3	0.9 (0.7)	SC/NC	61 /*[6,7]
Ta	4.48	dirty	2.12	CALC	103 Tab. 1 /*[3]
Sn	3.81	100	3.9	TI-n	62, 109 (p. 491)/*[7]
Sn, $Sn_{97.0}In_{3.0}$	~3.8, 3.86	280, 42	1.4 ± 0.2	TI-n	109/wt%, *[7]
Sn, Sn_xIn_{100-x}	-	-	1.5 ± 0.3	TI-n	105/x=3.0, 4.0wt%, *[7]
Sn	3.72	187	3.60	LVR	117
Sn	3.72	183	3.50	LVR	117
Sn	3.72	160	3.57	LVR	117
$Sn_{0.95}In_{0.05}$	3.69	37.5	3.33	LVR	117
$Sn_{0.95}In_{0.05}$	3.68	35.8	4.05	LVR	117
Sn, Sn In	-	-	3.6 ± 0.4	LVR	117 / averaged value
Sn	(3.722)	220	2.2	OPRL-c	125 /*[1,8]
Sn	(3.722)	122	1.9	OPRL-c	126 /*[1]
Sn	(3.722)	-	4.8	QPRK-m	132 Tab. 3 /*[1,9,10]
Sn	(3.722)	-	3.0	QPRK-c	132 Tab. 3 /*[1,10]
Sn	3.82	80	5.0	PSC-VI	84 / bridge 29 B
Sn	3.905	20	3.0	PSC-VI	84 / bridge 15 A
Sn	-	-	1 - 5	PSC-VI	84 / range
Sn	(3.722)	130	2.1 ±0.5 (1.8 ±0.4)	PSC-μ	70 /*[1,6,11]
Sn	(3.722)	180 (109)	1.4 ±0.3 (1.1 ±0.2)	PSC-μ	72 /*[1,6,12,13]
Sn	(3.722)	120 (102)	1.6 ±0.7 (1.3 ±0.6)	PSC-μ	72 /*[1,6,12,13]
Sn	(3.722)	80	1.9 ±0.8 (1.6 ±0.7)	PSC-μ	72 /*[1,6,12]
Sn	-	-	1.6 ± 0.4	PSC-μ	72 / averaged value
Sn	3.73	3550	2.6	SC/NC	93
$Sn_{0.99}In_{0.01}$	3.74	249	1.1	SC/NC	93
Sn	3.82	35	5.0	FF	140
Sn	~3.8	~100	2.8	Foot	141, 142
Sn	3.8	~115	1.4	HF	145
Sn	3.75	dirty	2.74	CALC	103 Tab. 1 /*[3]
Sn	3.722	pure bulk	3.56	CALC	*[4]
Sn	3.722	∞	402	CALC	$\tau_{\varepsilon\infty}$, *[5]
Sn	3.722	10	60.3	CALC	Fig. (A1), pres. work
In	3.405	1221	0.80	OPRL-a	124 /*[14]
In	3.367	1179	3.62 (2.98)	PSC-VI	40 /*[6]
In	3.356	926	3.31 (2.73)	PSC-VI	40 /*[6]

Material	T_{c0} / K	ℓ / nm	$\tau_E / 10^{-10}$ s	Method	Ref. / Remarks
In	3.350	738	2.73 (2.25)	PSC-VI	40 / *[6]
$In_{99.2}Pb_{0.8}$	3.406	333	1.45 (1.19)	PSC-VI	40, 14 / at%, *[6]
$In_{99.0}Pb_{1.0}$	3.410	292	1.09 (0.90)	PSC-VI	40, 14 / at%, *[6]
$In_{98.0}Pb_{2.0}$	3.477	145	0.95 (0.78)	PSC-VI	40, 14 / at%, *[6]
$In_{95.4}Pb_{4.6}$	3.702	67.6	2.57 (2.12)	PSC-VI	40, 14 / at%, *[6]
In	(3.405)	90	0.69 (0.57)	PSC-VI	33 / *[1, 6, 7, 15]
In	(3.405)	~375	1.5	PSC-te	137 / *[1]
In	3.41	1870	1.1	SC/NC	93
In	3.448	161	1.6	FF	140
In	3.422	145	1.4	FF	140
In	3.481	111	1.0	FF	140
In	3.486	80	0.8	FF	140
In	3.482	52	0.5	FF	140
In	3.40	dirty	0.95	CALC	103 Tab. 1 / *[3]
In	3.405	pure bulk	2.52	CALC	*[4]
In	3.405	∞	138	CALC	$\tau_{E\infty}$, *[5]
In	3.405	10	10.0	CALC	Fig. (A 1), pres. work
Al	1.219-2.113	409-1.0	143 - 10	TI-n	92 Tab. 1 / *[3, 16]
Al	< 1.3 (1.24)	> 49	119 ± 24	TI-n	92 eq. (4.1) / *[3, 17]
Al	-	-	119	TI-s	110 Fig. 22 / *[3]
Al	1.28	-	119 ± 17	TI-s	112 / *[3]
Al	2.4	-	2.7	TI-s	111 / *[3]
Al	1.321, 1.353	(45), (37)	98.8, 91.7	TEH	115 / *[3, 18]
Al	-	-	107	TEH	110 / *[3]
Al	1.211, 1.321	180, 70	160, 70	OPRL-i	118, 119
Al	1.245	(75)	76	OPRL-t	120 / *[18]
Al	1.30	(100)	44	OPRL-a	123 / *[18]
Al	-	(30)	520 ± 40	QPRK-m	135 / *[18, 19]
Al	(1.175)	-	500	QPRK-c	132 Tab. 3 / *[1, 10]
Al	1.20-1.31	10-95.4	435	PSC-VI	136 Fig. 2
Al	1.20-1.25	58.3-95.4	40 - 143	PSC-VI	136 Tab. 2 / *[16]
Al	1.25-1.31	27 - 52	40	PSC-μ	73 Fig. 6 / *[16]
Al	1.428, 1.947	13.5, 2.25	35, 8	FF	140
Al	1.19	-	100	Foot	143
Al	1.226-1.860	87.8-4.7	128 - 2.2	HF	146 Tab. 1 / *[16]
Al	1.19	dirty	521	CALC	103 Tab. 1 / *[3]
Al	1.175	pure bulk	464	CALC	*[4]
Al	1.175	∞	$9.63 \cdot 10^3$	CALC	$\tau_{E\infty}$, *[5]
Al	1.175	10	891	CALC	Fig. A 1, pres. work
Zn	0.9	-	(1.0)	PSC-VI	148 / *[20]
Zn, $Zn_{100-x}Al_x$	0.843-0.684	1950-280	320 ± 75	PSC-VI	20 / x = 1.2, 1.1 at%, *[21]
Zn	0.875	dirty	929	CALC	103 Tab. 1 / *[3]
Zn	0.875	pure bulk	1810	CALC	*[4]
Zn	0.875	∞	$3.30 \cdot 10^4$	CALC	$\tau_{E\infty}$, *[5]
Zn	0.875	10	6765	CALC	Fig. A 1, pres. work

Tab. A 2: *Material parameters for polycrystalline bulk material as used for the calculation of τ_E after Tinkham (see eq. (73) of the present work)*

Material	$\rho_\Theta \ell_\Theta$ [$10^{-3}\,\Omega\,\mu m^2$]	ρ_{298K} [$10^{-2}\,\Omega\,\mu m$]	v_F [10^6 m/s]	Θ [K]	T_{c0} [K]	References
Pb	0.70	21.2	0.61	104.0	7.23	14, 40
Sn	1.0 [a)]	11.3 [d)]	0.684 [c)]	200 [b)]	3.722 [a)]	a: 15, b: 26, c: 89, d: 349
In	1.48	9.06	0.72	111.2	3.405	14, 40
Al	0.49	2.84	1.3	400	1.175	136, 527
Zn	1.8	6.0	0.91	322.3	0.875	20

Tab. A 3: *Material parameters for the calculation of $\tau_{E\infty}$ and $\tau_E(\ell)$ from eqs. (87) and (88) of the present work, respectively. For T_{c0} and v_F see Tab. A 2. Values are valid for polycrystalline bulk material (except c_L and c_T for In which are given for a [100] and [001] direction in a single crystal).*

Material	ρ [10^3 kg/m^3]	c_L [10^3 m/s]	c_T [10^3 m/s]	References/Remarks
Pb	11.4	2.16	0.70	528
Sn	7.3	3.32	1.67	528
In	7.31	2.47	0.95	529, 530 / *22
Al	2.7	6.42	3.04	528
Zn	7.1	4.21	2.44	528

In Fig. A1 the dependence of τ_E on the electron mean free path is plotted for several materials as given by eq. (88) of the present work. Furthermore, a comparison with experiment is done in Fig. A2. For this purpose we took experimental values for τ_E from Tab. A1 and transformed the data to the critical temperature of pure bulk material as given in Tab. A2, using $\tau_E \sim T_{c0}^{-3}$. To calculate an experimental value of $\tau_{E\infty}$ at the critical temperature of pure bulk material it is then assumed that the experimental result for the largest ℓ is equal to the theoretical prediction. All other values are then divided through this experimental $\tau_{E\infty}$ and plotted into Fig. A2 at their electron mean free path.

For In we plotted all PSC-VI data (except sample $In_{95.4}Pb_{4.6}$), the PSC-te result, and all FF data. The data for ℓ and τ_E of the PSC-VI result of ref. 33 and the FF data of ref. 140 are transformed before plotting, because the material parameters $\rho_n \cdot \ell = 1.0 \cdot 10^{-3}\,\Omega\,\mu m$ and $v_F = 1.74 \cdot 10^6$ m/s used in that works are different from those given in ref. 40 and in Tab. A2 of the present work. While ℓ was calculated from the residual resistance and, thus, has simply to be multiplied by a factor 1.48, the kind of transformation used for τ_E depends on the procedure used for its evaluation from the experiments. In

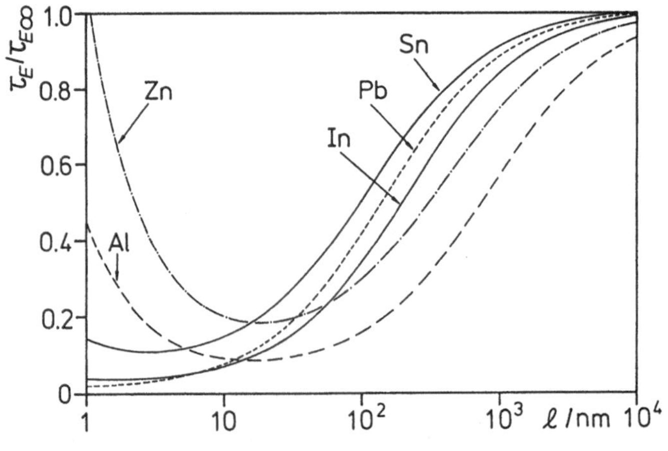

Fig. A1: *Electron mean free path dependence of the life-time τ_E of electrons in the normal state at the Fermi level due to electron-phonon collisions for Pb, Sn, In, Al, and Zn, respectively, as evaluated from the theory of Keck and Schmid (see eq. (88) of the present work).*

Values for $\tau_{E\infty}$ (that means for the limit of infinite electron mean free path ℓ) are given in Tab. A1 for each material. Here ℓ is the impurity induced electron mean free path as may be for instance calculated from the residual resistivity of a sample. Thus, $\tau_{E\infty}$ is the value of τ_E in a clean perfect infinite sample. The representation is semi-logarithmic.

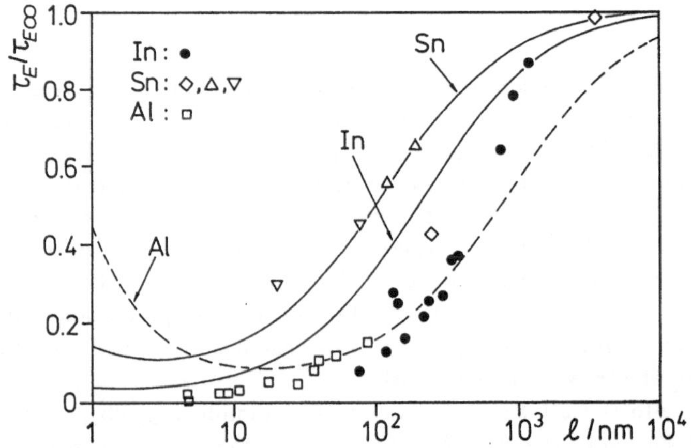

Fig. A2: *Comparison of the predicted $(\tau_E/\tau_{E\infty})(\ell)$ with experimental data from Tab. A1*
(●) In from refs. 33, 40, 137, and 140 with $\tau_{E\infty} = 4.05 \cdot 10^{-10}$ s
(◇, △, ▽) Sn from refs. 93, 125 and 126, 84 with $\tau_{E\infty}$ = 2.65, 3.38, 11.96 $\cdot 10^{-10}$ s
(□) Al from ref. 146 with $\tau_{E\infty} = 9.67 \cdot 10^{-10}$ s.
More details are given in the text.

ref. 33 τ_E is calculated from the quasiparticle diffusion length as concluded from the differential resistance. Thus, τ_E has to be multiplied by a factor $1.0 \cdot 1.74/1.48 \cdot 0.72 = 1/0.612$. In ref. 140 τ_E is calculated from the flux-flow behaviour (eq. (4) of ref. 140) and has, thus, to be multiplied by a factor of 0.612.

For Sn we plotted results from several references which, however, have to be normalized by different $\tau_{E\infty}$ calculated from the value with the larger ℓ of each pair of data. For Al we only plotted the results of ref. 146.

As can be seen from Fig. A 2, there seems to be some evidence for an electron mean free path dependence of τ_E. For In this dependence even seems to be stronger than calculated. For Al only a few measuring points join the theoretical curve. It is remarkable that these are just those points with a large sheet resistance, $R_\square$, which deviate from the theoretical curve for electron-electron scattering, as can be seen from Fig. 2 of ref. 146. For Sn only pairs of experimental results, each with its own $\tau_{E\infty}$ value, show the predicted behaviour.

However, there are several measurements given in Tab. A1 which were not plotted in Fig. A 2 and which do not follow the prediction for $(\tau_E/\tau_{E\infty})(\ell)$. In the case of In these are for instance the OPRL-a result of ref. 124 and the SC/NC result of ref. 93. For Al, for instance, none of the TI-n results of ref. 92 shows the predicted behaviour. For Sn the LVR results of ref. 117, the TI-n measurements of ref. 109, and the PSC-μ data of ref. 72 do not indicate any systematic dependence of τ_E on ℓ. For Pb the SC/NC results of ref. 93 indicate that there should not be any ℓ dependence of τ_E. For Zn there are not enough experimental values available to draw any conclusion.

A.4. Remarks on the Derivation of the Charge Imbalance Wave Equation

To evaluate the denominator in eq. (105), the actual nonequilibrium occupation probability of a $\underline{K}$ state with a quasiparticle, $\tilde{f}_K$, is split into two parts (as done in section 5.3), so that $\tilde{f}_{\underline{K}} = f_{\underline{K}}(E_{\underline{K}}) + \delta\tilde{f}_{\underline{K}}$. For a pure charge imbalance $\delta\tilde{f}_{\underline{K}}$ is odd in $\varepsilon_{\underline{K}}$. Since $E_{\underline{K}}$ is even in $\varepsilon_{\underline{K}}$, the contributions $(\mathbf{\Delta}^2/E_{\underline{K}}^3)\delta\tilde{f}_{\underline{K}}$ cancel out. The function $f_{\underline{K}}(E_{\underline{K}})$ may be expanded into a Taylor series at $E_{\underline{K}}/kT = \mathbf{\Delta}/kT$, yielding $f_{\underline{K}}(E_{\underline{K}}) \approx (1/2) - (1/4)(E_{\underline{K}}/kT - \mathbf{\Delta}/kT)$ for $E_{\underline{K}}/kT \ll 1$ which implies $\mathbf{\Delta}/kT \ll 1$. Since $f_{\underline{K}}(E_{\underline{K}})$ is weightened by $\mathbf{\Delta}^2/E_{\underline{K}}^3$ which rapidly decreases for $E_{\underline{K}} > \mathbf{\Delta}$, the approximation for $f_{\underline{K}}(E_{\underline{K}})$ may be used for all $\underline{K}$. Neglecting contributions $\sim \mathbf{\Delta}^3$ which should be allowed close to T_{c0} where $\mathbf{\Delta}$ is small, it follows

$$(1/\Omega N_0)\sum_{\underline{K}} \tilde{f}_{\underline{K}} \mathbf{\Delta}^2/E_{\underline{K}}^3 \approx (1/N_0)\int_{-\mu_{c,p}}^{\infty} (1/2 - E_{\underline{K}}/4kT)(\mathbf{\Delta}^2/E_{\underline{K}}^3) N_n(\varepsilon_{\underline{K}})\, d\varepsilon_{\underline{K}} \qquad \text{(A14)}$$

Introducing the quasiparticle density of states (see section A.2) the right hand side of eq. (A14) yields

$$2\int_{\mathbf{\Delta}}^{\infty} (1/2 - E_{\underline{K}}/4kT)(\mathbf{\Delta}^2/E_{\underline{K}}^2)(E_{\underline{K}}^2 - \mathbf{\Delta}^2)^{-1/2}\, dE_{\underline{K}} = 1 - \pi\mathbf{\Delta}/4kT \qquad \text{(A15)}$$

The integral has been evaluated by splitting it into two parts. Defining $x_1 := \Delta^2/E_K^2$, the first part is given by $-(1/2)\int_1^0 (1-x_1)^{-1/2} dx_1 = 1$. Defining $x_2 := \Delta/E_K$, the second part is given by $(\Delta/2kT)\int_1^0 (1-x_2^2)^{-1/2} dx = -(\Delta/2kT)\arcsin(1) = -\pi\Delta/4kT$.

A.5. Remarks on the Calculation of the Inelastic Electron-Phonon Scattering Time for In-Pb Alloy Whiskers

In section 9.3 it is remarked that in the case of In-rich In-Pb whiskers we did not strictly follow Tinkham's formula to calculate τ_E in ref. 40, although stated in that work. The formula contains a characteristic time τ_Θ which depends on the resistivity at room temperature (see eq. (73)). As τ_Θ denotes an electron-phonon scattering time, only the phonon-induced, temperature dependent part of the resistivity should be inserted. For pure samples this can be done with sufficient accuracy by using the measured total resistivity of pure bulk material with vanishing residual resistance ratio ρ^*. In the case of alloys with remarkable residual resistance ratio it is not allowed simply to insert the measured total resistivity of bulk material, as we have done in ref. 40.

The results for the correct calculation in this case are given below. It turns out that the conclusions of ref. 40 remain unchanged. Moreover, concerning the mean free path dependence of τ_E (Fig. A 2) the incorrect use of the resistivity is even of advantage. Since the total resistivity is larger than the temperature dependent part, its use leads to a lower value of τ_E. Thus, the reduction of τ_E with decreasing ℓ is somewhat better simulated although being still weaker than the theoretical prediction given in Fig. A 2. This can be seen by plotting into Fig. A 2 the results for τ_E after Tinkham's formula for the whiskers listed in Tab. 1 of ref. 40 (with the correct and the incorrect use of the resistivity). For this purpose it is assumed that the result for the pure In whisker with the largest ℓ (sample In 20) is in agreement with the theoretical curve after Keck and Schmid and then the normalization procedure of τ_E is applied analogous to the case of experimental values (see section A.3).

In the following we reevaluate τ_E for the whiskers of indium rich alloys mentioned in Tab. 1 and Tab. 8 of ref. 40 by strictly applying Tinkham's formula. Moreover, we reevaluate all parameters used in ref. 40 which depend on τ_E:

The total resistivities at room temperature used for our In-Pb alloy whiskers were obtained from the literature. However, the authors did not measure the residual resistance of their bulk material samples. Therefore, we take values for ρ^* of our samples to calculate the temperature dependent part of the resistivity. As pointed out in section 5.4, assuming the validity of Mathiessen's rule, this can be done by dividing the total resistivities from the literature (listed in the fifth column of Tab. 8 of ref. 40) by $(1+\rho^*)$. It is remarked that the value for τ_E of sample 2 given in Tab. 8 of ref. 40 should read $2.21 \cdot 10^{-10}$ s. The values for τ_E are then obtained by multiplying those given in Tab. 8 of ref. 40 by a factor $(1+\rho^*)$. The consequence is that the validity region ΔT_V of the TDGL theory of KR given in that table shrinks by a division through $(1+\rho^*)$ and the value $\Delta T(\gamma=100)$ by a division through $(1+\rho^*)^2$. To enhance accuracy, $\Delta T(\gamma=100)$ has now been evaluated directly from the equation of γ and not extrapolated from other γ values. The values for the pair-breaking parameter, γ, and also the ratios τ_E/τ_{0R} given in ref. 40 have to be enhanced by a factor $(1+\rho^*)$.

The reevaluated quantities are summarized in Tab. A4. The decrease of τ_E with decreasing electron mean free path is less strong. Nevertheless, the conclusions of ref. 40 are unchanged:

There remains an order of magnitude agreement between theoretical and experimental values for the forefactor $\tau_{Q^*}(0)$ of the charge imbalance relaxation time. The KR theory still could be applied for all measuring temperatures and all samples if the samples were dirty. For the normal-like length, L_{An1}, a $\Delta T^{-1/2}$ temperature dependence is observed now for γ between 9 and 24, still in agreement with the KR theory. Nearly all γ values for which $L_{An1} \sim \Delta T^{-1}$ range between 8 and 15. At the time when we wrote down ref. 40, this behaviour was not predicted by the TDGL theory of KR. Later Rangel also found this behaviour to be predicted by the TDGL theory (Fig. 20). The γ values for a $\Delta T^{-1/4}$ law range between 8 and 33 for samples In 20, 1, 2, 5, and 8 which only show this behaviour. The γ values for samples In 17, 'In', and 3 which should be homogeneous, range between 24 and 39. In this region of γ values the KR theory predicts a stronger temperature dependence than observed.

To summarize the results for dirty samples for which the KR theory can be strictly applied, the $\Delta T^{-1/2}$ dependence of L_{An1} is observed for γ between 11.6 and 22.4 as predicted by KR. The samples with a ΔT^{-1} behaviour have almost γ values in the lower part of the γ range for which KR predict a $\Delta T^{-1/2}$ law. Sample 8 with $L_{An1} \sim \Delta T^{-1/4}$ for γ between 7.8 and 21.8 is still far below the γ region for which KR predict this behaviour.

Furthermore, all conclusions concerning the hysteresis remain valid.

Tab. A4: *Re-evaluation of the τ_E values given in ref. 40 and the parameters which depend on this time, as summarized in Tab. 8 of ref. 40 (ρ_{298K}, τ_E, ΔT_V, $\Delta T(\gamma=100)$), Tab. 6 of ref. 40 (τ_E/τ_{OR}, γ_{Hy}), Tab. 1 of ref. 40 ($\tau_{Q^*\underline{In}Pb}(0)$) and Tabs. 2 - 4 of ref. 40 (γ_s, γ_f).*

It is remarked that in the present work ρ_{298K} denotes the temperature dependent phonon-induced part of the resistivity at room temperature while the values summarized in Tab. 8 of ref. 40 are measured total bulk material resistivities. The definition of all quantities is in detail discussed in ref. 40. Most quantities are also introduced in the present work. It may be added that γ_{Hy} characterizes the onset of hysteresis and γ_s and γ_f the limits within which a certain temperature law for the normal-like length is observed.

Sample	$1/\rho^*$	ρ_{298K} [$10^{-2}\,\Omega\,\mu m$]	τ_E [10^{-10} s]	ΔT_V [mK]	$\Delta T(\gamma=100)$ [mK]	τ_E/τ_{OR}
1	21.21	9.36	2.36	12.7	82.1	1.1 - 1.1
2	18.60	9.30	2.33	12.9	84.0	-
3	19.08	9.31	2.30	13.0	85.6	-
5	9.48	9.50	2.09	14.4	102.3	2.0 - 2.2
6.1	5.40	9.79	1.74	17.2	142.4	0.6 - 0.7
7	3.98	9.75	1.58	19.0	169.9	1.9 - 2.1
8	4.85	10.20	1.48	20.2	190.5	2.1 - 2.3
11	2.89	9.96	1.27	23.6	249.8	1.3 - 1.4
13	3.04	10.61	1.09	27.6	333.8	-
14	2.90	10.48	1.08	27.8	335.5	1.1 - 1.2
18	1.80	11.64	0.41	73.4	1858.5	0.8 - 0.9

Sample	γ_{Hy}	Theory $\tau_{Q^*\underline{In}Pb}(0)$ [10^{-10} s]	Tab. II		Tab. III		Tab. IV	
			γ_s	γ_f	γ_s	γ_f	γ_s	γ_f
1	17.1 - 17.5	0.992	-	-	-	-	28.2	11.3
2	-	0.977	-	-	-	-	19.8	11.9
3	-	-	23.5	12.3	-	-	29.0	23.5
5	22.1 - 23.2	0.877	-	-	-	-	32.8	14.7
6.1	11.5 - 12.4	-	15.6	12.7	12.7	9.1	-	-
7	19.3 - 20.1	-	> 22.4	11.6	-	-	-	-
8	19.4 - 20.5	0.621	-	-	-	-	21.8	7.8
11	14.5 - 15.3	-	14.7	12.0	20.9	8.7	-	-
13	-	-	21.9	14.9	14.9	12.2	-	-
14	12.4 - 12.8	-	-	-	14.4	10.3	-	-
18	7.2 - 7.9	-	-	-	9.9	8.4	-	-

Final Note

During the print of the present work we developed additional ideas for the interpretation of our ongoing work mentioned in chap. 13 (for details see ref. 531):

1. The low temperature behaviour of a phase-slip center (sec. 13.1) seems to be governed by charge imbalance waves: We found that the maximum of the differential resistance appears at a temperature for which the decay length of the charge imbalance wave, $\lambda_{d,KSS}$, in the high frequency limit (see p. 71 of the present work) has its maximum value.

2. We roughly estimated the dimensions of our microcontacts (sec. 13.3). The length of the microcontact is of the order 0.2 μm, while the diameters of the contacts range between 0.01 and 0.10 μm.

As far as known to us there is not any phenomenon reported in the literature which would be able to explain the shape of the V-I characteristics of our microcontacts.

We assume that the core region of a phase-slip center establishes in the microcontact and that this phase-slip center shows discrete excited states. For increasing current the phase-slip center changes from one state to the other. Since the voltage is detected by superconducting parts of the whiskers which extend to the borders of the microcontact, the voltage should be a measure for differences of the electrochemical pair potential across the core region of the phase-slip center. As far as we know, excited states of phase-slip centers have not been observed up to now, and there is not any theoretical prediction of these states.

3. We interpreted the interaction of neighbouring phase-slip centers in Zn whiskers in more detail. For this purpose the shifts of the critical current shown in Figs. 102 and 104 were split into an asymmetric and a symmetric part (as discussed on p. 147 of the present work). We found that the asymmetric part is well described by time averaged quasiparticle currents generated by the influencing phase-slip center as given by the SBT model.

References

1. W. Buckel, Supraleitung, 1st ed. (Physik Verlag, Weinheim, 1972)
2. M. Tinkham, Introduction to Superconductivity (McGraw Hill, New York, 1975)
3. R.P. Hübener, Magnetic Flux Structures in Superconductors (Springer Verlag, Berlin, 1979)
4. R.P. Hübener, Phys. Rep. 13, 143 (1974)
5. D. Saint-James, E.J. Thomas, and G. Sarma, Type II, Superconductivity (Pergamon Press, London, 1969), chapters 2 and 5
6. K.E. Gray (Editor), Nonequilibrium Superconductivity, Phonons, and Kapitza Boundaries (Plenum Press, New York, 1981)
7. D.N. Langenberg and A.I. Larkin (Editors), Nonequilibrium Superconductivity (North Holland, Amsterdam, 1986)
8. M. Tinkham, Festkörperprobleme (Adv. Solid State Phys.) XIX, 363 (1979)
9. J.A. Pals, K. Weiss, P.M.T.M. van Attekum, R.E. Horstman, and J. Wolter, Phys. Rep. 89, 323 (1982)
10. K.-Th. Wilke, Methoden der Kristallzüchtung, 1st ed. (Deutsch, Frankfurt/Main, 1963), pp. 434 - 474
11. J. Franks, Acta Met. 6, 103 (1958)
12. D.R. Overcash, E.P. Stillwell, M.J. Skove, and J.H. Davis, Phil. Mag. 25, 1481 (1972)
13. J.D. Eshelby, Phys. Rev. 91, 755 (1953)
14. Th. Werner, R. Tidecks, and B.D. Johnston, J. Cryst. Growth 73, 467 (1985)
15. R. Tidecks and J.D. Meyer, Z. Physik B 32, 363 (1979)
16. R. Tidecks and G. Slama, Z. Physik B 37, 103 (1980)
17. G. Slama and R. Tidecks, Solid State Commun. 44, 425 (1982)
18. U. Schulz, P.J. Wilbrandt, and R. Tidecks, J. Cryst. Growth 85, 472 (1987)
19. J.D. Meyer, Appl. Phys. 2, 303 (1973)
20. U. Schulz and R. Tidecks, J. Low Temp. Phys. 71, 151 (1988)
21. J.D. Meyer and R. Tidecks, Solid State Commun. 24, 639 (1977)
22. R. Tidecks, J. Low Temp. Phys. 58, 183 (1985); Errata 60, 347 (1985)
23. R. Tidecks, Z. Physik B - Condensed Matter 57, 127 (1984)
24. R. Tidecks, Nichtgleichgewichtssupraleitung in stromtragenden Haarkristallen. Thesis, Universität Göttingen (1980)
25. T. Werner and R. Tidecks, Cryogenics 26, 556 (1986); Erratum 27, 220 (1987)
26. C. Kittel, Einführung in die Festkörperphysik, 2nd/3rd ed. (Oldenbourg, München, 1969/1973)
27. T. Werner, Stromerzwungener Zusammenbruch der Supraleitung nahe

der kritischen Temperatur in Whiskern aus Indium-Blei-Legierungen und reinem Blei. Thesis, Universität Göttingen (1986)
28. R. Tidecks, Stromerzwungener Phasenübergang Supraleitung/Normalleitung von Zinn-Whiskern mit Indium-Verunreinigungen. Diplomarbeit, Universität Göttingen (1975)
29. J.D. Meyer, Spannungsstufen in den U(T)-Übergangskurven und U(I)-Kennlinien stromtragender Zinn-Whisker. Thesis, Universität Köln (1973)
30. R. Tidecks and T. Werner, Cryogenics 25, 366 (1985)
31. U. Schulz and R. Tidecks, Cryogenics 25, 700 (1985)
32. W.J. Skocpol, M.R. Beasley, and M. Tinkham, J. Low Temp. Phys. 16, 145 (1974)
33. D.W. Jillie, Interactions Between Coupled Thin-Film Microbridge Josephson Junctions. Thesis, State University of New York at Stony Brook (1976)
34. T.M. Klapwijk and J.E. Mooij, Phys. Lett. 57 A, 97 (1976)
35. T.M. Klapwijk, M. Sepers, and J.E. Mooij, J. Low Temp. Phys. 27, 801 (1977)
36. J. Meyer and G.v. Minnigerode, Phys. Lett. 38 A, 529 (1972)
37. J. Meyer and G.v. Minnigerode, in Low Temperature Physics LT 13. K.D. Timmerhaus, W.J. O'Sullivan, and E.F. Hammel, eds. (Plenum Press, New York, 1974) Vol. 3, pp. 701 - 704
38. W.W. Webb and R.J. Warburton, Phys. Rev. Lett. 20, 461 (1968)
39. J.D. Meyer and R. Tidecks, Solid State Commun. 18, 305 (1976)
40. R. Tidecks and T. Werner, J. Low Temp. Phys. 65, 151 (1986)
41. U. Schulz and R. Tidecks, Solid State Commun. 57, 829 (1986)
42. R.J. Warburton and W.W. Webb, Fluctuations Near the Phase Transition in 'One-Dimensional' Superconductors, in: Critical Phenomena in Alloys, Magnets, and Superconductors. R.E. Mills, E. Asher, and R.J. Jaffee, eds. (McGraw Hill, New York, 1970), pp. 451 - 469
43. J.E. Lukens, R.J. Warburton, and W.W. Webb, Phys. Rev. Lett. 25, 1180 (1970)
44. R.S. Newbower, M.R. Beasley, and M. Tinkham, Phys. Rev. B5, 864 (1972)
45. J.R. Schrieffer, Theory of Superconductivity. 1st ed. (W.A. Benjamin, New York, 1964)
46. G. Lüders and K.D. Usadel, The Method of Correlation Function in Superconductivity Theory. Springer Tracts in Modern Physics, Vol. 56 (Springer-Verlag, Berlin, 1971)
47. N.R. Werthamer, in Superconductivity, R.D. Parks, ed., 1st ed. (Marcel Dekker, Inc., New York, USA (1969)), chap. 6
48. J. Bardeen, L.N. Cooper, and J.R. Schrieffer, Phys. Rev. 108, 1175 (1957)
49. L.N. Cooper, Am.J. Phys. 28, 91 (1960)
50. N.N. Bogoljubov, Il Nuovo Cimento 7, 794 (1958)
51. J.G. Valentin, Il Nuovo Cimento 7, 843 (1958)
52. M. Tinkham, Superconductivity, chaps. IV and V, in Documents on Modern Physics. E.W. Montrell, G.H. Vineyard, and M. Lévy, eds. (Gordon and Breach Science Publishers, New York, 1965)
53. M. Tinkham, Superconductivity, chaps. IV and V, in Low Temperature Physics (Physique des basses temperatures). C. De Witt, B. Dreyfuß,

and P.G. De Gennes, eds. (Presses Universitaires de France, Paris; Gordon and Breach Science Publishers, New York, 1961)
54. J.M. Ziman, Prinzipien der Festkörpertheorie, 1st ed. (Akademie Verlag, Berlin, 1974), chap. 11
55. P.G. De Gennes, Superconductivity of Metals and Alloys, 1st ed. (W. Benjamin Inc., 1966), chap. 4
56. J. Clarke, in Nonequilibrium Superconductivity, Phonons, and Kapitza Boundaries. E.E. Gray, ed. (Plenum Press, New York, 1981), chap. 13
57. W.W. Parker, Phys. Rev. B12, 3667 (1975)
58. R. Becker, Theorie der Wärme (Springer Verlag, Berlin, 1966)
59. J.T. Rieger, D.J. Scalapino, and J.E. Mercereau, Phys. Rev. Lett. 27, 1787 (1971)
60. M.L. Yu and J.E. Mercereau, Phys. Rev. Lett. 28, 1117 (1972)
61. M.L. Yu and J.E. Mercereau, Phys. Rev. B 12, 4909 (1975)
62. J. Clarke, Phys. Rev. Lett. 28, 1363 (1972)
63. M. Tinkham and J. Clarke, Phys. Rev. Lett. 28, 1366 (1972)
64. J.R. Waldram, Proc. R. Soc. Lond. A 345, 231 (1975)
65. C.J. Pethick and H. Smith, Ann. Phys. (N.Y.) 119, 133 (1979)
66. C.J. Pethick and H. Smith, J. Phys. C: Solid St. Phys. 13, 6313 (1980)
67. C.J. Pethick and H. Smith, in Nonequilibrium Superconductivity, Phonons, and Kapitza Boundaries, K.E. Gray, ed. (Plenum Press, New York, 1981), chap. 15
68. J.R. Waldram, Rep. Prog. Phys. 39, 751 (1976)
69. J.E. Mercereau, in: SQUID, Superconducting Quantum Interference Devices and their Applications. H.D. Hahlbohm and H. Lübbig, eds. (Walter de Gruyter, Berlin, 1977), pp. 101 - 131
70. G.J. Dolan and L.D. Jackel, Phys. Rev. Lett. 39, 1628 (1977)
71. W.J. Skocpol, A.M. Kadin, and M. Tinkham, J. Phys. (Paris) 39 (Suppl. 8), C 6 - 1421 (1978)
72. J.M. Aponte and M. Tinkham, J. Low Temp. Phys. 51, 189 (1983)
73. M. Stuivinga, C.L.G. Ham, T.M. Klapwijk, and J.E. Mooij, J. Low Temp. Phys. 53, 633 (1983)
74. S.N. Artemenko and A.F. Volkov, Sov. Phys. JETP 43, 548 (1977)
75. C.M. Falco, Phys. Rev. Lett. 39, 660 (1977)
76. C.J. Pethick and H. Smith, Phys. Rev. Lett. 43, 640 (1979)
77. J. Clarke, B.R. Fjordbøge, and P.E. Lindeloff, Phys. Rev. Lett. 43, 642 (1979)
78. M. Tinkham, Phys. Rev. B 6, 1747 (1972)
79. A. Schmid and G. Schön, J. Low Temp. Phys. 20, 207 (1975)
80. J. Clarke, U. Eckern, A. Schmid, G. Schön, and M. Tinkham, Phys. Rev. B 20, 3933 (1979)
81. A. Schmid, in Nonequilibrium Superconductivity, Phonons, and Kapitza Boundaries, K.E. Gray, ed. (Plenum Press, New York, 1981), chap. 14
82. J.J. Chang, in Nonequilibrium Superconductivity, Phonons, and Kapitza Boundaries, K.E. Gray, ed. (Plenum Press, New York, 1981), chap. 9
83. T.R. Lemberger and J. Clarke, Phys. Rev. B 23, 1088 (1981)
84. A.M. Kadin, W.J. Skocpol, and M. Tinkham, J. Low Temp. Phys. 33, 481 (1978)
85. T.R. Lemberger and J. Clarke, Phys. Rev. B 23, 1100 (1981)
86. M. Stuivinga, J.E. Mooij, and T.M. Klapwijk, J. Low Temp. Phys. 46, 555 (1982)

87. M. Tinkham, in Nonequilibrium Superconductivity, Phonons, and Kapitza Boundaries, K.E. Gray, ed. (Plenum Press, New York, 1981), chap. 8
88. J.T. Anderson and A.M. Goldman, Phys. Rev. Lett. 25, 743 (1970)
89. D. Markowitz and L.P. Kadanoff, Phys. Rev. 131, 563 (1963)
90. P.W. Anderson, J. Phys. Chem. Solids 11, 26 (1959)
91. J. Bardeen and D. Mattis, Phys. Rev. 111, 412 (1958)
92. C.C. Chi and J. Clarke, Phys. Rev. B 19, 4495 (1979)
93. T.Y. Hsiang and J. Clarke, Phys. Rev. B 21, 945 (1980)
94. J. Clarke, in Nonequilibrium Superconductivity, D.N. Langenberg and A.I. Larkin, eds. (North Holland, Amsterdam, 1986), chap. 1
95. O. Entin-Wohlman and R. Orbach, Phys. Rev. B 24, 1177 (1981)
96. A. Schmid, Z. Physik 271, 251 (1974)
97. E. Abrahams, P.W. Anderson, P.A. Lee, and T.V. Ramakrishnan, Phys. Rev. B 24, 6783 (1981)
98. J.M. Gordon, C.J. Lobb, and M. Tinkham, Phys. Rev. B 28, 4046 (1983)
99. P. Santhanam and D.E. Prober, Phys. Rev. B 29, 3733 (1984)
100. C.C. Chi and J. Clarke, Phys. Rev. B 21, 333 (1980)
101. J.J. Chang, Phys. Rev. B 19, 1420 (1979)
102. M. Tinkham, Phys. Rev. B 6, 1747 (1972)
103. S.B. Kaplan, C.C. Chi, D.N. Langenberg, J.J. Chang, S. Jafarey, and D.J. Scalapino, Phys. Rev. B 14, 4854 (1976); Erratum Phys. Rev. B 15, 3567 (1977)
104. D.N. Langenberg, in Proceedings of the 14th International Conference on Low Temperature Physics, Vol. 5, M. Krusius and M. Vuorio, eds. (North Holland Publishing Company, Amsterdam; American Elsevier Publishing Company, New York, 1975), pp. 223 - 263
105. M.V. Moody and J.L. Paterson, J. Low Temp. Phys. 34, 83 (1979)
106. J. Bardeen, Rev. Mod. Phys. 34, 667 (1962)
107. P. van den Hamer, T. M. Klapwijk, and J. E. Mooij, J. Low Temp. Phys. 54, 607 (1984)
108. J.L. Levine, Phys. Rev. Lett. 15, 154 (1965)
109. J. Clarke and J.L. Paterson, J. Low Temp. Phys. 15, 491 (1974)
110. K.E. Gray, in Nonequilibrium Superconductivity, Phonons, and Kapitza Boundaries, K.E. Gray, ed. (Plenum Press, New York, 1981), chap. 5
111. J.R. Kirtley, D.S. Kent, D.N. Langenberg, S.B. Kaplan, J.J. Chang, and C.C. Yang, Phys. Rev. B 22, 1218 (1980)
112. M.V. Moody and J.L. Paterson, Phys. Rev. B 23, 133 (1981)
113. K.E. Gray, Solid State Commun. 26, 633 (1978)
114. J.J. Chang, Phys. Rev. B 17, 2137 (1978)
115. C.C. Chi and J. Clarke, Phys. Rev. B 20, 4465 (1979)
116. T.R. Lemberger, Y. Yen, and S.G. Lee, Phys. Rev. B 35, 6670 (1987)
117. Y. Yen and T.R. Lemberger, Phys. Rev. B 37, 3324 (1988)
118. I. Schuller and K.E. Gray, Phys. Rev. Lett. 36, 429 (1976)
119. I. Schuller and K.E. Gray, Solid State Commun. 23, 337 (1977)
120. J. Wolter, P.M.Th.M. van Attekum, R.E. Horstman, and M.C.H.M. Wouters, Solid State Commun. 40, 433 (1981)
121. J. Wolter, P.M.Th.M. van Attekum, R.E. Horstman, and M.C.H.M. Wouters, Physica 108 B, 781 (1981)

122. I.F. Oppenheim and S. Frota-Pessôa, Phys. Rev. B 25, 4495 (1982)
123. J.A. Pals, J.A. Geurst, and J.J. Ramekers, Phys. Rev. B 23, 6184 (1981)
124. H. Weissbrod, R. Gross, and R.P. Hübener, Solid State Commun. 60, 147 (1986)
125. R. Peters and H. Meissner, Phys. Rev. Lett. 30, 965 (1973)
126. A.J. Ritger and H. Meissner, J. Low Temp. Phys. 40, 495 (1980)
127. R. Rangel and L. Kramer, J. Low Temp. Phys. 68, 85 (1987)
128. A.J. Ritger, H. Meissner, R. Rangel, and L. Kramer, J. Low Temp. Phys. 73, 221 (1988)
129. W. Eisenmenger, in Tunneling Phenomena in Solids, E. Burstein and S. Lundqvist, eds. (Plenum Press, New York, 1969), chap. 26
130. W. Eisenmenger, in Nonequilibrium Superconductivity, Phonons, and Kapitza Boundaries, K.E. Gray, ed. (Plenum Press, New York), chap. 3
131. W. Eisenmenger, K. Laßmann, H.J. Trumpp, and R. Krauß, Appl. Phys. 11, 307 (1976)
132. W. Eisenmenger, K. Laßmann, H.J. Trumpp, and R. Krauß, Appl. Phys. 12, 163 (1977)
133. P.W. Epperlein, K. Lassmann, and W. Eisenmenger, Z. Physik B 31, 377 (1978)
134. P.W. Epperlein and W. Eisenmenger, Z. Physik B 32, 167 (1979)
135. L.N. Smith and J.M. Mochel, Phys. Rev. Lett. 35, 1597 (1975)
136. O. Liengme, A. Baratoff, and P. Martinoli, J. Low Temp. Phys. 65, 113 (1986)
137. D.J. Frank and M. Tinkham, Phys. Rev. B 28, 5345 (1983)
138. T.Y. Hsiang, Phys. Rev. B 21, 956 (1980)
139. V.V. Ryazanov, V.V. Schmidt, and L.A. Ermolaeva, J. Low Temp. Phys. 45, 507 (1981)
140. W. Klein, R.P. Hübener, S. Gauss, and J. Parisi, J. Low Temp. Phys. 61, 413 (1985)
141. M. Octavio, W.J. Skocpol, and M. Tinkham, Phys. Rev. B 17, 159 (1978)
142. A. Schmid, G. Schön, and M. Tinkham, Phys. Rev. B 21, 5076 (1980)
143. S.G. Wang and P.E. Lindeloff, in Proceedings of the 17th International Conference on Low Temperature Physics, LT 17, U. Eckern, A. Schmid, W. Weber, and H. Wühl, eds. (North Holland, Amsterdam, 1984), pp. 805 - 806
144. J.E. Mooij, in Nonequilibrium Superconductivity, Phonons, and Kapitza Boundaries, K.E. Gray, ed. (Plenum Press, New York, 1981), chap. 7
145. Y.I. Latyshev and F.Y. Nad', Sov. Phys. JETP 44, 1136 (1976)
146. P.C. van Son, J. Romijn, T.M. Klapwijk, and J.E. Mooij, Phys. Rev. B 29, 1503 (1984)
147. R. Tidecks and T. Werner, J. Low Temp. Phys. 67, 225 (1987)
148. J.E. Mooij and T.M. Klapwijk, in Localization, Interaction, and Transport Phenomena (Springer-Verlag, Berlin, 1984), pp. 233 - 244
149. P. Santhanam, S. Wind, and D.E. Prober, Phys. Rev. B 35, 3188 (1987)
150. H. Fukuyama and E. Abrahams, Phys. Rev. B 27, 5976 (1983)
151. G. Bergmann, Phys. Rep. 107, 1 (1984)
152. J.M.B. Lopes dos Santos, Phys. Rev. B 28, 1189 (1983)
153. A. Schmid, Z. Physik 259, 421 (1973)
154. B. Keck and A. Schmid, J. Low Temp. Phys. 24, 611 (1976)

155. H. Takayama, Z. Physik 263, 329 (1973)
156. A. A. Abrikosov, L.P. Gorkov, and I. Ye. Dzyaloskinskii, Quantum Field Theoretical Methods in Statistical Physics, 2nd ed. (Pergamon Press, Oxford, 1965)
157. A.B. Pippard, Phil. Mag. 46, 1104 (1955)
158. A.M. Kadin, L.N. Smith, and W.J. Skocpol, J. Low Temp. Phys. 38, 497 (1980)
159. T.R. Lemberger, Phys. Rev. B 24, 4105 (1981)
160. A.F.G. Wyatt, V.M. Dmitriev, W.S. Moore, and F.W. Sheard, Phys. Rev. Lett. 16, 1166 (1966)
161. A.H. Dayem and J.J. Wiegand, Phys. Rev. 155, 419 (1967)
162. T.M. Klapwijk and J.E. Mooij, Physica 81 B, 132 (1976)
163. T. M. Klapwijk, J. N. van den Bergh, and J. E. Mooij, J. Low Temp. Phys. 26, 385 (1977)
164. T.M. Klapwijk, H.B. van Linden van den Heuvel, and J.E. Mooij, Journal de Physique (Suppl. 8) 39 (I), C 6 - 525 (1978)
165. J.A. Pals, Phys. Lett. 61 A, 275 (1977)
166. J.A. Pals and J. Dobben, Journal de Physique (Suppl. 8) 39 (I), C 6 - 523 (1978)
167. J.A. Pals and J. Dobben, Phys. Rev. Lett. 42, 270 (1979)
168. J.A. Pals and J. Dobben, Phys. Rev. B 20, 935 (1979)
169. Y.I. Latychev and F.Y. Nad', JETP Letters 19, 380 (1974)
170. Y.I. Latychev and F.Y. Nad', JETP Letters 29, 50 (1979)
171. S.A. Peskovatskiǐ and L.P. Stritzhko, JETP Letters 29, 54 (1979)
172. T. Kommers and J. Clarke, Phys. Rev. Lett. 38, 1091 (1977)
173. R.L. Dahlberg, E.D. Orbach, and I. Schuller, J. Low Temp. Phys. 36, 367 (1979)
174. J.T. Hall, L.B. Holdeman, and R.J. Soulen jr., Phys. Rev. Lett. 45, 1011 (1980)
175. C.M. Falco, T.R. Werner, and I.K. Schuller, Solid State Commun. 34, 535 (1980)
176. J.E. Mooij, N. Lambert, and T.M. Klapwijk, Solid State Commun. 36, 585 (1980)
177. J.A. Pals and J. Dobben, Phys. Rev. Lett. 44, 1143 (1980)
178. O. Entin-Wohlmann, Phys. Rev. B 23, 2428 (1981)
179. J.A. Pals, P.M.T.M. van Attekum, and J.J. Ramekers, Physica 108 B, 831 (1981)
180. R.E. Horstman and J. Wolter, Phys. Lett. 82 A, 43 (1981)
181. O. Entin-Wohlman, J. Low Temp. Phys. 43, 91 (1981)
182. P.M.T.M. van Attekum and J.J. Ramekers, Solid State Commun. 43, 735 (1982)
183. P.M.T.M. van Attekum, J.J. Ramekers, J.A. Pals, and A.A.M. Hoeben, Phys. Rev. B 27, 1623 (1983)
184. R. Escudero and H.J.T. Smith, Phys. Rev. B 31, 2725 (1985)
185. T.J. Tredwell and E.H. Jacobson, Phys. Rev. Lett. 35, 244 (1975)
186. T.J. Tredwell and E.H. Jacobson, Phys. Rev. B 13, 2931 (1976)
187. N.D. Miller and J.E. Rutledge, Phys. Rev. B 26, 4739 (1982)
188. D. Seligson and J. Clarke, Phys. Rev. B 28, 6297 (1983)
189. V.M. Dmitriev, V.N. Gubankov, and F.Ya. Nad', in Nonequilibrium Superconductivity, D.N. Langenberg and A.I. Larkin, eds. (North Holland, Amsterdam, 1986), chap. 5
190. U. Eckern, A. Schmid, M. Schmutz, and G. Schön, J. Low Temp. Phys. 36, 643 (1979)
191. G.M. Eliashberg, Sov. Phys. JETP Lett. 11, 114 (1970)

192. G.M. Eliashberg and B.I. Ivlev, in Nonequilibrium Superconductivity, D.N. Langenberg and A.I. Larkin, eds. (North Holland, Amsterdam, 1986), chap. 6
193. J.J. Chang and D.J. Scalapino, Phys. Rev. B 15, 2651 (1977)
194. J.J. Chang and D.J. Scalapino, J. Low Temp. Phys. 29, 477 (1977)
195. J.J. Chang and D.J. Scalapino, J. Low Temp. Phys. 31, 1 (1978)
196. P. van den Hamer, E. A. Montie, J. E. Mooij, and T. M. Klapwijk, J. Low Temp. Phys. 69, 265 (1987)
197. P. van den Hamer, E.A. Montie, P.B.L. Meijer, J.E. Mooij, and T.M. Klapwijk, J. Low Temp. Phys. 69, 287 (1987)
198. J. Fuchs, P.W. Epperlein, M. Welte, and W. Eisenmenger, Phys. Rev. Lett. 38, 919 (1977)
199. J.T.C. Yeh and D.N. Langenberg, Phys. Rev. B 17, 4303 (1978)
200. T.V. Rajevakumar, J.J. Chang, and J.T. Chen, J. Low Temp. Phys. 37, 77 (1979)
201. I. Iguchi and K. Hara, Phys. Lett. 59 A, 313 (1976)
202. I. Iguchi, Phys. Rev. B 16, 1954 (1977); Erratum, Phys. Rev. B 17, 3023 (1978)
203. I. Iguchi, Phys. Lett. 64 A, 415 (1978)
204. I. Iguchi, J. Low Temp. Phys. 31, 605 (1978)
205. I. Iguchi, J. Low Temp. Phys. 33, 439 (1978)
206. T. Wong, J.T.C. Yeh, and D.N. Langenberg, Phys. Rev. Lett. 37, 150 (1976)
207. R.C. Dynes, V. Narayanamurti, and J.P. Garno, Phys. Rev. Lett. 39, 229 (1977)
208. K.E. Gray and H.W. Willemsen, J. Low Temp. Phys. 31, 911 (1978)
209. I. Iguchi and D.N. Langenberg, Phys. Rev. Lett. 44, 486 (1980)
210. I. Iguchi, D. Kent, H. Gilmartin, and D.N. Langenberg, Phys. Rev. B 23, 3240 (1981)
211. I. Iguchi, S. Kotani, Y. Yamaki, Y. Suzuki, M. Manabe, and K. Harada, Phys. Rev. B 24, 1193 (1981)
212. H. Akoh and K. Kajimura, Phys. Rev. B 25, 4467 (1982)
213. I. Iguchi, A. Nishiura, and V.S. Tomar, Phys. Rev. B 28, 4037 (1983)
214. M. Sugahara, J. Phys. Soc. Japan 46, 410 (1979)
215. C.H. Wang and X.Y. Zhu, J. Low Temp. Phys. 42, 277 (1981)
216. G. Schön and A.M. Trembley, Phys. Rev. Lett. 42, 1086 (1979)
217. A.M. Trembley, in Nonequilibrium Superconductivity, Phonons, and Kapitza Boundaries, K.E. Gray, ed. (Plenum Press, New York, 1981), chap. 11
218. G. Schön, Physica 109 and 110 B, 1677 (1982)
219. V.M. Galitskii, V.F. Elesin, and Y.V. Kopaev, in Nonequilibrium Superconductivity, D.N. Langenberg and A.I. Larkin, eds. (North Holland, Amsterdam, 1986), chap. 9
220. V.F. Elesin, Sov. Phys. JETP 49, 1121 (1979)
221. I. Iguchi and H. Konno, Phys. Rev. B 28, 4040 (1983)
222. I. Iguchi and Y. Suzuki, Phys. Rev. B 28, 4043 (1983)
223. R. Gross, B.D. Schmid, and R.P. Hübener, J. Low Temp. Phys. 62, 245 (1986)
224. G. Dharmadurai and B.A. Ratnam, Phys. Lett. 68 A, 371 (1978)
225. G. Dharmadurai and B.A. Ratnam, Cryogenics 18, 553 (1978)
226. G. Dharmadurai and B.A. Ratnam, Journal de Physique (Suppl. 8) 39 (II), C 6-1234 (1978)

227. G. Dharmadurai and B.A. Ratnam, Phys. Rev. B 19, 5711 (1979)
228. G. Dharmadurai and B.A. Ratnam, Phys. Rev. B 20, 4633 (1979)
229. G. Dharmadurai and B.A. Ratnam, Solid State Commun. 32, 1293 (1979)
230. G. Dharmadurai and B.A. Ratnam, J. Low Temp. Phys. 37, 149 (1979)
231. L.R. Testardi, Phys. Rev. B 4, 2189 (1971)
232. W.H. Parker and W.D. Williams, Phys. Rev. Lett. 29, 924 (1972)
233. J.T.C. Yeh and D.N. Langenberg, Appl. Phys. Lett. 32, 191 (1978)
234. L.N. Smith, J. Low Temp. Phys. 38, 553 (1980)
235. C. Vanneste, A. Gilabert, P. Sibillot, and D.B. Ostrowski, Appl. Phys. Lett. 38, 941 (1981)
236. A.F. Görlach and R.P. Huebener, J. Low Temp. Phys. 53, 633 (1983)
237. A.D. Smith, W.J. Skocpol, and M. Tinkham, Phys. Rev. B 21, 3879 (1980)
238. C.S. Owen and D.J. Scalapino, Phys. Rev. Lett. 28, 1559 (1972)
239. J.J. Chang and D.J. Scalapino, Phys. Rev. B 9, 4769 (1974)
240. W.H. Parker, Phys. Rev. B 12, 3667 (1975)
241. J.J. Chang, W.Y. Lai, and D.J. Scalapino, Phys. Rev. B 20, 2739 (1979)
242. V.F. Elesin, V.E. Kondrashov, and A.S. Sukhikh, Sov. Phys. Solid State 21, 1861 (1979)
243. V.F. Elesin, Sov. Phys. JETP 44, 780 (1976)
244. V.F. Elesin, Sov. Phys. Solid State 19, 1744 (1977)
245. H.W. Willemsen and K.E. Gray, Phys. Rev. Lett. 41, 812 (1978)
246. U. Eckern and G. Schön, J. Low Temp. Phys. 32, 821 (1978)
247. P.C. Martin, in Superconductivity, R.D. Parks, ed. (Marcel Decker, New York, 1961), chap. 7
248. C.P. Enz, Rev. Mod. Phys. 46, 705 (1974)
249. A.M. Goldman, in Nonequilibrium Superconductivity, Phonons, and Kapitza Boundaries, K.E. Gray, ed. (Plenum Press, New York, 1981), chap. 17
250. G. Schön, in Nonequilibrium Superconductivity, D.N. Langenberg and A.I. Larkin, eds. (North Holland, Amsterdam, 1986), chap. 13
251. R.V. Carlson and A.M. Goldman, Phys. Rev. Lett. 31, 880 (1973)
252. R.V. Carlson and A.M. Goldman, Phys. Rev. Lett. 34, 11 (1975)
253. R.V. Carlson and A.M. Goldman, J. Low Temp. Phys. 27, 67 (1976)
254. A. Schmid and G. Schön, Phys. Rev. Lett. 34, 941 (1975)
255. S.N. Artemenko and A.F. Volkov, Sov. Phys. JETP 42, 896 (1975)
256. I.O. Kulik, O. Entin-Wohlman, and R. Orbach, J. Low Temp. Phys. 43, 591 (1981)
257. P.B. Littlewood and C.M. Varma, Phys. Rev. B 26, 4883 (1982)
258. J. Wilks, The Properties of Liquid and Solid Helium, 1st ed. (Clarendon Press, Oxford, 1967)
259. A. Schmid, Phys. kondens. Materie 8, 129 (1968)
260. W.J. Skocpol, in Nonequilibrium Superconductivity, Phonons, and Kapitza Boundaries, K.E. Gray, ed. (Plenum Press, New York, 1981), chap. 17
261. A.M. Kadin and A.M. Goldman, in Nonequilibrium Superconductivity, D.N. Langenberg and A.I. Larkin, eds. (North Holland, Amsterdam, 1986), chap. 13

262. I.O. Kulik, Sov. J. Low Temp. Phys. 5, 656 (1979)
263. B.D. Josephson, Phys. Lett. 1, 251 (1962)
264. J.S. Langer and V. Ambegoakar, Phys. Rev. 164, 498 (1967)
265. T.J. Rieger, D.J. Scalapino, and J.E. Mercereau, Phys. Rev. B 6, 1734 (1972)
266. R.K. Kirschmann, H.A. Notarys, and J.E. Mercereau, Phys. Lett. 34A, 209 (1971)
267. J.D. Franson and J.E. Mercereau, J. Appl. Phys. 47, 3261 (1976)
268. L.D. Jackel, R.A. Buhrman, and W.W. Webb, Phys. Rev. B 10, 2782 (1974)
269. L.D. Jackel, J.M. Warlaumont, T.D. Clark, J.C. Brown, R.A. Buhrman, and M.T. Levinsen, Appl. Phys. Lett. 28, 353 (1976)
270. L.D. Jackel, W.H. Henkels, J.M. Warlaumont, and R.A. Buhrman, Appl. Phys. Lett. 29, 214 (1976)
271. S.S. Pei, J.E. Lukens, and R.D. Sandell, Appl. Phys. Lett. 36, 88 (1980)
272. K.K. Likarev, Rev. Mod. Phys. 51, 101 (1979)
273. M. Tinkham, Rev. Mod. Phys. 46, 587 (1974)
274. H.A. Notarys and J.E. Mercereau, Physica 55, 424 (1971)
275. A.B. Pippard, J.G. Shepherd, and D.A. Tindall, Proc. Roy. Soc. Lond. A 324, 17 (1971)
276. M. Tinkham, J. Low Temp. Phys. 35, 147 (1979)
277. A.G. Aronov and V.L. Gurevich, Sov. Phys. Solid State 16, 1722 (1975)
278. L. Solymar, Superconductive Tunneling and Applications (J. Wiley and Sons, New York, 1972)
279. A. Barone and G. Paternò, Physics and Applications of the Josephson Effect (J. Wiley and Sons, New York, 1982)
280. L.G. Aslamasov and A.F. Volkov, in Nonequilibrium Superconductivity, D.N. Langenberg and A.I. Larkin, eds. (North Holland, Amsterdam, 1986), chap. 2
281. M. Octavio and W.J. Skocpol, Physica 107 B, 173 (1981)
282. T.R. Lemberger, Physica 107 B, 163 (1981)
283. A.M. Kadin, C. Varmazis, and J.E. Lukens, Physica 107 B, 159 (1981)
284. W.J. Skocpol and L.D. Jackel, Physica 108 B, 1021 (1981)
285. L.P. Gorkov, Sov. Phys. JETP 34, 505 (1958)
286. B.D. Josephson, Rev. Mod. Phys. 46, 251 (1974)
287. P.W. Anderson, N.R. Werthamer, and J.M. Luttinger, Phys. Rev. 138, A1157 (1965)
288. M.J. Stephen and H. Suhl, Phys. Rev. Lett. 13, 797 (1964)
289. M. Cyrot, Rep. Prog. Phys. 36, 103 (1973)
290. A. Schmid, Phys. kondens. Materie 5, 302 (1966)
291. E. Abrahams and T. Tsusento, Phys. Rev. 152, 416 (1966)
292. L.P. Gorkov and G.M. Eliashberg, Sov. Phys. JETP 27, 328 (1968)
293. A.A. Abrikosov and L.P. Gorkov, Sov. Phys. JETP 12, 1243 (1961)
294. R.S. Thompson and C.R. Hu, Phys. Rev. Lett. 20, 1352 (1971)
295. C.R. Hu and R.S. Thompson, Phys. Rev. B 6, 110 (1972)
296. G.M. Eliashberg, Sov. Phys. JETP 28, 1298 (1969)
297. C.R. Hu, Phys. Rev. B 14, 4834 (1976)
298. A.I. Larkin and Y.N. Ovchinikov, Sov. Phys. JETP 41, 960 (1976)
299. A.I. Larkin and Y.N. Ovchinikov, Sov. Phys. JETP 46, 155 (1978)
300. Y.N. Ovchinikov, J. Low Temp. Phys. 28, 43 (1977)

301. L. Kramer and R.J. Watts-Tobin, Phys. Rev. Lett. 40, 1041 (1978)
302. G. Schön and V. Ambegaokar, Phys. Rev. B 19, 3515 (1979)
303. A.L. de Lozanne and M.R. Beasley in: Nonequilibrium Superconductivity, D.N. Langenberg and A.I. Larkin, eds. (North-Holland, Amsterdam, 1986), chap. 3
304. C.R. Hu, Phys. Rev. B 21, 2775 (1980)
305. R.J. Watts-Tobin, Y. Krähenbühl, and L. Kramer, J. Low Temp. Phys. 42, 459 (1981)
306. K. Maki, in Superconductivity, R.D. Parks, ed. (Marcel Dekker, Inc., New York, 1969), chap. 18
307. A. Baratoff, Phys. Rev. Lett. 48, 434 (1982)
308. A. Baratoff, Physica 109 and 110 B, 2058 (1982)
309. Y. Krähenbühl, Z. Physik B - Condensed Matter 52, 219 (1983)
310. R. Rangel and L. Kramer, J. Low Temp. Phys. 74, 163 (1989)
311. R. Rangel, On Dissipative States in Narrow Current-Carrying Superconducting Filaments. Thesis, Universität Bayreuth (1987)
312. H.J. Fink, Phys. Lett. 42 A, 465 (1973)
313. H.J. Fink, Phys. Stat. Sol. (b) 60, 843 (1973)
314. H.J. Fink and R.S. Poulsen, Phys. Rev. Lett. 32, 762 (1974)
315. H.J. Fink and R.S. Poulsen, Phys. Rev. B 11, 1870 (1975)
316. L. Kramer and R. Rangel, J. Low Temp. Phys. 57, 391 (1984)
317. A. Baratoff and L. Kramer, in SQUID, Superconducting Quantum Interference Devices and their Applications, H.D. Hahlbohm and H. Lübbig, eds. (Walter de Gruyter, Berlin, 1977), pp. 51 - 62
318. L. Kramer and A. Baratoff, Phys. Rev. Lett. 38, 518 (1977)
319. A. Baratoff and L. Kramer, Journal de Physique (Suppl. 8) 39 (I), C 6 - 548 (1978)
320. R.J. Watts-Tobin and L. Kramer, Journal de Physique (Suppl. 8) 39 (III), C 6 - 554 (1978)
321. L. Kramer, personal communication (1985)
322. O. Liengme, A. Baratoff, and P. Martinoli, J. Low Temp. Phys. 65, 113 (1986)
323. V.P. Galaiko, J. Low Temp. Phys. 26, 483 (1977)
324. I.O. Kulik, Solid State Commun. 35, 383 (1980)
325. B.I. Ivlev and N.B. Kopnin, J. Low Temp. Phys. 39, 137 (1980)
326. B.I. Ivlev and N.B. Kopnin, J. Low Temp. Phys. 44, 453 (1981)
327. B.I. Ivlev and N.B. Kopnin, Adv. Phys. 33, 47 (1984)
328. L.P. Gorkov, Sov. Phys. JETP 36, 1364 (1959)
329. L.P. Gorkov, Sov. Phys. JETP 37, 998 (1960)
330. J.J. Hauser, Phys. Rev. B 10, 2792 (1974)
331. D.J. Scalapino, in Superconductivity, R.D. Parks, ed. (Marcel Dekker Inc., New York, 1969), chap. 10
332. B. Damaschke, X. Yang, and R. Tidecks, J. Low Temp. Phys. 70, 131 (1988)
333. D.E. Mapother, Phys. Rev. B 126, 2021 (1962)
334. M.A. Biondi, A.T. Forrester, M.P. Garfunkel, and C.B. Satterthwaite, Rev. Mod. Phys. 30, 1109 (1958)
335. D.L. Dekker, D.E. Mapother, and R.W. Shaw, Phys. Rev. 112, 1888 (1958)

336. N.E. Phillips, Phys. Rev. 114, 676 (1959)
337. D.K. Finnemore, D.E. Mapother, and R.W. Shaw, Phys. Rev. 118, 127 (1960)
338. R.W. Shaw, D.E. Mapother, and D.C. Hopkins, Phys. Rev. 120, 88 (1960)
339. D.K. Finnemore and D.E. Mapother, Phys. Rev. 140, A 507 (1965)
340. J. Romijn, T.M. Klapwijk, M.J. Renne, and J.E. Mooij, Phys. Rev. B 26, 3648 (1982)
341. J. Romijn, T.M. Klapwijk, and J.E. Mooij, Physica 108 B, 981 (1981)
342. M.Y. Kupriyanov and V.F. Lukichev, Sov. J. Low Temp. Phys. 6, 210 (1980)
343. L.G. Neumann and Y.H. Kao, J. Low Temp. Phys. 48, 321 (1982)
344. W.J. Skocpol, Phys. Rev. B 14, 1045 (1976)
345. F.V. Burckbuchler and C.A. Reynolds, Phys. Rev. 175, 550 (1968)
346. F.V. Burckbuchler, D. Markowitz, and C.A. Reynolds, Phys. Rev. 175, 543 (1968)
347. C.M. Hurd, Advan. Phys. 23, 315 (1973)
348. M.L. Lea, D.D. Llewellyn, D.R. Peck, and E.R. Dobbs, Proc. Roy. Soc. Lond. A 334, 357 (1973)
349. B.N. Aleksandrov, Sov. Phys. JETP 16, 286 (1963)
350. D.P. Almond, M.J. Lea, and E.R. Dobbs, Proc. Roy. Soc. Lond. A 343, 537 (1975)
351. E.R. Dobbs, M.J. Lea, and D.R. Peck, Proc. Roy. Soc. Lond. A 334, 379 (1973)
352. C.R. Cleavelin and B.I. Marshall, Phys. Rev. B 10, 1902 (1974)
353. E.H. Breashears, C.R. Cleavelin, and B.I. Marshall, Phys. Rev. B 13, 4801 (1976)
354. S. Anluk, Phys. Rev. B 18, 522 (1978)
355. D. Farrel, J.G. Park, and B.R. Coles, Phys. Rev. Lett. 13, 328 (1964); Errata 13, 650 (1964)
356. Y.P. Gaidukov and J. Kadlecovå, J. Low Temp. Phys. 2, 131 (1970)
357. G. Chamin, E.A. Lynton, and B. Serin, Phys. Rev. 114, 719 (1959)
358. D.P. Seraphim, C. Chion, and D.J. Quinn, Acta Met. 9, 861 (1961)
359. B. Damaschke and R. Tidecks, Phase-Slip Centers in Superconducting Whiskers of Pb with Small Concentrations of In and Bi. To be published in J. Low Temp. Phys. 79, No. 3/4, pp. 117-134 (1990)
360. E. Nembach, J. Phys. Chem. Solids 29, 1205 (1968)
361. M. Satō, N. Kumasaka, and M. Mitani, J. Phys. Soc. Japan 21, 1617 (1966)
362. H. Gamari-Seale and B.R. Coles, Proc. Phys. Soc. (Lond.) 86, 1199 (1985)
363. D.U. Gubser, Phys. Rev. B 6, 827 (1972)
364. J.P. Carbotte, in Anisotropy Effects in Superconductors, H.W. Weber, ed. (Plenum, New York, 1977), chap. 5.R-5
365. J.L. Bostock and M.L.A. MacVicar, in Anisotropy Effects in Superconductors, H.W. Weber, ed. (Plenum, New York, 1977), chap. 5. R-6
366. W.D. Gregory, A.J. Grekas, S. Horn, and L. Morelli, in Anisotropy Effects in Superconductors, H.W. Weber, ed. (Plenum, New York, 1977), chap. 6. C-14
367. R. Tidecks, J. Low Temp. Phys. 58, 233 (1985); Erratum 60, 457 (1985)
368. E. Meyer and R. Pottel, Physikalische Grundlagen der Hochfrequenztechnik, 1st ed. (Vieweg, Braunschweig, 1969), chap. 6.2.
369. H. Weissbrod, R.P. Huebener, and W. Clauss, J. Low Temp. Phys. 69, 77 (1987); Erratum 73, 171 (1988)

370. G.v.Minnigerode and U.Schulz, Einfluß thermischer Effekte auf den stromerzwungenen Zusammenbruch der Supraleitung in Haarkristallen aus Zink, in Nachrichten der Akademie der Wissenschaften in Göttingen, II. Mathematisch-Physikalische Klasse, Jahrgang 1988, Nr. 3 (Vandenhoek and Ruprecht, Göttingen, 1988)
371. R.Tidecks and G.v.Minnigerode, Phys.Stat.Sol.(a) 52, 421 (1979)
372. B.Damaschke and R.Tidecks, Z. Phys. B 77, 17 (1989)
373. S.Shapiro, Phys.Rev.Lett. 11, 80 (1963)
374. D.N.Langenberg, in Tunneling Phenomena in Solids, E.Burstein and S.Lundquist, eds. (Plenum Press, New York, 1969), chap. 33
375. W.C.Danchi, F.Habbal, and M.Tinkham, Appl.Phys.Lett. 41, 883 (1982)
376. H.Happ and U.Kaiser-Diekhoff, Phys.Lett. 29 A, 161 (1969)
377. D.A.Weitz, W.J.Skocpol, and M.Tinkham, Phys.Rev.Lett. 40, 253 (1978)
378. P.W.Anderson and A.H.Dayem, Phys.Rev.Lett. 13, 195 (1964)
379. T.M.Klapwijk and T.B.Veenstra, Phys.Lett. 47 A, 351 (1974)
380. S.Kuriki and K.H.Gundlach, J.Appl.Phys. 50, 3514 (1979)
381. M.Mück, H.Rogalla, B.David, and C.Heiden, Z.Physik B - Condensed Matter 61, 81 (1985)
382. R.B.Laibowitz, A.N.Broers, J.T.C.Yeh, and J.M.Viggiano, Appl.Phys.Lett. 35, 891 (1979)
383. R.K. Kirschman, Proceedings of the Applied Superconductivity Conference, pp. 707 - 708, Annapolis, Maryland (1972)
384. H.Seifert and R.P.Huebener, J.Low Temp.Phys. 41, 275 (1980)
385. R.Gross, H.Seifert, R.P.Huebener, and K.Yoshida, J.Low Temp.Phys. 54, 277, (1984)
386. D.W.Palmer and J.E.Mercereau, Appl.Phys.Lett. 25, 467 (1974)
387. J.Niemeyer, J.H.Hinken, and W.Meier, IEEE Trans.Instrum.Meas. IM-33, 311 (1984)
388. J.Niemeyer, J.H.Hinken, and R.L.Kautz, Appl.Phys.Lett. 45, 478 (1984)
389. J.Niemeyer, L.Grimm, W.Meier, J.H.Hinken, and E.Vollmer, Appl.Phys. Lett. 47, 1222 (1985)
390. K.H.Gundlach and J.Kadlec, J.Low Temp.Phys. 26, 603 (1977)
391. J.Kadlec and K.H.Gundlach, J.Low Temp.Phys. 27, 887 (1977)
392. J.D.Franson and J.E.Mercereau, J.Appl.Phys. 47, 3261 (1976)
393. B.Damaschke, Phasenschlupfzentren und deren hysteretisches Verhalten in supraleitenden Whiskern. Thesis, Universität Göttingen (1989)
394. M.Abramowitz and I.A.Stegun, Handbook of Math. Functions, 7th ed. (Dover Publ. Inc., New York, 1970), pp. 358 - 361
395. D.W.Jillie, J.Lukens, and Y.H.Kao, IEEE Trans. MAG-11, 671 (1975)
396. M.Octavio and W.J.Skocpol, J.Appl.Phys. 50, 3505 (1979)
397. Y.D.Dai, W.J.Yeh, and Y.H.Kao, J.Low Temp.Phys. 48, 373 (1982)
398. M.L.Yu and J.E.Mercereau, Phys.Rev.Lett. 37, 1148 (1976)
399. D.W.Jillie, J.E.Lukens, Y.H.Kao, and G.J.Dolan, Phys.Lett. 55 A, 381 (1976)
400. P.E.Lindeloff and J.Bindslev-Hansen, J.Low Temp.Phys. 29, 369 (1977)
401. M.A.H.Nerenberg, J.A.Blackburn, and D.W.Jillie, Phys.Rev.B 21, 118 (1980)
402. D.W.Jillie, M.A.H.Nerenberg, J.A.Blackburn, Phys.Rev.B 21, 125 (1980)

403. J. Bindslev-Hansen and P.E. Lindeloff, Rev. Mod. Phys. 56, 431 (1984)
404. A.K. Jain, K.K. Likarev, J.E. Lukens, and J.E. Sauvageau, Phys. Rep. 109, 311 (1984)
405. Y.D. Dai, W.J. Yeh, and Y.H. Kao, J. Low Temp. Phys. 52, 99 (1983)
406. B.Y. Shi, L. Zhang, Y.D. Dai, and Y.H. Kao, J. Low Temp. Phys. 54, 519 (1984)
407. P.E. Lindelof, Rep. Prog. Phys. 44, 949 (1981)
408. N.F. Pedersen, O.H. Soerensen, J. Mygind, P.E. Lindelof, M.T. Levinsen, and T.D. Clark, Appl. Phys. Lett. 28, 562 (1976)
409. T.F. Finnegan and S. Wahlsten, Appl. Phys. Lett. 21, 541 (1972)
410. D.W. Palmer and J.E. Mercereau, Phys. Lett. 61 A, 135 (1977)
411. P.E. Lindelof, J. Bindslev-Hansen, J. Mygind, N.F. Pedersen, and O.H. Sørensen, Phys. Lett. 60 A, 451 (1977)
412. C. Varmazis, R.D. Sandell, A.K. Jain, and J.E. Lukens, Appl. Phys. Lett. 33, 357 (1978)
413. P.E. Lindelof and J. Bindslev-Hansen, in Nonequilibrium Superconductivity, Phonons, and Kapitza Boundaries, K.E. Gray, ed. (Plenum Press, New York, 1981), chap. 19
414. J.A. Blackburn, J. Low Temp. Phys. 50, 475 (1983)
415. G.W. Frank, A.S. Deakin, M.A.H. Nerenberg, and J.A. Blackborn, Phys. Rev. B 35, 3138 (1987)
416. H. Seifert, J. Low Temp. Phys. 37, 595 (1979)
417. C. Vanneste, A. Gilabert, P. Sibillot, and D.B. Ostrowsky, J. Low Temp. Phys. 45, 517 (1981)
418. R. Sobolewski and C.V. Stancampiano, Phys. Rev. B 31, 6063 (1985)
419. C. Vanneste, A. Gilabert, P. Sibillot, and D.B. Ostrowsky, Apply. Phys. Lett. 38, 941 (1981)
420. G.Y. Logvenov, M. Osherov, and V.V. Ryazanov, Solid State Commun. 57, 99 (1986)
421. B.R. Fjordbøge and P.E. Lindelof, J. Low Temp. Phys. 31, 83 (1978)
422. F.J. Rachford, C.Y. Huang, S.A. Wolf, and M. Niesenhoff, Solid State Commun. 17, 1493 (1975)
423. P. Bergé, Y. Pomeau, and C. Vidal, Order within Chaos, 1st ed. (John Wiley and Sons, New York; Hermann publishers, Paris, 1984)
424. J.P. Eckmann and D. Ruelle, Rev. Mod. Phys. 57, 617 (1985); Addendum 57, 1115 (1985)
425. B.A. Huberman, J.P. Crutchfield and N.H. Packard, Appl. Phys. Lett. 37, 750 (1980)
426. K. Okuyama, H.J. Hartfuss, and K.H. Gundlach, J. Low Temp. Phys. 44, 283 (1981)
427. R.L. Kautz, J. Appl. Phys. 52, 6241 (1981)
428. M.R. Beasley and B.A. Huberman, Comments on Solid State Phys. 10, 155 (1982)
429. M. Civillo and N.F. Pedersen, Phys. Lett. 90 A, 150 (1982)
430. D. D'Humieres, M.R. Beasley, B.A. Huberman, and A. Libchaber, Phys. Rev. A 26, 3483 (1982)
431. H. Seifert, Phys. Lett. 98 A, 213 (1983)
432. H. Seifert and C. Nöldeke, Proceedings of the 17th International Conference on Low Temperature Physics, LT-17. U. Eckern, A. Schmid,

W. Weber, and W. Wühl, eds. (North Holland, Amsterdam, 1984), pp. 1135 - 1136
433. P. Alstrøm, M.T. Levinsen, and M.H. Jensen, in Proceedings of the 17th International Conference on Low Temperature Physics, LT-17. U. Eckern, A. Schmid, W. Weber, and W. Wühl, eds. (Nort Holland, Amsterdam, 1984), pp. 1133 - 1134
434. I. Goldhirsch, Y. Irmy, G. Wasserman, and E. Ben-Jacob, Phys. Rev. B 29, 1218 (1984)
435. M. Octavio, Phys. Rev. B 29, 1231 (1984)
436. D.R. He, W.Y. Yeh, and Y.H. Kao, Phys. Rev. B 30, 172 (1984)
437. K. Sakai and Y. Yamaguchi, Phys. Rev. B 30, 1219 (1984)
438. M. Octavio and C.R. Nasser, Phys. Rev. B 30, 1586 (1984)
439. M. Jansiti, Q. Hu, R.M. Westervelt, and M. Tinkham, Phys. Rev. Lett. 55, 746 (1985)
440. C. Nöldecke, Phys. Lett. 112 A, 178 (1985)
441. D.C. Gronemeyer, C.C. Chi, A. Davidson, and N.F. Pedersen, Phys. Rev. B 31, 2667 (1985)
442. Y.H. Kao, J.C. Huang, and Y.S. Gon, Phys. Rev. A 34, 1628 (1986)
443. Y.H. Kao, J.C. Huang, and Y.S. Gon, J. Low Temp. Phys. 63, 287 (1986)
444. C. Noeldeke, R. Gross, M. Bauer, G. Reiner, and H. Seifert, J. Low Temp. Phys. 64, 235 (1986)
445. J. Kuznik and M. Odehnal, J. Low Temp. Phys. 65, 353 (1986)
446. J. Kuznik, M. Odehnal, and P. Seidel, J. Low Temp. Phys. 72, 391 (1988)
447. N.F. Pedersen, O.H. Soerensen, B. Dueholm, and J. Mygind, J. Low Temp. Phys. 38, 1 (1980)
448. M.T. Levinsen, J. Appl. Phys. 53, 4294 (1982)
449. M. Bartucelli, P. L. Christiansen, N. F. Pedersen, and M. P. Soerensen, Phys. Rev. B 33, 4686 (1986)
450. C.R. Nasser and M. Octavio, in Proceedings of the 17th International Conference on Low Temperature Physics, LT-17. U. Eckern, A. Schmid, W. Weber, and W. Wühl, eds. (North Holland, Amsterdam, 1984), pp. 1129 - 1130
451. R. Sobolewski, D. R. Dykaar, T. Y. Hsiang, C. Vanneste, and C. C. Chi, Jap. J. of Appl. Phys. 26 (Supplemment 26-3), 1595 (1987)
452. R.F. Miracky, J. Clarke, and R.H. Koch, Phys. Rev. Lett. 50, 856 (1983)
453. K. Wiesenfeld, E. Knoblauch, R.F. Miracky, and J. Clarke, Phys. Rev. A 29, 2102 (1984)
454. M.A.H. Nerenberg, J.A. Blackburn, and S. Vik, Phys. Rev. B 30, 5084 (1984)
455. W. Binruo, Y. Zhon-jing, J. A. Blackburn, S. Vik, H.J. T. Smith, and M.A.H. Nerenberg, Phys. Rev. B 37, 3349 (1988)
456. M.A.H. Nerenberg, J.H. Baskey, and J.A. Blackburn, Phys. Rev. B 36, 8333 (1987)
457. P. Hadley and M.R. Beasley, Appl. Phys. Lett. 50, 621 (1987)
458. O.H. Olsen and M.R. Samuelsen, Appl. Phys. Lett. 47, 1007 (1985)
459. M. Cirillo, J. Appl. Phys. 60, 338 (1986)
460. J.D. Meyer and R. Tidecks, Solid State Commun. 24, 643 (1977)
461. W.J. Yeh, Y.D. Dai, and Y.H. Kao, Physica 108 B, 1025 (1981)

462. W.J. Yeh, Y.D. Dai and Y.H. Kao, J. Low Temp. Phys. 52, 249 (1983)
463. J.H. Davis, M.J. Skove, and E.P. Stillwell, Solid State Commun. 4, 597 (1966)
464. D.R. Overcash, M.J. Skove, and E.P. Stillwell, Phys. Rev. B 3, 3765 (1971)
465. J.W. Cook, W.T. Davis, J.H. Chandler, and M.J. Skove, Phys. Rev. B 15, 1357 (1977)
466. B. Rothberg-Bibby, D.S. MacLachlen, and F.R.N. Nabarro, in Proceedings of the 14th International Conference of Low Temperature Physics, Vol. 2, M. Krusius and M. Vuorio, eds. (North Holland Publishing Company, Amsterdam; American Elsevier Publishing Company, New York, 1975), pp. 117 - 120
467. C.L. Watlington, J.W. Cook jr., and M.J. Skove, Phys. Rev. B 15, 1370 (1977)
468. E.P. Stillwell, M.J. Skove, D.R. Overcash, and W.B. Gettys, Phys. kondens. Materie 9, 183 (1969)
469. D.W. Jillie, J.E. Lukens, and Y.H. Kao, Phys. Rev. Lett. 38, 915 (1977)
470. Y.S. Way, K.S. Hsu, and Y.H. Kao, Phys. Rev. Lett. 39, 1684 (1977)
471. L.G. Neuman, Y.D. Dai, and Y.H. Kao, Appl. Phys. Lett. 39, 648 (1981)
472. L.G. Neuman, Y.D. Dai, and Y.H. Kao, J. Low Temp. Phys. 49, 457 (1982)
473. F. Kober, W. Clauss, and R.P. Huebener, J. Low Temp. Phys. 74, 215 (1989)
474. R. Tidecks, J. Low Temp. Phys. 58, 439 (1985); Erratum 60, 459 (1985)
475. P. Haasen, Physikalische Metallkunde, 1st ed. (Springer Verlag, Berlin, 1974), chap. 6.3.3.
476. M. Hansen, Constitution of Binary Alloys, 2nd ed. (McGraw Hill, New York, 1958), pp. 854 - 856
477. U. Schulz and R. Tidecks, J. Low Temp. Phys. 71, 63 (1988)
478. G. Slama, Der Zusammenbruch der Supraleitung in stromtragenden Indium-Haarkristallen. Diplomarbeit, Universität Göttingen (1979)
479. J.D. Meyer, Appl. Phys. 7, 127 (1975)
480. T. Werner, R. Tidecks, H.J. Schulze, and K. Keck, Cryogenics 25, 705 (1985); Errata 26, 198 (1986)
481. U. Schulz and R. Tidecks, Cryogenics 28, 161 (1987)
482. U. Schulz and R. Tidecks, Solid State Commun. 66, 59 (1988)
483. B. Damaschke and R. Tidecks, Hysteretic Behaviour of a Phase-Slip Center in a Quasi-One-Dimensional Superconductor. To be published. See also ref. 393.
484. W.J. Skocpol, M.R. Beasley, and M. Tinkham, J. Appl. Phys. 45, 4054 (1974)
485. S.K. Decker and D.W. Palmer, J. Appl. Phys. 48, 2043 (1977)
486. G. Dharmadurai and B.A. Ratnam, J. Low Temp. Phys. 33, 395 (1978)
487. G. Dharmadurai and B.A. Ratnam, Physica 96 B, 134 (1979)
488. G. Dharmadurai and N.S. Satya Murthy, J. Low Temp. Phys. 37, 269 (1979)
489. G. Dharmadurai, Phys. Stat. Sol. (a) 62, 11 (1980)
490. H.J. Schulze and K. Keck, Z. Phys. B - Condensed Matter 51, 215 (1983)
491. H.J. Schulze and K. Keck, Appl. Phys. A 34, 2690 (1984)
492. H.J. Schulze and K. Keck, Solid State Commun. 55, 509 (1985)
493. R. Eichele, R.P. Huebener, H. Pavlicek, and H. Seifert, Physica 108 B, 1029 (1981)
494. R. Eichele, H. Seifert, and R.P. Huebener, Appl. Phys. Lett. 38, 383 (1981)
495. R. Eichele, L. Freytag, M. Seifert, R.P. Huebener, and J.R. Clem, J. Low Temp. Phys. 52, 449 (1983)

496. L. Freytag, R.P. Huebener, and H. Seifert, J. Low Temp. Phys. 60, 365 (1985)
497. L. Freytag and R.P. Huebener, J. Low Temp. Phys. 60, 377 (1985)
498. A.V. Gurevich and R.G. Mints, Rev. Mod. Phys. 59, 941 (1987)
499. A.F. Andreev, Sov. Phys. JETP 19, 1228 (1964)
500. A.J. Walton, Proc. Roy. Soc. London A 289, 377 (1965)
501. G.L. Pollack, Rev. Mod. Phys. 41, 48 (1969)
502. L.J. Challis, J. Phys. C 7, 481 (1974)
503. L.J. Challis and J.D.N. Cheeke, Proc. Roy. Soc. London A 304, 479 (1968)
504. O.V. Lounasmaa, Experimental Principles and Methods Below 1 K (Academic Press, London, 1974), chap. 9.6
505. A.F.G. Wyatt, in Nonequilibrium Superconductivity, Phonons, Kapitza Boundaries, K.E. Gray, ed. (Plenum Press, New York, 1981), chap. 2
506. W.A. Little, Can. J. Phys. 37, 334 (1959)
507. A.C. Anderson and W.L. Johnson, J. Low Temp. Phys. 7, 1 (1972)
508. T. Nakayama, J. Phys. C 10, 3273 (1977)
509. N.S. Snyder, Cryogenics 10, 89 (1970)
510. R.E. Jones and W.B. Pennebaker, Cryogenics 3, 215 (1963)
511. L.J. Challis, Phys. Lett. 26 A, 105 (1968)
512. N.W. Ashcroft and N.D. Mermin, Solid State Physics (Holt Saunders International Editions, Philadelphia, 1976)
513. T.B. Massalski, U. Mizutani, and S. Noguchi, Proc. Roy. Soc. London A 343, 363 (1975)
514. J.T. Folinsbee and A.C. Anderson, Phys. Rev. Lett. 27, 1580 (1973)
515. M. Stuivinga, T.M. Klapwijk, J.E. Mooij, and A. Bezuijen, J. Low Temp. Phys. 53, 673 (1983)
516. A. Rotwarf, G.A. Sai-Halàsz and D.N. Langenberg, Phys. Rev. Lett. 33, 212 (1974)
517. S.B. Kaplan, J. Low Temp. Phys. 37, 343 (1979)
518. L. Kramer and R. Rangel, J. Low Temp. Phys. 75, 65 (1989)
519. A. Gilabert, D.B. Ostrowsky, C. Vanneste, M. Papuchon, and B. Puech, Appl. Phys. Lett. 31, 590 (1977)
520. S.B. Kaplan, J. Appl. Phys. 51, 1682 (1980)
521. Y. Okabe, P. Anprung, and K. Fukuoka, Jap. J. Appl. Phys. 25, 1342 (1986)
522. K. Kojima, S. Nara, and K. Hamanaka, Jap. J. Appl. Phys. 26, 81 (1987)
523. K. Kojima, S. Nara, and K. Hamanaka, Jap. J. Appl. Phys. 26, 216 (1987)
524. K. Kojima, S. Nara, and K. Hamanaka, Extended Abstracts of the 18th (1986 International) Conference on Solid State Devices and Materials, Tokyo, 1986, pp. 455 - 458
525. J.A. Pals and L.H.J. Graat, in Proceedings of the 14th International Conference on Low Temperature Physics, Vol. 2, M. Krusius and M. Vuorio, eds. (North Holland Publishing Company, Amsterdam; American Elsevier Publishing Company, New York, 1975), pp. 251 - 254
526. Y.G. Bevza, V.I. Karamushko, and I.M. Dimitrenko, Journal de Physique (Suppl. 8) 39 (I), C 6 - 537 (1978)
527. B.W. Roberts, Properties of Selected Superconductive Materials, U. S. Department of Commerce, National Bureau of Standards, Washington D. C. (1972), p. 10

528. R.C. Weast (editor), Handbook of Chemistry and Physics, 56th ed. (Chemical Rubber Company Press, 1974/1975), pp. E 47, E 84
529. C.A. Hampel (editor), Rare Metals Handbook, 2nd ed. (Reinhold, London, 1961), p. 225
530. D.E. Gray (editor), American Institute of Physics Handbook, 3rd ed. (McGraw Hill, New York, 1972), pp. 2-54, 3-99, 3-100
531. X. Yang, Intrinsische und steuerbare Phasenschlupfzentren in quasieindimensionalen Supraleitern, Thesis, Universität Göttingen (1990)

List of Abbreviations

BCS	Bardeen, Cooper, Schrieffer
GL	Ginsburg, Landau
KR	Kramer, Rangel
KSS	Kadin, Smith, Skocpol
MKSA	Meter, kilogram, second, ampere
PSC	Phase-slip center
RSM	Rieger, Scalapino, Mercereau
SBT	Skocpol, Beasley, Tinkham
SC/NC	Superconductor/normalconductor
SEM	Scanning electron microscope
TDGL	Time-dependent Ginsburg-Landau
TEM	Transmission electron microscope
TWL	Tunable weak link

List of Symbols

General Remarks

1. The equations are usually written in a 'linear form', that means, for instance,

$$1 + ab/cd \qquad \text{instead of} \qquad 1 + \frac{ab}{cd}$$

2. The vector nature of a quantity is indicated by underlining (for instance, $\underline{A}(\underline{r})$).

Symbols

A	Cross-sectional area of a sample
$\underline{A}(\underline{r})$	Vector potential of the magnetic field at locus $\underline{r}$
A_s	Boundary surface for heat transfer (sec. 11.2)
a	Factor (sec. 7.3) equal to I_0/I_c
$\tilde{a}$	Constant used in the SBT model
$\langle a^2 \rangle$	Normalized mean-square deviation of the gap, $\langle a^2 \rangle_0$ value for clean bulk material
$\underline{B}$	Magnetic field
$B_{c\,th}$	Thermodynamical critical magnetic field
$B_{T,T*}$	Weighted average of Bose distributions b_{T*} and b_T (sec. 5.5)
$\mathbb{B}^*_{\underline{K},\sigma}$, $\mathbb{B}_{\underline{K},\sigma}$	Creation and annihilation operator, respectively, for a quasiparticle excitation with wave number vector $\underline{K}$ and spin σ (Bogolubov operators)
b_m	Coefficient (sec. 7.3)
C	1. A constant used in the discussion of the electron mean free path dependence of τ_E 2. A constant used in sec. 6.2 3. $C = C(T)$ Coefficient characterizing the heat transfer properties through the surface (sec. 11.2)
$C(\gamma)$	A function defined in the TDGL theory
$\tilde{C}$	Leakage capacitance (KSS model)
C_K	A constant introduced in sec. 7.6
$\tilde{C}_{Ku}$	Leakage capacitance (Kulik, sec. 7.2)

c_{CG}	Bare wave phase velocity of the Carlsson-Goldman mode
c_{KSS}	Bare wave phase velocity of the charge imbalance wave
c_{Ku}	Bare wave phase velocity of the collective mode according to Kulik
c_L, c_T	Longitudinal and transversal sound velocity, respectively
$c_{\tilde{m}}(\omega_J)$	Coefficients (KSS model)
c_{Pb}	Lead concentration
c_p, $c_{\dot{\varphi}}$	Constants used in the discussion of μ_p (sec. 5.4)
$c^*_{\underline{K}\uparrow}$, $c_{\underline{K}\uparrow}$	Creation and annihilation operator, respectively, for an electron in the state $\underline{K}\uparrow$
D	Diffusion constant
D(T)	Deviation of the critical magnetic field from a parabolic law
d	Film thickness
d_p	Periodic length (distance of phase-slip centers in an array, TDGL theory)
$\underline{E}$	Electrical field
$E_{\underline{K}}$	Quasiparticle excitation energy
e	Absolute value of the electron charge (it is $e > 0$, the electron charge is denoted by $-e$)
e'	Twice the electron charge ($e' = -2e$)
F	Damping factor of the Carlsson-Goldman mode
F^*	Dimensionless factor introduced in the discussion of charge imbalance relaxation times (sec. 5.4)
F_{KSS}	Damping factor of the charge imbalance wave
F_{Ku}	Damping factor of the collective mode according to Kulik
δF_0	Free energy barrier according to Langer and Ambegaokar
$f_{\underline{K}}$	Fermi function
$\tilde{f}_{\underline{K}}$	Occupation probability of a $\underline{K}$ state by a quasiparticle excitation
$f_1(x)$, $f_2(x)$	Functions introduced in sec. 7.6
$f(\varphi_{+-}(t))$	Current-phase relation (KSS model)
$\langle f \rangle_{\varphi_{+-}}$	Phase angle average of the function f (KSS model)
G	1. Free enthalpy 2. Abbreviation used in the discussion of the Carlsson-Goldman mode (sec. 5.6)
$\tilde{G}$	Leakage conductance (KSS model)
G_{KSS}	Abbreviation used in the discussion of the charge imbalance waves
G_{Ku}	Abbreviation used in the discussion of the collective mode according to Kulik (sec. 5.7)
$\tilde{G}_{Ku}$	Leakage conductance (Kulik, sec. 7.2)
G_s	Ginsburg-Landau free enthalpy

g_L	A constant used in the discussion of the electron mean free path dependence of τ_E
$\hbar$	It is $\hbar = h/2\pi$, where h is Planck's constant
I	Total current
ΔI	1. Width of a current step (sec. 7.3) 2. Difference between the onset currents of the first two voltage steps in the V-I characteristics of a zinc whisker (sample Zn 21, Fig. 45) 3. Current range over which the first linear portion in the V-I characteristics of a Pb whisker exists (sec. 9.2.3) 4. Distance in current between the first two voltage steps in sample WWZn 7 (sec. 13.4)
I_{AB}	A conventional current flowing from contact A to contact B (sec. 7.6, p. 156)
I_{AF}	Current flowing from contact A to contact F (sample WWZn7, sec. 13.4)
$I_{AF,c}$	Critical current, where the current flows from contact A to contact F (sec. 13.4)
I_c	Critical current
$I_c(T^*)$	Critical current at the temperature T^* (sec. 8.3)
$I_{c,PSC2}$	Critical current at which the phase-slip center PSC 2 appears in the V-I characteristics of sample WWZn 7 (sec. 13.4)
$I_{c,TWL}$	Critical current of the TWL (chap. 12)
$\Delta I_{c,PSC3}^{PSC2}$	Shift of the critical current of phase-slip center PSC 3 under the influence of phase-slip center PSC 2 in sample WWZn 7 (sec. 13.4)
ΔI_{c1}	Shift of the critical current of bridge 1 in the closely spaced two-microbridge system (sec. 7.5)
ΔI_{c1}^a	Asymmetric part of ΔI_{c1}
ΔI_{c1}^s	Symmetric part of ΔI_{c1}
I_F	Fixed current (chap. 11)
I_{F1}, I_{F2}	Fixed currents introduced in sec. 11.4.1 for the description of the fitting procedure in the case of the hysteresis model with two fitting parameters
I_G	Current in a generalized resistivity shunted junction model (KSS model)
I_H	Extrapolated zero hysteresis width intercept (sec. 11.2)
$\tilde{I}_1(\tilde{\chi}_1)$	A function characterizing the anisotropy effect (sec. 6.2)
I_k	Control current (sec. 7.5)
I_m	Measuring current (sec. 7.5)
I_{max}	Maximum current for the existence of $\tilde{m}$ phase-slip centers (SBT model)
I_{min}	Limit of the hysteresis in the KSS model
$I_n(x)$	Normal current at locus x, ⟨...⟩ time average

I_R	Jump-back current (sec. 6.1, 11.1 and Fig. 25)
$I_{R,TDGL}$	Jump-back current as predicted by the TDGL theory (sec. 11.5)
I_s	Supercurrent
$I_s(x)$	Supercurrent at locus x, ⟨...⟩ time average
$\overline{I}_s(t)$	Spatial average of the supercurrent in the weak link in the RSM model
$I_s(X_{PSC}, t)$	Supercurrent in the core of a phase-slip center, ⟨...⟩ time average
I_t	Control (or tuning) current through the controlling whisker of a tunable weak link arrangement (chap. 12)
I_{tA}	Control current needed until the phase-slip center at the TWL appears before the onset of the intrinsic V-I characteristic (chap. 12)
I_0	Zero voltage intercept obtained by back extrapolation of the V-I characteristic after the first voltage step to $V = 0$
$(I_0/I_c)_{0InW}$	'Low temperature value' of the ratio I_0/I_c for In whiskers (sec. 7.2)
$(I_0/I_c)_{0\underline{In}PbW}$	'Low temperature value' of the ratio I_0/I_c for <u>In</u>-Pb whiskers (sec. 9.2.1)
$(dI_c^{2/3}/dT_c)_0$	Slope of the $I_c^{2/3}(T_c)$ straight line for the first intrinsic voltage step of the measuring whisker in a tunable weak link arrangement (chap. 12)
$J_n(...)$	Bessel functions of the first kind of integer order (sec. 7.3)
$\hat{\underline{j}}$	Normalized current (TDGL theory)
j_c	Critical current density ($j_c = I_c/A$)
$\hat{j}_c$	Ginsburg-Landau critical current (normalized units, TDGL theory)
$\hat{j}_{max}$, $\hat{j}_{min}$	Upper and lower border, respectively, of the current range in which phase-slip solutions exist (normalized units, TDGL theory)
$\underline{j_n}$	'Normal' current density (i.e. quasiparticle current density), $j_n(x)$ space dependent absolute value, ⟨...⟩ time average
$\hat{j}_n$	Absolute value of the normal current (normalized units, TDGL theory)
$\underline{j_s}$	Supercurrent density, $j_s(x)$ space dependent absolute value, ⟨...⟩ time average
j_0	Normalization current (TDGL theory)
$\hat{j}_0$	Extrapolated zero voltage intercept of a voltage-current characteristic (normalized units, TDGL theory)
$\underline{K}$	Wave number vector of an electron
K_F	Fermi wave number
K_{GL}	Constant appearing in the Ginsburg-Landau critical current density
K^i	A parameter characterizing all effects besides the anisotropy effect (sec. 6.2)

$2\delta K$	Range of nonzero interaction potential in the BCS theory
L	Sample length
$\tilde{L}$	Series inductance per unit length (KSS model)
L_{An1}	Normal-like length related to the differential resistance of the first phase-slip center of a V-I characteristic. The subscripts '0 In W', '0 Pb W', '0 In Pb W', and '0 Pb Bi W' denote the 'low temperature values' of the normal-like length L_{An1} for whiskers of In, Pb, In-Pb, and Pb-Bi, respectively (sec. 7.2, 9.2.4, 9.2.1, and 9.2.4, respectively)
L_{An2}	Normal-like length related to the increase of the differential resistance, $\Delta(dV/dI)_2$, after the appearance of the second voltage step in a V-I characteristic. The subscript '0 In W' denotes the 'low temperature value' of the normal-like length L_{An2} for whiskers of In (sec. 7.2)
L_{An3}	Normal-like length as calculated from the differential resistance of the first linear portion of the V-I characteristic of a Pb whisker generated by 3 active phase-slip centers (sec. 9.2.3)
$\tilde{L}_{Ku}$	Series inductance (Kulik, sec. 7.2)
L_{spot}	Length of the hot spot (sec. 11.2)
ℓ	Electron mean free path
m	Electron mass (except in sec. 7.3, where m denotes an index)
m'	Twice the electron mass ($m' = 2m$)
N	Number of particles in a system (for instance electrons)
N_{max}	Maximum quasiparticle density (sec. 11.4.1)
$N_n(\varepsilon_{\underline{K}})$	Normal density of states for one spin direction
N_{op}	Quasiparticle overpopulation, $N_{op}(x, t=0)$: spatial dependence at the moment of the phase-slip event, $N_{op}(x=0, t)$: time dependence at the core of the phase-slip center
$N_q(E_{\underline{K}})$	Quasiparticle density of states in a superconductor (sec. A.2)
N_s	Abbreviation used in the discussion of Kulik's work in sec. 5.7
N_T	Number of quasiparticles per volume at the bath temperature T
N_0	Number of electronic states (in a free electron model) for one spin direction per volume and energy interval at the Fermi energy
n	Electron density (except in sec. 7.3, where n denotes an index)
$\tilde{n}$	1. Effective number of electrons added to the quasiparticle system if a quasiparticle with wave number K is excited (sec. 5.2) 2. Positive integer or fractional number (sec. 7.3)
n_{ar}	Number of proximity effect bridges in an array (sec. 7.3)
$n_s(\underline{r})$	Local density of the superconducting charge carriers in the Ginsburg-Landau theory

n_{se}	Superelectron density (i.e. number of electrons bound to Cooper pairs per volume)
n_{s0}	Density of particles described by ψ in the absence of currents or magnetic fields (Ginsburg-Landau theory)
n_0	Maximum possible density of superconducting particles (Cooper pairs) as present at T = 0 K (Ginsburg-Landau theory)
P1, P2, P3, P4	Potential probes
p_F	Fermi momentum
Q	Branch imbalance
$\underline{Q}$	Abbreviation (TDGL theory)
$\dot{Q}$	Heat dissipation in a sample (chap. 11)
Q^*	Charge imbalance of the quasiparticle system per unit volume
$Q_{\underline{K}}$	Quasiparticle charge
$Q_{\bar{P}}$	Charge of the condensate
$\underline{q}$	Phonon wave number vector
$\tilde{q}$	Gradient of the phase in the SBT model
q_{CG}	Wave number of the Carlsson-Goldman mode
$\underline{q}_{KSS}$	Wave number vector of the charge imbalance wave (absolute value: q_{KSS})
$q_{\tilde{m}}$	Abbreviation used in the KSS model
$\tilde{R}$	Series resistance per unit length (KSS model)
R_{ac}	Impedance of the transmission line in the high-frequency limit (KSS model)
R_{eff}	Normal resistance of length $2\Lambda_{Q^*in}$ of the transmission line (KSS model)
R_K	Kapitza resistance (sec. 11.2)
$\tilde{R}_{Ku}$	Series resistance (Kulik, sec. 7.2)
R_n	Residual resistance
R_{298K}	Resistance at room temperature
$R_\square$	Sheet resistance (sec. 5.4)
$\underline{r}$	Space vector
r_w	Radius of a whisker assuming a circular cross-sectional area
S	Surface of the warm region around a phase-slip center through which the dissipated energy is transmitted into the surrounding helium bath (sec. 11.4.1)
S_u	Abbreviation (sec. A.2)
s_B	Factor accounting for the nonexponential charge imbalance decay in the matching regions (sec. 5.9)
T	Absolute temperature (bath temperature, sec. 11.2)
ΔT	Temperature difference of the actual helium bath temperature relative to the critical temperature ($\Delta T = T_{c0} - T$)
T^*	Effective quasiparticle temperature

δT^*	Enhancement of the effective quasiparticle temperature, T^*, above the bath temperature, T, i. e. $\delta T^* = T^* - T$
T_c	Transition temperature at current I_c
$T_c(I_F)$	Transition temperature at current I_F (chap. 11)
T_{c0}	Critical temperature
$T_{c0,\,part1}$	Critical temperature of part 1 of sample WWZn 7 (sec. 13.4)
T_{c0r}	Critical temperature of the clean matrix (sec. 6.2)
T_E	Upper temperature border for the local equilibrium approximation (TDGL theory)
ΔT_E	It is $\Delta T_E = \lvert T_{c0} - T_E \rvert$
T_{GL}	Temperature down to which the simple (i. e. gapless) version of the TDGL theory may be used
ΔT_{GL}	It is $\Delta T_{GL} = \lvert T_{c0} - T_{GL} \rvert$
$\delta T_H(I_F)$	Width of the temperature hysteresis at current I_F
T_{Hy}^{exp}	Measured onset temperature of hysteretic behaviour (sec. 11.1)
ΔT_{Hy}^{exp}	Measured onset of hysteretic behaviour with $\Delta T_{Hy}^{exp} = T_{c0} - T_{Hy}^{exp}$ (sec. 11.1)
ΔT_{Hy}^{KSS}	Prediction for the onset of hysteresis after the KSS model (sec. 11.1)
T_{Hy}^{op}	Onset temperature of the quasiparticle overpopulation induced hysteresis (sec. 11.4.1)
$\delta T_{H1}, \delta T_{H2}, \delta T_{H2}'$	Quantities introduced in sec. 11.4.1 for the description of the fitting procedure in the case of the hysteresis model with two fitting parameters
δT_{KSS}	Charge imbalance wave induced hysteresis (KSS mechanism, sec. 11.4.1)
T_L	Lattice temperature (sec. 11.4.1)
δT_L	Enhancement of the lattice temperture above the bath temperature T, i. e. $\delta T_L = T_L - T$
$T_R(I_F)$	Jump-back temperature of the helium bath at current I_F (chap. 11)
ΔT_V	The same as ΔT_E (see footnote p. 195)
T_W	Sample temperature (sec. 11.2)
δT_W	Enhancement of the sample temperature above the bath temperature, i. e. $\delta T_W = T_W - T$
T_1	A constant used in the calculation of τ_{ee} (sec. 5.4)
t	Time
$\hat{t}$	Normalized time (TDGL theory)
t_{sb}	Time defined in sec. 5.8.3 (SBT model)
t_0	Normalization time (TDGL theory)
u	Normalization constant (TDGL theory)
$\lvert u_{\underline{K}} \rvert^2$	Occupation probability of the state with wave number vector $\underline{K}$ by a hole in a BCS superconductor ($= u_{\underline{K}}^2$)
$u_s(\underline{r}, t)$	Displacement amplitude of the superfluid part in the Carlsson-Goldman mode, $\langle \ldots \rangle$ time average (sec. 7.6)

U_r	Constant used in the SBT model
V	Voltage, ⟨...⟩ time average
$\hat{V}$	Normalized voltage (TDGL theory), ⟨...⟩ time average
v_a	Radiation induced voltage (sec. 7.3)
V_{AB}	A conventional voltage measured between contacts A and B (sec. 7.6, p. 156)
V_{AF}	Voltage measured between contacts A and F (sample WWZn7, sec. 13.4)
V_c	Scaling voltage (KSS model)
V_{CBf}	Voltage of the voltage foot (sec. 7.6)
$V_{eff}(\underline{K}, \underline{q})$	Effective electron-electron interaction potential
$\langle V \rangle_G$	Time averaged voltage in a generalized resistivity shunted junction model (KSS model)
$V_{J\tilde{n}}$	Voltage of harmonic and subharmonic current steps (sec. 7.3)
V_{J1}	Voltage of the main current step (sec. 7.3)
$V_{\tilde{m}}$	Voltage developed by $\tilde{m}$ phase-slip centers (SBT model)
V_R	Jump-back voltage
V_{RL}	Voltage between the points x_R and x_L (SBT model)
V_{RS}	Saturating value of the jump-back voltage, V_R (sec. 11.1)
V_t	Voltage across the controlling (or 'tuning') whisker of a tunable weak link arrangement (chap. 12)
$V_1(I_c)$	Height of the first voltage step
V_{12}	Voltage (i. e. electrostatic potential difference) between $\underline{r}_1$ and $\underline{r}_2$: $V_{12} = \Phi_{\dot{\varphi}}(\underline{r}_1) - \Phi_{\dot{\varphi}}(\underline{r}_2)$
$V_{-+}(V_{+-})$	Voltage between the left (right) and right (left) side of the Josephson oscillator (KSS model)
$V_{\mp\infty}$	Voltage measured by probes far away from the phase-slip center (KSS model)
$(dV/dI)_1$	Differential resistance of the characteristics after the first voltage jump (or of the first linear portion) in a V-I characteristic at constant temperature
v_F	Fermi velocity
$\lvert v_{\underline{K}} \rvert^2$	Occupation probability of the state with wave number vector $\underline{K}$ by an electron in a BCS superconductor ($= v_{\underline{K}}^2$)
v_{KSS}	Phase velocity of the charge imbalance wave
$\underline{v}_s$	Superfluid velocity (absolute value, v_s)
X_G, X_G^*	Abbreviations used in eq. (89) of sec. 5.5
X_{PSC}	Locus of the core of the phase-slip center
$x = x(V)$	Abbreviation (sec. 11.4.1) with $x_R = x(V_R)$
$\tilde{x}$	Distance from the core of the phase-slip center (TDGL theory)
$\hat{\underline{x}}$	Unit vector along the axis of the filament (SBT model)
x_L, x_R	Point left and right from the phase-slip center where $\mu = \mu_p$ (SBT model)
x_l, x_r	Point left and right from the phase-slip center

$Z(\omega_{KSS})$	Impedance of the transmission line (KSS model)
$Z(T)$	Abbreviation (sec. A.2)
$Z_{Ku}(\omega_{Ku})$	Impedance of the transmission line (Kulik, sec. 7.2)
α	Constant of the Ginsburg-Landau theory
α_K	Heat transfer coefficient (sec. 11.2)
$\tilde{\alpha}^2 F(\omega)$	Electron-phonon coupling function of the Eliashberg theory
β	Constant of the Ginsburg-Landau theory
β_{KR}	Ratio of the 'excess current' I_0 and the critical current (TDGL theory)
β_H	Fitting parameter of the phenomenological hysteresis model (sec. 11.4.1)
β_{SBT}	Ratio $\langle I_s(X_{PSC})\rangle / I_C$ of the SBT model
β_{TWL}	A constant, relating the quasiparticle overpopulation in a tunable weak link (TWL) with the control current (sec. 12.2)
Γ	Pair-breaking parameter (sec. 5.4), except in sec. 7.3, where Γ denotes the Gamma function
Γ_0	Abbreviation used in the discussion of the damping of the Carlsson-Goldman mode
γ	Pair-breaking parameter (TDGL theory)
$\tilde{\gamma}$	Euler's constant
γ_c	Value of the pair-breaking parameter for which the upper current border of the phase-slip solution is equal to the Ginsburg-Landau critical current
γ_{Hy}^{exp}	Value of the pair-breaking parameter, γ, at the onset temperature of hysteretic behaviour (sec. 11.1)
γ_{max}	Upper limit of the pair-breaking parameter (concerning the local equilibrium approximation, TDGL theory)
γ_R	Abbreviation (TDGL theory)
$\langle\boldsymbol{\Delta}\rangle$	Average value of the gap (angle average)
$\boldsymbol{\Delta}(\underline{r})$	Space dependent order parameter of the microscopic theory of superconductivity (in the absence of currents and magnetic fields: $\boldsymbol{\Delta}_0(\underline{r})$)
$\boldsymbol{\Delta}(T)$	Energy gap of the superconductor at temperature T ($\boldsymbol{\Delta}(0) = \boldsymbol{\Delta}(T=0)$)
$\hat{\boldsymbol{\Delta}}_s$	Normalized order parameter in the presence of magnetic impurities (TDGL theory)
$\boldsymbol{\Delta}_0$	Equilibrium value of the order parameter (TDGL theory)
$\boldsymbol{\Delta}_{0s}$	Equilibrium order parameter in the presence of magnetic impurities (TDGL theory)
$\langle(\delta\boldsymbol{\Delta})^2\rangle$	Mean-square deviation of the gap from the average value (angular average)

ε_H	Fitting parameter of the phenomenological hysteresis model (sec. 11.4.1)
$\varepsilon_{\underline{K}}$	Plane wave energy of a free electron state measured relative to the chemical potential, μ_F
ε_0	Influence constant
η_F	Fermi energy
$\eta_{\underline{K}}$	Plane wave energy of a free electron state
Θ	Debye temperature
$\tilde{\Theta}$	Phase of the order parameter (TDGL theory)
$\varkappa$	Ginsburg-Landau parameter
Λ	Quasiparticle diffusion length (SBT model)
$\Lambda(t)$	Quasiparticle diffusion length, $\Lambda(t)=((1/3)\ell\, v_F t)^{1/2}$, sec. 11.4.3
Λ_E	Diffusion length related to τ_E (TDGL theory)
Λ_{KR}	Normal-like length (TDGL theory)
Λ_{Ku}	Decay length of the scalar potential or penetration depth of the electric field into a superconductor (according to Kulik)
Λ_L	London's constant
Λ_{Q^*}	Charge imbalance relaxation length
Λ_{Q^*in}	Charge imbalance relaxation length due to inelastic electron-phonon collisions
$\lambda(T)$	Magnetic penetration depth
λ_d	Decay length of the Carlsson-Goldman mode
$\lambda_{d,KSS}$	Decay length of the charge imbalance wave
$\lambda_L(0)$	London penetration depth at $T=0\,K$
μ	Electrochemical potential of quasiparticles, $\langle ... \rangle$ time average
$\hat{\mu}$	Normalized electrochemical quasiparticle potential (TDGL theory), $\langle ... \rangle$ time average
μ^*	Chemical potential in the quasi-thermal Owen-Scalapino model for quasiparticle overpopulation
μ_{chem}	Chemical potential
$\mu_{c,p}$	Chemical potential of Cooper pairs
$\mu_{c,q}$	Chemical potential of quasiparticles
μ_{elchem}	Electrochemical potential
μ_F	Chemical potential of electrons in the normal state (often called 'Fermi energy' although rigorously $\eta_F=\mu_F(T=0)$)
$\langle\mu\rangle_K$	Time averaged value of the contribution of collective excitations to the electrochemical potential of quasiparticles (sec. 7.6)
μ_p	Electrochemical potential of Cooper pairs, $\langle ... \rangle$ time average
$\langle\Delta\mu_p\rangle$	Time averaged difference of the electrochemical pair potentials far away from the core of the phase-slip center (sec. 7.3)

$\mu_{p,a}$	Slope of the electrochemical pair potential during the accelleration part of the phase-slip cycle (SBT model)
$\mu_{p,d}$	Slope of the electrochemical pair potential during the decelleration part of the phase-slip cycle (SBT model)
$\langle\mu_p\rangle_l$, $\langle\mu_p\rangle_r$	Time averaged electrochemical potentials left and right of the phase-slip center (SBT model)
$\mu_{p,12}$	Difference of the electrochemical pair potential between the locus $\underline{r}_1$ and $\underline{r}_2$ (sec. 7.3: between the core region of a phase-slip center), i. e. $\mu_{p,12} = \mu_p(\underline{r}_1) - \mu_p(\underline{r}_2)$, $\langle ... \rangle$ time average
$\langle\mu\rangle_{P1}$	Time averaged electrochemical quasiparticle potential at the locus of potential probe P1 (Fig. 36)
$\langle\mu\rangle_{SBT}$	Time averaged electrochemical quasiparticle potential as proposed by SBT (sec. 7.6)
μ_0	Induction constant, $\mu_0 = 4\pi \cdot 10^{-7}$ Vs A^{-1} m^{-1}
ν	Frequency of the HF radiation field
ν_c	Low frequency border of HF stimulated superconductivity (sec. 7.3)
$\nu_{c,min}$	Lowest frequency that can lead to an enhancement of superconductivity by HF radiation (sec. 7.3)
ν_J	Josephson frequency
ν_{PSC}	Phase-slip rate (TDGL theory)
$\xi(T)$	Ginsburg-Landau coherence length
ξ_D	Ginsburg-Landau coherence length in the dirty limit
ξ_0	BCS coherence length (typical decay length of the probability density for a Cooper pair in a weak-coupling superconductor at T = 0 K)
$\xi_0(T_{c0})$	Coherence length (typical decay length of the probability density for a Cooper pair in a weak-coupling superconductor at $T = T_{c0}$)
ρ	Resistivity (Density, in g_L^2, eq. (85), and Tab. A3)
ρ^*	Residual resistance ratio
ρ_n	Residual resistivity
$\rho_{nSnIn\perp}$	Residual resistivity of tin with indium impurities perpendicular to the tetrad axis of the tetragonal elementary cell of 'white tin'
$\rho_\Theta \ell_\Theta$	Product $\rho \cdot \ell$ at the Debye temperature, Θ
ρ_{298K}	Phonon induced temperature dependent part of the resistivity at room temperature (298 K)
$\rho_{298K,tot}$	Total resistivity at room temperature (sometimes called ρ_{RT})
σ	Conductivity in the normal state (in sec. 5.2: spin index)
σ_0	Effective cross-sectional area for screened coulomb interaction
τ	Impurity scattering time in the normal metal ($\tau = \ell / v_F$)

τ_E	Inelastic electron-phonon collision time, i.e. lifetime of electrons in the normal state at the Fermi level and at T_{c0} due to electron-phonon collisions
$\tau_{E\infty}$	The value of τ_E for pure materials after Keck and Schmid (sec. 5.4)
τ_{eb}	Electron-bogolon scattering rate
τ_{eb0}	Abbreviation used in the discussion of Kulik's work in sec. 5.7
τ_{eff}	Effective quasiparticle recombination time, including phonon trapping effects (sec. 11.4.3)
τ_{ee}	Inelastic electron-electron scattering time
τ_{es}	Phonon escape time
τ_{GL}	Ginsburg-Landau order parameter relaxation time (gap relaxation time in the gapless limit)
τ_{PSC}	Period of the phase-slip process
τ_Q	Branch imbalance relaxation time
τ_{Qel}	Branch imbalance relaxation time due to gap anisotropy
τ_{Qin}	Branch imbalance relaxation time due to inelastic electron-phonon scattering
τ_{Q^*}	Charge imbalance relaxation time
τ_{Q^*ee}	Charge imbalance relaxation time due to inelastic electron-electron scattering
τ_{Q^*el}	Charge imbalance relaxation time due to elastic scattering processes in the presence of gap anisotropy
τ_{Q^*g}	Charge imbalance relaxation time due to a spatial dependence of the gap
τ_{Q^*in}	Charge imbalance relaxation time due to inelastic electron-phonon processes only
$\tau_{Q^*in}(0)$	Forefactor in the expression of the charge imbalance relaxation time due to inelastic electron-phonon scattering in the case of a weak-coupling superconductor
τ_{Q^*in,j_c}	Charge imbalance relaxation time due to inelastic electron-phonon collisions and a supercurrent
$\tau_{Q^*in,m}$	Charge imbalance relaxation time due to inelastic electron-phonon collisions and magnetic impurities
τ_R	Relaxation time of the order parameter (TDGL theory)
τ_r	Intrinsic recombination time (between two recombination events) for the recombination of a given quasiparticle (sec. 11.4.3)
τ_r'	Intrinsic quasiparticle recombination time characterizing the decay of a quasiparticle population (sec. 11.4.3)
τ_s	Elastic spin-flip scattering time
τ_s^{ph}	Scattering time for a phonon with a quasiparticle
τ_α	Abbreviation used in the charge imbalance wave equation
τ_Δ	Order parameter relaxation time
τ_Θ	Electron-phonon scattering time extrapolated back to the Debye temperature from the high temperature limit where the resistivity is proportional to the temperature (sec. 5.4)

τ_0	Unit time of Kaplan et al. [103]
$\tau_0(\dots)$	Unit time of Kaplan et al. [103] for the material mentioned, $(\dots) = (Sn), (In)$, sec. 7.2, $(\dots) = (Pb)$, sec. 9.3
τ_{0InW}	'Low temperature value' of the quasiparticle relaxation time τ_2 for In whiskers (sec. 7.2)
$\tau_{0\underline{In}PbW}$	'Low temperature value' of the quasiparticle relaxation time τ_2 for $\underline{In}$-Pb whiskers (sec. 9.3)
$\tau_{0\underline{Pb}In,BiW}$	Temperature independent quasiparticle relaxation time τ_2 evaluated for Pb, $\underline{Pb}$-In, and $\underline{Pb}$-Bi whiskers (sec. 9.3)
τ_{0PbW}	Temperature independent quasiparticle relaxation time τ_2 evaluated for pure Pb whiskers only (sec. 9.3)
τ_{0R}	Supercurrent response time
τ_{0SnW}	Temperature independent relaxation time τ_2 evaluated for Sn and $\underline{Sn}$-In whiskers (sec. 7.2)
τ_{0ZnW}	Temperature independent relaxation time τ_2 evaluated for Zn whiskers (sec. 8.3)
τ_2	Quasiparticle relaxation time (SBT model)
$\tilde{\Phi}$	Potential introduced in the TDGL theory
$\tilde{\Phi}_0$	Normalization potential (TDGL theory)
Φ	Electrostatic potential
$\Phi(\underline{r}, t)$	Scalar potential of the electromagnetic field
Φ_{Q^*}	Charge imbalance induced part of the electrostatic potential
$\Phi_{\dot{\varphi}}$	Order parameter induced part of the electrostatic potential
Φ_L, Φ_T	Functions used in the discussion of the electron mean free path dependence of τ_E
$\varphi(\underline{r})$	Space dependent phase of the order parameter of the Ginsburg-Landau theory
φ_a	Phase angle of the radiation of a HF field (sec. 7.3)
φ_{12}	Phase difference between $\underline{r}_1$ and $\underline{r}_2$, i.e. $\varphi_{12} = \varphi(\underline{r}_1) - \varphi(\underline{r}_2)$, in sec. 7.3: across the core region of a phase-slip center
$\tilde{\varphi}_{12}(0)$	Abbreviation used in sec. 7.3
$\varphi_{+-}(t)$	Difference between the phases of the order parameter at both sides of the Josephson oscillator (KSS model)
χ	'Impurity function' of the Ginsburg-Landau theory (sec. 5.1)
$\tilde{\chi}_i$	A quantity which is proportional to the density of impurities (sec. 6.2)
Ψ_{BCS}	Wave function of the superconducting ground state as proposed by BCS
$\psi(\underline{r})$	Complex order parameter of the Ginsburg-Landau theory at locus $\underline{r}$
ψ_0	Equilibrium value of the order parameter in the absence of currents or magnetic fields

$\lvert\psi_0\rvert^2$	Density of particles described by ψ in the absence of currents or magnetic fields (Ginsburg-Landau theory)
$\lvert\psi_T\rvert$	Absolute value of the equilibrium value of the order parameter far away from the phase-slip center
$\hat{\psi}$	Normalized order parameter (TDGL theory)
$\langle\lvert\hat{\psi}\rvert\rangle$	Time average of the absolute value of the normalized order parameter (TDGL theory)
Ω	Volume of the superconductor
ω	Phonon frequency
ω_a	Angular frequency of the radiation of a HF field (sec. 7.3)
ω_c	Scaling angular frequency (KSS model)
ω_{CG}	Angular frequency of the Carlsson-Goldman mode
$\omega_{c,\,min}$	Lowest angular frequency that can lead to an enhancement of superconductivity by HF radiation (sec. 7.3)
ω_D	Debye (angular) frequency
ω_J	Josephson (angular) frequency (KSS model)
ω_{KSS}	Angular frequency of the charge imbalance wave
ω_{Ku}	Angular frequency of the collective mode according to Kulik
ω_{OP}	Frequency of order parameter oscillations
ω_{PSC}	Repetition (angular) frequency of the phase-slip process
$\uparrow$, $\downarrow$	Direction of electron spin (spin up, spin down)

Subject Index

V. Kose, Physikalisch-Technische Bundesanstalt, Braunschweig (Ed.)

Superconducting Quantum Electronics

Foreword by W. Buckel

With contributions by numerous experts

1989. XV, 299 pp. 180 figs. Hardcover DM 98,- ISBN 3-540-51176-8

This unique collection of papers by leading German scientists reviews recent accomplishments, presents new results and discusses possible future developments of superconducting quantum electronics and high T_c superconductivity. The three main parts of the book deal with fundamentals, sensitive detectors, and precision metrology. The book will be valuable to researchers and students at academic and industrial institutions who are interested in theory and precision experiments.

M. A. Savchenko, A.V. Stefanovich, Moscow

Fluctuational Superconductivity of Magnetic Systems

Translated from the Russian by R. S. Wadhwa, N. Wadhwa

1990. IX, 258 pp. 63 figs. (Research Reports in Physics) Softcover DM 98,- ISBN 3-540-50561-X

The fluctuational theory of superconductivity predicted the existence of high-temperature superconductivity in compounds of rare-earth metals, in ceramic systems, and polymers before its experimental discovery by Bednorz und Müller. This prediction was published in the 1986 (Russian) edition of this monograph. The theory is based on the enhancement of conductivity by fluctuations in the electron spin that arise due to the exchange nature of electron-phonon interactions. This text is intended for physicists and chemists investigating and synthesizing new high temperature superconducting materials. Methods of increasing the critical temperature of high-temperature superconductors are proposed. Macroscopic properties of high-temperature superconductors are investigated both in the superconducting and normal phases by using the theory of gauge fields, differential geometry, and topology. In view of these methods, the book should also be of interest to mathematicians working in nonlinear differential equations.

Springer-Verlag
Berlin
Heidelberg
New York
London
Paris
Tokyo
Hong Kong
Barcelona

Springer Tracts in Modern Physics

90* **Jets of Hadrons**
By W. Hofmann 1981. 165 figs. VIII, 215 pages

91 **Structural Studies of Surfaces**
By K. Heinz, K. Müller, T. Engel, and K. H. Rieder
1982. 120 figs. VII, 180 pages

92 **Single-Particle Rotations in Molecular Crystals**
By W. Press 1981. 53 figs. IX, 129 pages

93 **Coherent Inelastic Neutron Scattering in Lattice Dynamics**
By B. Dorner 1982. 47 figs. VIII, 96 pages

94 **Exciton Dynamics in Molecular Crystals and Aggregates**
By V. M. Kenkre and P. Reinecker 1982. 37 figs. IX, 226 pages

95 **Projection Operator Techniques in Nonequilibrium Statistical Mechanics**
By H. Grabert 1982. 4 figs. X, 164 pages

96 **Hyperfine Structure in 4d- and 5d-Shell Atoms**
By S. Büttgenbach 1982. 14 figs. VIII, 97 pages

97 **Elements of Flow and Diffusion Processes in Separation Nozzles**
By W. Ehrfeld 1983. 73 figs. VIII, 140 pages

98 **Narrow-Gap Semiconductors**
With contributions by R. Dornhaus, G. Nimtz, and B. Schlicht
1983. 2nd printing 1985. 244 figs, X, 309 pages

99 **Dynamical Properties of IV–VI Compounds**
With contributions by H. Bilz, A. Bussmann-Holderer, W. Jantsch, and P. Vogl
1983. 47 figs. VIII, 101 pages

100* **Quarks and Nuclear Forces**
Edited by D. C. Fries and B. Zeitnitz 1982. 69 figs. XI, 223 pages

101 **Neutron Scattering and Muon Spin Rotation**
With contributions by R. E. Lechner, D. Richter, and C. Riekel
1983. 118 figs. IX, 229 pages

102 **Theory of Jets in Electron-Positron Annihilation**
By G. Kramer 1984. 86 figs. VII, 140 pages

103 **Rare Gas Solids**
With contributions by H. Coufal, E. Lüscher, H. Micklitz, and R. E. Norberg
1984. 40 figs. IX, 99 pages

104 **Surfaces Enhanced Raman Vibrational Studies at Solid/Gas Interfaces**
By I. Pockrand 1984. 60 figs. IX, 164 pages

105 **Two-Photon Physics at e^+e^- Storage Rings**
By H. Kolanoski 1984. 123 figs. IX, 187 pages

106 **Polarized Electrons at Surfaces**
By J. Kirschner 1985. 57 figs. VIII, 158 pages

---

* denotes a volume which contains a Classified Index starting from Volume 36